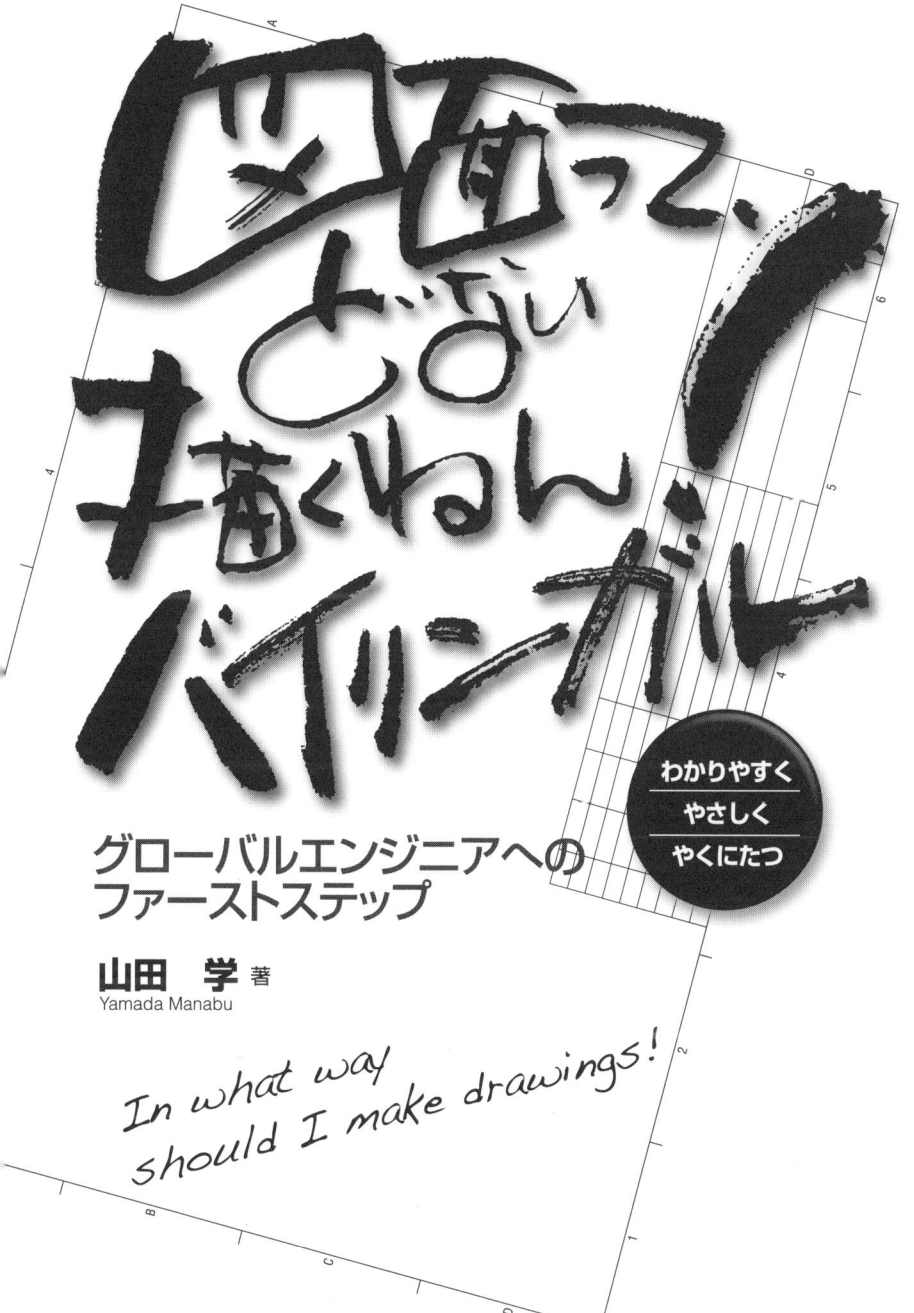

図面って、どない描くねん！バイリンガル

グローバルエンジニアへの
ファーストステップ

わかりやすく
やさしく
やくにたつ

山田 学 著
Yamada Manabu

In what way should I make drawings!

日刊工業新聞社

誇り高き日本人エンジニアに！

　「モノづくり」に国境がなくなると、日本人エンジニアが海外で活動しなければいけなくなります。そのため、文化や言語の違うエンジニアと意思の疎通を図ることが重要になります。

　海外のエンジニアとアイデアを共有するときに、いかにコミュニケーション能力が重要であるかを思い知ることになるでしょう。幸いにも、エンジニアは図面やポンチ絵という意思を伝えるための手段を持っています。

　本書は、JIS（日本工業規格）の定める製図のルールについて、日本語と英語で解説しており、機械設計業務の中でこの知識を使用してもらうことを期待しています。(^O^)／

For a proud Japanese engineer!

　Since "Manufacturing" has no border, Japanese engineers have to work overseas. Therefore, it becomes important to communicate with the engineer whose culture and languages are different.

　Sharing ideas with overseas engineers will teach you how important communication competence is. Fortunately, engineers have drawings and caricatures (PONCHI-E in Japanese) to communicate design intention.

　This textbook explains in Japanese and English about the rule of technical drawing by JIS (Japanese Industrial Standards).

　So, I hope to use this knowledge in your mechanical design. ;-)

機械製図のルールとして、暗黙の了解による解釈があります。
代表的な例を下記に示します。
As a rule of engineering drawings, there are interpretations of implied annotation. Typical example is shown as follows:

誤解の恐れがない限り、以下の状況において寸法が指示されなくても暗に寸法を示します。
A dimension may be implied and not indicated on a drawing in the following situations, so long as there is no risk of misunderstanding.

・2つの形体が同一平面上に整列している場合、距離寸法の0、あるいは角度寸法の0°は示しません。
- Where two features are aligned in the same plane, there is no requirement to indicate a linear dimension of 0 or an angular dimension of 0°.

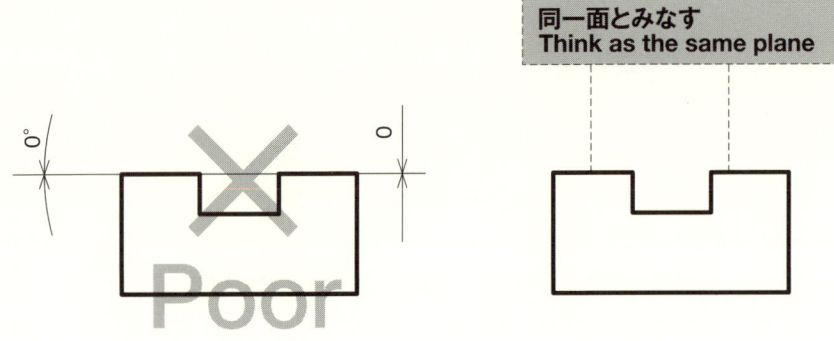

- 2つの形体が互いに平行である場合、0°または180°の角度寸法は示しません。
- Where two features are parallel to each other, there is no requirement to indicate an angular dimension of 0° or 180°.

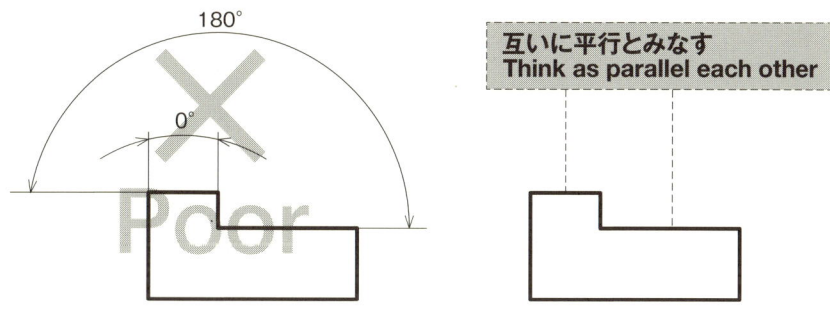

- 2つの形体が互いに垂直である場合、90°の角度寸法は示しません。
- Where two features are perpendicular to each other, there is no requirement to indicate an angular dimension of 90°.

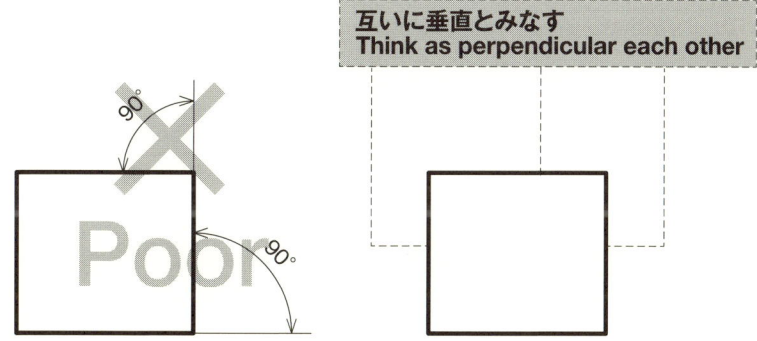

製図のルールを知るだけでは図面を描くことはできません。
図面の目的は「意思の疎通」です。機能や加工のことを考え、論理的に寸法を指示するテクニックを習得しましょう！

問題点のフィードバックなど、ホームページを通して紹介しています。

「Labnotes by 六自由度」
書籍サポートページ
http://www.labnotes.jp/

本書の執筆にあたり、お世話いただいた日刊工業新聞社出版局の方々にお礼を申し上げます。

2014年9月

山田 学

You cannot make drawings only by knowing the rule of engineering drawings.
Drawing purpose is "communication of intention". Let's learn technique to indicate dimensions logically, considering the function and processing!

We introduce feedback of problems in our website.

"Labnotes by ROKUJIYUUDO(Six Degree of Freedom in English)"
Books support page
http://www.labnotes.jp/

I would like to give special thanks to the publishers of NIKKAN KOGYO SHIMBUN.

Sep.2014

m. yamach

目次 CONTENTS

誇り高き日本人エンジニアに! ……………………………………………… i

第1章 図面の基本って、なんやねん! ……………………………………… 1
- 1-1 製図の目的 ……………………………………………………………… 2
- 1-2 用紙のサイズと様式 …………………………………………………… 4
- 1-3 表題欄に記入する記号 ………………………………………………… 10
- 1-4 線、文字および文章、尺度 …………………………………………… 20

第2章 投影法って、なんやねん! ………………………………………… 31
- 2-1 対象物の投影 …………………………………………………………… 32
- 2-2 第三角法 ………………………………………………………………… 36
- 2-3 正面図の表し方と向き ………………………………………………… 44
- 2-4 わかりやすい投影図の表し方 ………………………………………… 50

第3章 投影図のテクニックって、なんやねん! …………………………… 59
- 3-1 対象物の一部のみを表す投影図 ……………………………………… 60
- 3-2 断面図 …………………………………………………………………… 72
- 3-3 図形の省略 ……………………………………………………………… 92
- 3-4 特殊な図示 ……………………………………………………………… 96

第4章 寸法記入のルールって、なんやねん! ……………………………… 101
- 4-1 寸法記入の基本 ………………………………………………………… 102
- 4-2 寸法補助記号 …………………………………………………………… 112

For a proud Japanese engineer! ································· i

Chapter 1 **What are the fundamentals of drawings!** ················ 1

    1-1 Purpose of the technical drawings ·················· 3
    1-2 Size and format of the sheet ······················ 5
    1-3 Signs to fill in the title block ······················ 11
    1-4 Lines, Characters and texts, Scale ·················· 21

Chapter 2 **What is the projection method!** ····················· 31

    2-1 Projection of the object ·························· 33
    2-2 Third angle projection ···························· 37
    2-3 Expression and direction of front view ·············· 45
    2-4 Expression of the simply views ···················· 51

Chapter 3 **What is the technique of the projection views!** ········· 59

    3-1 Views expressing only a part of the object ·········· 61
    3-2 Sectional view ·································· 73
    3-3 Omission of view ································ 93
    3-4 Special expression of view ························ 97

Chapter 4 **What is the rule of dimensioning!** ··················· 101

    4-1 Fundamentals of dimensioning ···················· 103
    4-2 General symbols for dimensioning ················ 113

目次 CONTENTS

第5章 設計意図って、どない伝えんねん！ ……………………… 151
- 5-1 設計意図を伝える表現 …………………………… 152
- 5-2 寸法記入のコツ ………………………………… 162
- 5-3 寸法公差 ………………………………………… 172
- 5-4 はめあい公差の指示とその意味 ………………… 176
- 5-5 公差解析 ………………………………………… 182
- 5-6 面の肌記号の指示 ……………………………… 186

第6章 特殊な記号って、どない使うねん！ ……………………… 193
- 6-1 ねじの指示 ……………………………………… 194
- 6-2 溶接記号の理解 ………………………………… 200
- 6-3 幾何公差記号の理解 …………………………… 218

第7章 図面って、どない描くねん！ ……………………………… 239
- 7-1 寸法指示の論理的思考（治具の例） …………… 240
- 7-2 寸法指示の論理的思考（ドリルプレスバイスの例） ……… 256

参考文献 …………………………………………………………… 272

Chapter 5 In what way should I communicate my design intentions! ... 151

- 5-1 Expression to communicate design intentions ... 153
- 5-2 Point of dimensioning ... 163
- 5-3 Dimension tolerances ... 173
- 5-4 Indication and meaning of fit tolerance ... 177
- 5-5 Tolerance analysis ... 183
- 5-6 Indication of surface texture symbols ... 187

Chapter 6 In what way should I use special symbols! ... 193

- 6-1 Indication of screw thread ... 195
- 6-2 Understanding weld symbols ... 201
- 6-3 Understanding geometrical tolerance symbols ... 219

Chapter 7 In what way should I make drawings! ... 239

- 7-1 Logical mind of the dimensions indication (Example of Jig) ... 240
- 7-2 Logical mind of the dimensions indication (Example of Drilling-press vice) ... 256

Reference literature ... 272

第1章

図面の基本って、なんやねん！

Chapter1

What are the fundamentals of drawings!

この章を学習すると、次の知識を得ることができます。
1）製図の目的
2）用紙のサイズと様式
3）表題欄に記入する記号
4）線、文字および文章、尺度

After studying this chapter, you will be able to get knowledge as follows:
-1. Purpose of technical drawings
-2. Size and format of the sheet
-3. Signs to fill in the title block
-4. Lines, Characters and texts, Scale

第1章 1 製図の目的

> JISとは、Japan Industrial Standard（日本工業規格）の略です。
> JISとは、工業標準化法に基づき、すべての工業製品について定められる日本の国家規格です。
> 製図のルールは、「JIS B 0001（機械製図）」に規定されています。

　図面とは、設計者の意思を言葉ではなく、対象物を文字や記号、投影図を使って正確に相手に伝えるための仕様書です。

　そのためには、以下のことに注意しなければいけません。

①正確さ
　図面には、下記の事項を記入します。
- いつ（日時）
- どこで（企業名）
- 誰が（担当者）
- なにを（投影対象物）
- どのように（基準や加工方法、精度、仕上げなど）

②簡潔さ
　形状や寸法を理解できない投影図を描いてはいけません。
　製図のルールに従わない投影図のレイアウトや記号を使ってはいけません。

③解りやすさ
　断面図や特殊な図示法を使います。
　加工や機能を考慮して、寸法を整列させます。

> 機械製図のルールは、JIS B 0001に制定されてるんや。この番号は覚えとかなアカンで！

Chapter1 1 Purpose of the technical drawings

> JIS stands for Japan Industrial Standard.
> JIS is the national standard of Japan applied about all the industrial products based on the Industrial Standardization Law.
> The rule of technical drawings is specified in "JIS B 0001 (Technical drawings for Mechanical Engineering)".

No words can communicate a designer's intention, drawings are specification of objects which communicate correctly using texts and symbols, signs, and projection views.

So, drawing has to be careful of the following things:

-1. Accuracy

The following items are written in the drawing.
- When (Date)
- Where (Company Name)
- Who (In charge)
- What (Object)
- How (Datum, Processing, Accuracy, Finishing, etc.)

-2. Simplicity

Do not draw the projection view which can understand neither shape nor size.
Do not use the arrangement of projection views or symbols which are not in the rule of technical drawing.

-3. Plainness

Use a sectional view, and a special expression of view.
Align the dimensions in consideration of processing or function.

> The rule of technical drawings is specified in JIS B 0001. Keep this number in your mind!

Chapter1   What are the fundamentals of drawings!

第1章 2 用紙のサイズと様式

> 用紙のサイズや様式はJIS Z 8311に規定されています。
> 図面は長辺を横向きにして使用しますが、A4サイズは長辺を縦向きにして使用することができます。
> 図面には図面の輪郭線、表題欄、中心マークを描きます。

1-2-1 用紙のサイズ

用紙の大きさは、A0〜A4を標準として使用します（**表1-1**）。ただし、投影図がこれらの用紙に入りきらない場合には、別途規定された延長用紙を用います。

用紙は、投影図を見やすい大きさで描ける最小の用紙を選ぶとええで！

A列用紙とはどのくらいの大きさなのでしょうか？
本書の大きさはA5で、A4サイズの半分になります。A4サイズを2枚並べたものがA3サイズ、同様にA2サイズ、A1サイズと2倍ずつ大きくなりA0サイズが最大となります（**図1-1**）。

表1-1　A列サイズ（第一優先）

用紙の大きさ	短辺×長辺
A0	841×1189
A1	594×841
A2	420×594
A3	297×420
A4	210×297

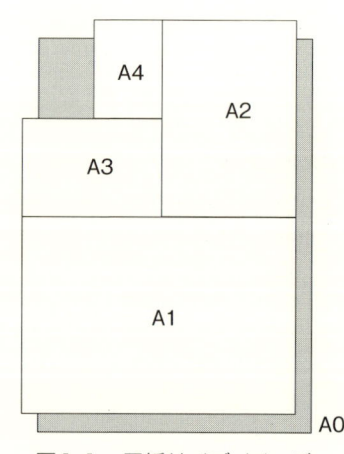

図1-1　用紙サイズイメージ

| Chapter1 | 2 | # Size and format of the sheet |

> Size and format of the sheet is specified in JIS Z 8311.
> The drawings are used with its long side in the horizontal direction, but A4 can be used with its long side in the vertical direction.
> On a drawing, the borderline, the title block, and the centering mark are drawn.

1-2-1 Size of the sheet

From A0 to A4 sheets are used for the drawings as a standard. Refer to Table 1-1. However, when projection views are protruded from these sizes, the extended sheets specified separately can be used.

You had better choose minimum sheet that can be drawn in a legible size!

What is A series paper size?
This textbook is A5, and is half of A4. A3 is two times A4.
Similarly, A2 is two times A3. A1 is two times A2. A0 is two times A1, and is max size. Refer to Figure 1-1.

Table1-1 Row A size (primary)

Drawing sheet size	Dimension a x b
A0	841×1189
A1	594×841
A2	420×594
A3	297×420
A4	210×297

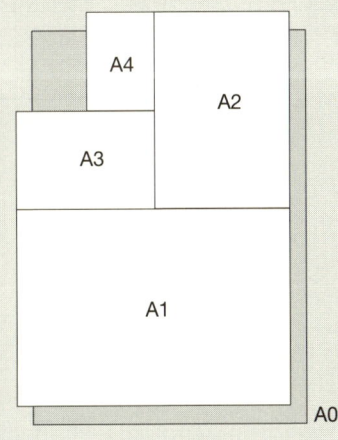

Figure 1-1 Image of the sheet size

Chapter1 What are the fundamentals of drawings!

1-2-2 輪郭線と中心マーク

　用紙の縁は破れなど破損する可能性があるため、輪郭線を描かなければいけません。

　また、図面をコピーするときに便利なよう、図面の各辺の中央に太い実線で中心マークをつけ、輪郭線の約5mm内側まで描きます。

　A0とA1用紙は、輪郭線を用紙の縁から内側に20mmオフセットして描き、A2～A4用紙は輪郭線を用紙の縁から内側に10mmオフセットして描きます（図1-2）。

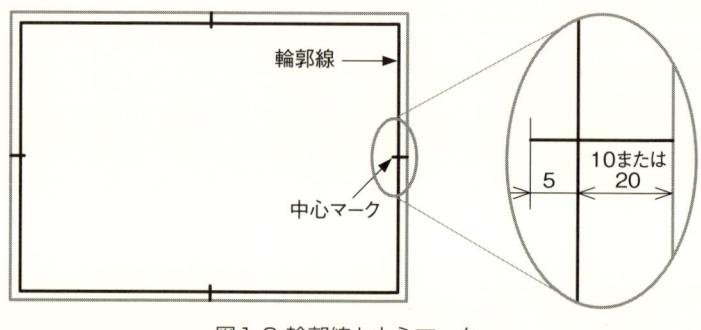

図1-2 輪郭線と中心マーク

1-2-3 格子参照方式

　輪郭線を偶数等分して区域を分けることができます（図1-3）。

　例えば電話で図面について説明する場合、「A-2エリアにある直径16mmの穴」のように使うことができます。

　この方式は道路地図にも用いられています。

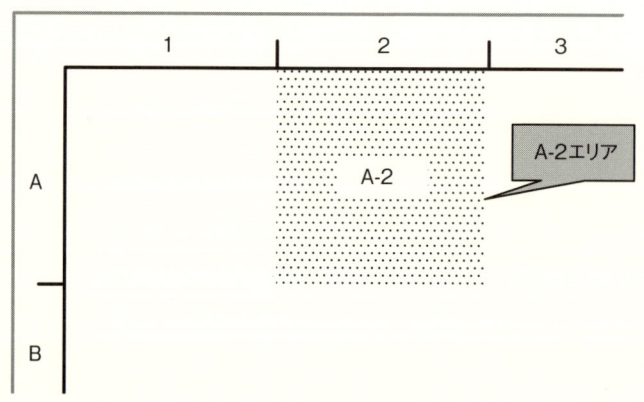

図1-3 格子参照方式

1-2-2 Border line and the centering mark

The border line has to draw because the edge of sheet may be broken.

Also, the centering marks are attached in the center of all side as a thick continuous line for convenience when copying sheets. The centering mark is drawn to the about 5mm inside of the border line.

In A0 and A1 sheet, the border line is drawn 20mm inside of the sheet edge. And in A2-A4 sheet, the border line is drawn 10mm inside of the sheet edge. Refer to Figure 1-2.

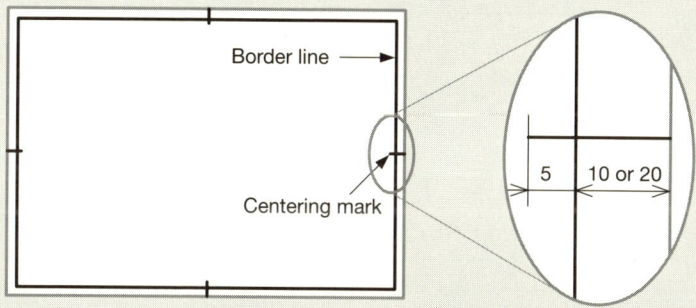

Figure 1-2 Border line and centering mark

1-2-3 Grid reference system

The border line can divide equally by even number. Refer to Figure 1-3.

For example, you can explain by telephone as follows:

"a 16 mm hole in area A-2."

This system is used also for the road map.

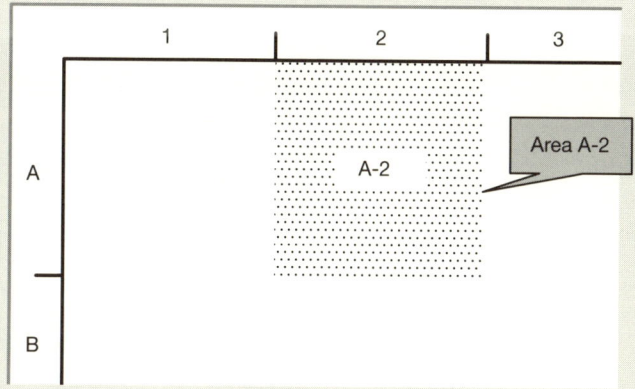

Figure1-3 Grid reference system

Chapter1 What are the fundamentals of drawings!

1-2-4 表題欄

表題欄は図面の管理事項（図面番号や品名、材質記号など）を記入するための枠です。

表題欄の位置は作図スペースの右下に位置します。ただし、A4用紙を縦向きに使用するときは、下部に配置されます（図1-4）。

図面情報が多い場合、表題欄は右上隅にも追加できます。

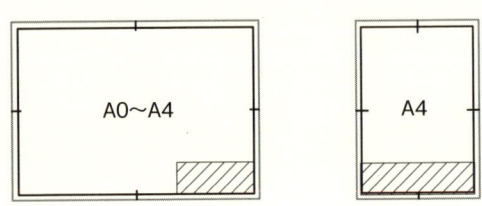

図1-4 表題欄

一般的に表題欄には、次のような項目や記号を記載します。

- 投影法
- 尺度
- 版数
- 部品名
- 図面番号（品番）
- 材料名
- 表面処理
- 製図者
- 設計担当者
- 検図者
- 承認者
- 日付
- 一般許容差
- 企業名

表題欄の項目については、決まりごとはないやで！

マジで〜？ そやから会社ごとにちゃうんですね？

1-2-4 Title block

The title block is a frame for filling in the management items (drawing number, part name, material, etc.) of a drawing.

Location of the title block shall be situated in the bottom right-hand corner of the drawing space. But title block is arranged in the bottom when A4 is used its long side vertically. Refer to Figure 1-4.

When there is much drawing information, title block can add in an upper right corner.

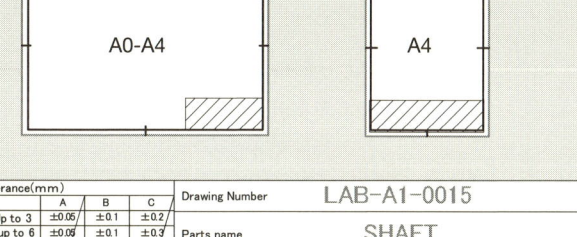

Tolerance(mm)			
Dimension	A	B	C
Up to 3	±0.05	±0.1	±0.2
Over 3 up to 6	±0.05	±0.1	±0.3
Over 6 up to 30	±0.1	±0.2	±0.5
Over 30 up to 120	±0.15	±0.3	±0.8
Over 120 up to 400	±0.2	±0.5	±1.2
Over 400	±0.3	±0.8	±2.0
Screw location	±0.1	±0.2	±0.5
Angle Tolerance (°) bending	±1.0	±1.5	±2.0
machining	±0.3	±0.5	±1.0

Labnotes Co., Ltd.

Drawing Number	LAB-A1-0015			
Parts name	SHAFT			
APPROVALS	CHECK	IN CHARGE	DRAWN	
Tanaka	Satou	Suzuki	Yamada	
Date	30/09/2014	30/09/2014	29/09/2014	26/09/2014
Size A3	Material S45C	Finish induction hardening		
Scale 1:1	Projection	Sheet 1/1	Revision 0	

Figure 1-4 Title block

Generally the following text documents or signs are written in title block.

Projection	Scale	Revision	Part name
Drawing No. Parts.No.	Material	Finishing	Drawn
In charge	Check	Approval	Date
Tolerance	Company name		

"For management items of title block, there are no requirements!"

"No way! So, it is different for every company, isn't it?"

Chapter1 What are the fundamentals of drawings!

第1章 3 表題欄に記入する記号

1-3-1 JISの材質記号

表題欄には材質記号を記入します。

1) 金属材料

金属材料は、図1-5のように分類されます。

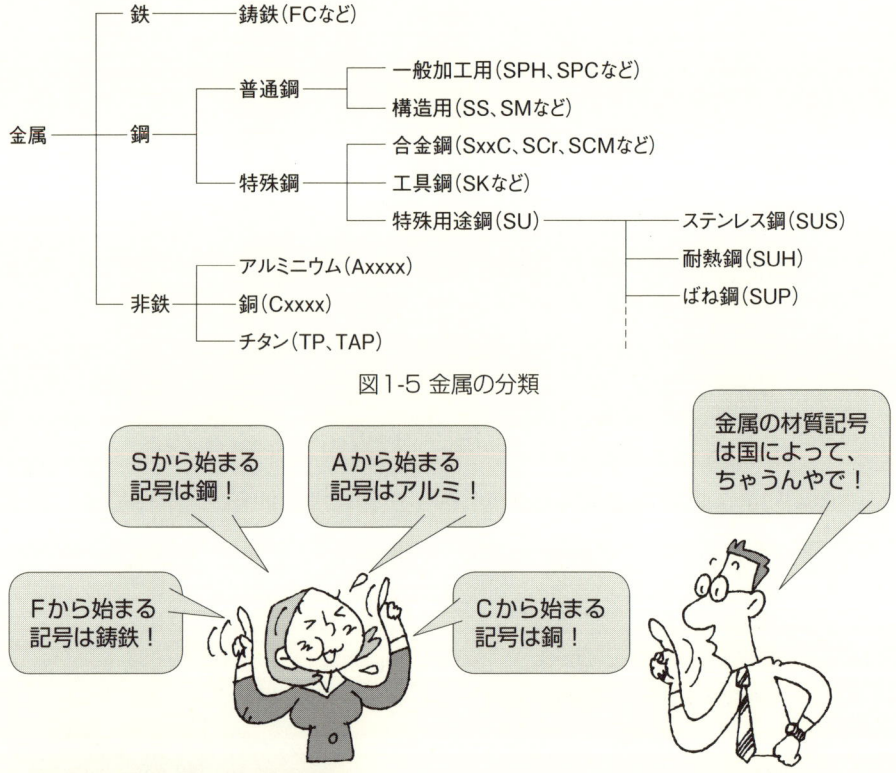

図1-5 金属の分類

φ(@°▽°@) メモメモ

鉄と鋼の違いは炭素量です。
　炭素含有量が0.04%未満の金属を純鉄、0.04%以上2.1%未満の金属を鋼（はがね）と呼び、2.1%以上入った金属を鋳鉄（または銑鉄）と呼びます。
　※炭素含有量の値は、文献によって多少異なります。

| Chapter1 | 3 | **Signs to fill in the title block** |

1-3-1 Material signs of JIS

The material sign is filled in the title block.

-1 Metallic material

Metal is classified as shown in Figure 1-5.

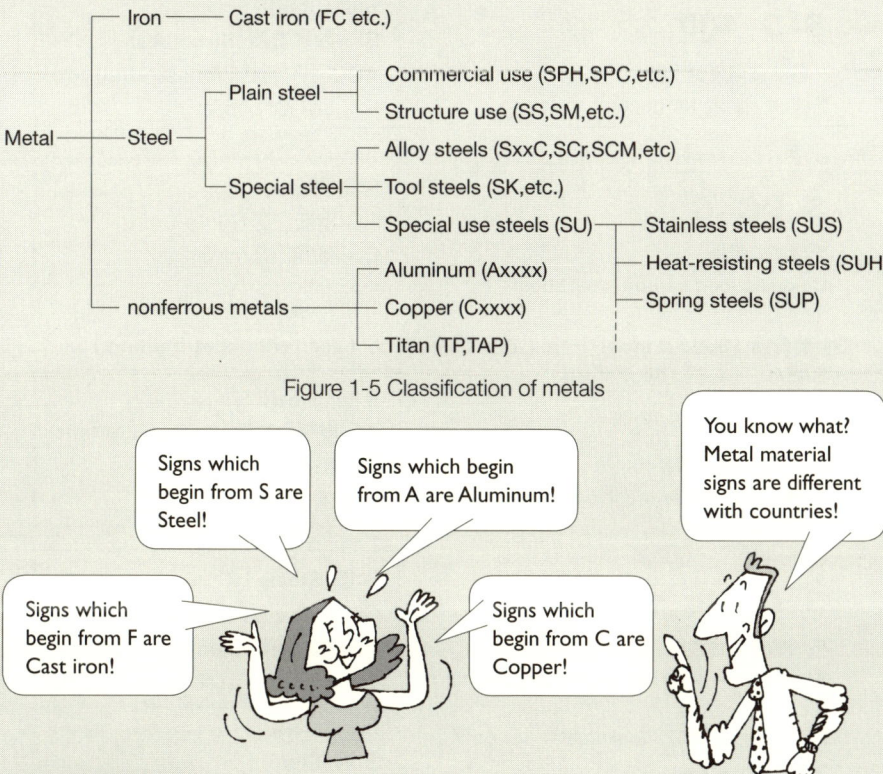

Figure 1-5 Classification of metals

> Signs which begin from S are Steel!
>
> Signs which begin from A are Aluminum!
>
> You know what? Metal material signs are different with countries!
>
> Signs which begin from F are Cast iron!
>
> Signs which begin from C are Copper!

Note ;-)

The difference between iron and steel is carbon content.
The metal of under 0.04% carbon content is called pure iron, 2.1% to 0.04% is called steel, and 2.1% or more is called cast iron (or pig iron).
 * The value of carbon content changes somewhat with literature.

φ(@ﾟ▽ﾟ@) メモメモ Note ;-)

代表的な金属材料（Typical metal materials）

1）鋳鉄（Cast Iron）

・FC250　鋳鉄（Cast Iron）

F C 250

ねずみ鋳鉄
Gray iron castings

> F:鉄（Ferrum）
> C:鋳造（Casting）
> 250:引張強さ（Tensile strength）
> 　　　$250N/mm^2$

2）普通鋼（Plain steel）…熱処理せずに使用する。（Used without heat-treating.）

・SS400　一般構造用（Structure use）

S S 400

一般構造用圧延鋼材
Rolled steels for general structure

> S:鋼（Steel）
> S:一般構造用（Structure）
> 400:引張強さ（Tensile strength）
> 　　　$400N/mm^2$

・SPCC　一般加工用（Commercial use）

S PC C

冷間圧延鋼板
Cold-reduced carbon steel sheet

> S:鋼（Steel）
> PC:冷間圧延（Plate Cold）
> C:一般用（Commercial）

3）特殊鋼（Special steel）…熱処理して使用する。（Used with heat-treating.）

・S45C　合金鋼（Alloy steels）

S 45 C

機械構造用炭素鋼
Carbon steels for machine structural use

> S:鋼（Steel）
> 45:炭素含有量（Carbon content）
> 　　　0.45（％）
> C:炭素（Carbon）

・SCM415　合金鋼（Alloy steels）

S CM 4 15

機械構造用合金鋼
Alloyed steels for machine structural use

> S:鋼（Steel）
> CM:クロム、モリブデン（Chromium, Molybdenum）
> 4:元素コード（Element cord）
> 15:炭素含有量（Carbon content）0.15（％）

・SK140　工具鋼（Tool steels）

S K 140

炭素工具鋼
Carbon tool steels

> S:鋼（Steel）
> K:工具（Kougu）* Kougu means tool in English.
> 140:炭素含有量（Carbon content）1.4（％）

φ(@°▽°@) メモメモ Note ;-)

4) 特殊用途鋼 (Special use steels)

・SUS304

S U S 304

ステンレス鋼
Stainless steels

```
S:鋼 (Steel)
U:用途 (Use)
S:ステンレス (Stainless)
304:種類番号 (Grade)
```

・SUP9

S U P 9

ばね鋼
Spring steels

```
S:鋼 (Steel)
U:用途 (Use)
P:スプリング (Spring)
9:種類番号 (Grade)
```

・SWPA

S W P A

ピアノ線
Piano wires

```
S:鋼 (Steel)
W:線 (Wire)
P:ピアノ (Piano)
A:種類番号 (Grade)
```

・SUM22

S U M 22

硫黄及び硫黄複合快削鋼
Free machining steels

```
S:鋼 (Steel)
U:用途 (Use)
M:切削性 (Machinability)
22:種類番号 (Grade)
```

5) 非鉄 (Nonferrous metal)

・A5052　アルミニウム (Aluminum)

A 5052

アルミニウム合金
Aluminum alloy

```
A:アルミニウム (Aluminium)
5052:種類番号 (Grade)
```

・C2800　銅 (Copper)

C 2800

銅合金
Copper alloy

```
C:銅 (Copper)
2800:種類番号 (Grade)
```

2) 樹脂材料

　樹脂成型品の場合、リサイクル性を考慮して部品に材質記号を表示しなければいけません。

　対象物に表示する場合は、図1-6のように指示することができます。

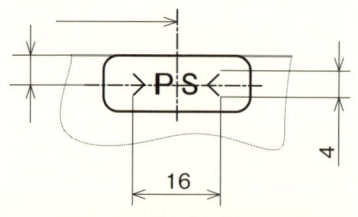

図1-6　材質表記例

材料記号の指示を忘れたら、金型修正せなあかんで！

表1-2　代表的な樹脂材料

材質記号	材料名
ABS	ABS樹脂
AS	AS樹脂
PA6	ポリアミド6（6ナイロン）
PC	ポリカーボネート
PE	ポリエチレン
PET	ポリエチレンテレフタレート（ペット）
PF	フェノール樹脂
POM	アセタール樹脂（ポリアセタール）
PP	ポリプロピレン（PPシート）
PS	ポリスチレン（スチロール樹脂）
PUまたはPUR	ポリウレタン
-HI	耐衝撃性〜
（例）PS-HI	耐衝撃性ポリスチレン
-P、-U	軟質〜、硬質〜
（例）PVC-U	硬質塩化ビニル
-GF	ガラス繊維〜％混入する〜
（例）PS-GF	ガラス繊維〜％混入するポリスチレン

樹脂の材質記号は、金属と違って世界的に共通なんや！

-2 Resin materials

When using resin molding part, do not forget to indicate the material sign for recycling.

When displaying material sign on the object, it can indicate as shown in Figure 1-6.

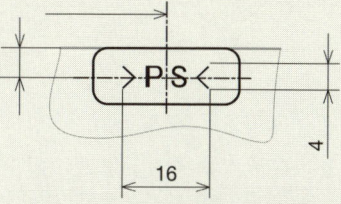

Figure 1-6 Example of resin material sign

Take care! Don't forget to indicate material sign. Mold should be revised if you've forgotten!

Table1-2 Typical resin materials

Material signs	Material names
ABS	Acrylonitrile butadiene styrene
AS	Styrene acrylonitrile
PA6	Poly amide6 (nylon6)
PC	Poly carbonate
PE	Poly ethylene
PET	Poly ethylene terephthalate (PET)
PF	Phenol-formaldehyde
POM	Poly oxymethylene
PP	Poly propylene
PS	Poly styrene
PU or PUR	Poly urethane
-HI	High-impact modified~
e.g. PS-HI	High-impact modified polystyrene
-P、-U	Plasticized~、Unplasticized~
e.g. PVC-U	Unplasticized poly vinyl chloride
-GF~	Grass-fiber reinforced~
e.g. PS-GF	Grass-fiber reinforced polystyrene

Resin material signs are common worldwide unlike metal!

Chapter1 What are the fundamentals of drawings!

1-3-2 表面処理記号

表面処理は、綺麗に見せるための「装飾性」と、錆を抑えたり硬くしたりという「機能性」を目的とします。
　自動車や電気製品に使われる部品の防錆には、電気めっきが一般的に使われます。

電気めっきの記号

Ep-Fe/Zn 8/CM 2:B
(電気めっき、鉄素地、亜鉛めっき8μm以上、有色クロメート処理、通常の屋外での使用)

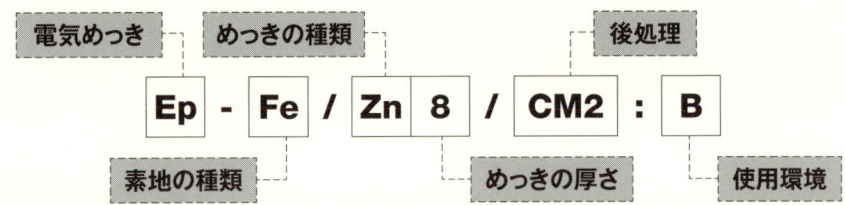

電気めっき	Ep:電気めっき　　ELp:無電解めっき
素地の種類	Fe:鉄　　Cu:銅・銅合金　　Zn:亜鉛合金　など
めっきの種類	Ni:ニッケル　Cr:クロム　ICr:工業用クロム　Cu:銅　Zn:亜鉛　など
めっきの厚さ	μm厚さ
後処理	CM1:光沢クロメート　　CM2:有色クロメート　＊黒色クロメートは記号なし
使用環境	A:腐食性の高い屋外　B:通常の屋外　C:湿気の高い屋内　D:通常の屋内

φ(＠°▽°＠)　メモメモ

めっきの旧記号

古いめっきの記号は、次のように表されます。
※MFZ nⅢ-Cは、MFZ n8-Cと同じ解釈となります。

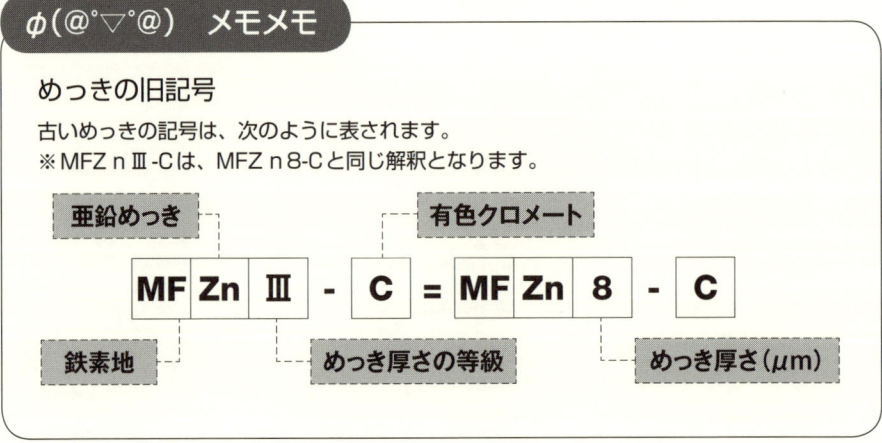

1-3-2 Signs of surface treatment

The surface treatment aims at the "decorativeness" for showing beautiful, and the "functionality" for rust prevention or hardening.

For cars or electric appliances parts, electroplating is used for the rust prevention generally.

The signs of electroplating

Ep-Fe/Zn 8/CM 2:B

(electroplating, ferrum foundation, 8μm or more of galvanization, colored chromate, usual outdoor use)

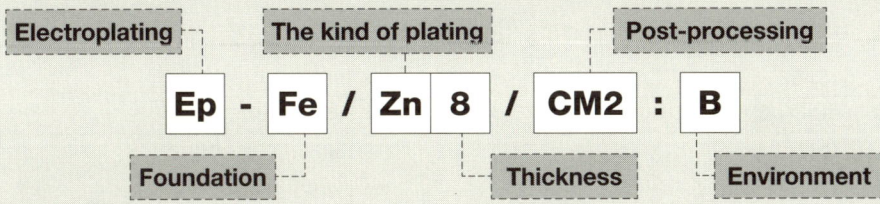

Electroplating	Ep: electroplating ELp: electroless plating
Foundation	Fe: ferrum Cu: copper or copper alloy Zn: zinc alloy etc.
The kind of plating	Ni: nickel Cr: chrome ICr: Industrial chrome Cu: copper Zn: zinc etc.
Thickness	μm thickness
Post-processing	CM1: bright chromate CM2: colored chromate *Black chromate has no sign
Environment	A: high corrosiveness outdoor use B: usual outdoor use C: high humidity indoor use D: usual indoor use

Note ;-)

Old signs of electroplating

The old signs of electroplating are expressed as follows:
* MFZn Ⅲ -C is the same as MFZn8-C.

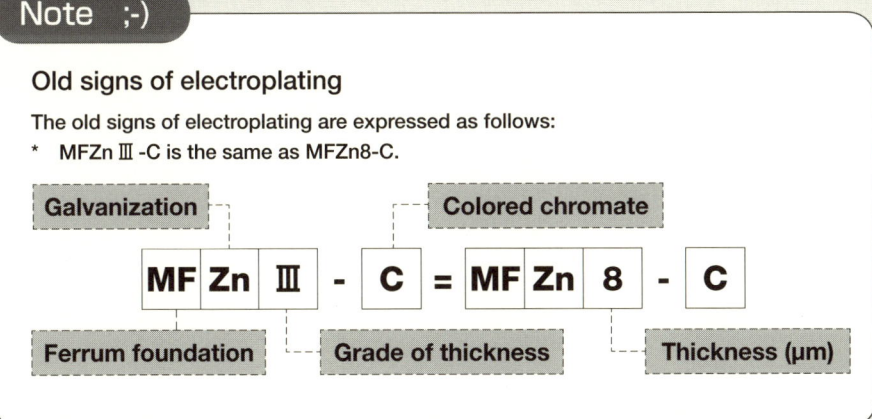

＼(°o°;)／ エンジニアリング テクノロジー

環境問題と材質、表面処理

設計の際に、環境負荷について配慮しなければいけません。

・RoHS指令

　　RoHSは、欧州連合内（EU）における製造メーカと供給元にとって非常に重要な指令です。

　　RoHSの文字は、Restriction of Hazardous Substances（危険物質の制限）の略です。

　　日常生活で遭遇したり生態系に入り込んだりする危険物質を減少することが目的です。そのため、使用が制限される6物質があります。

物質	製品内許容濃度	含有物質
Pb（鉛）	0.1wt%	はんだ、合金など
Hg（水銀）	0.1wt%	スイッチなど
Cd（カドミウム）	0.01wt%	電池、塗料など
Cr^{6+}（六価クロム）	0.1wt%	亜鉛めっきなど
PBB（ポリ臭化ビフェニール）	0.1wt%	コンデンサなど
PBDE（ポリ臭化ジフェニルエーテル）	0.1wt%	樹脂の難燃材など

※0.1Wt%とは、1kgの製品に対して1gが許容されることを意味する。

・材料使用上の注意点

　　特殊鋼のSUM22LやSUM31Lは、最大0.35%鉛を含んでいます。

　　したがって、SUM22LやSUM31Lを100g使用した場合、製品に0.35gの鉛を含有してしまい影響が大きくなります。

・表面処理使用上の注意点

　　従来の亜鉛めっきには六価クロムが含有されていましたが、RoHS対応によって三価クロムが代用されています。

　　亜鉛めっきをする場合は、六価クロムを使用していないか確認した方がよいでしょう。

:0 Engineering Technology

An environmental problem and materials, surface treatment
In designing, consider about an environmental impact.

-RoHS Regulations
   RoHS is a very important directive to manufacturers and suppliers within European Union (EU).
   The letters stand for Restriction of Hazardous Substances.
   Its aim is to reduce the hazardous substances that are encountered in everyday life and also enter the ecosystem.
   Therefore, there are six substances whose use is restricted:

substances	allowable level in a product	contained material
Pb (Lead)	0.1wt%	Solder, Alloy, etc.
Hg (Mercury)	0.1wt%	switch, etc.
Cd (Cadmium)	0.01wt%	battery, paint, etc.
Cr^{6+} (Chromium VI or hexavalent chromium)	0.1wt%	galvanization, etc.
PBB (Polybrominated biphenyl)	0.1wt%	electrostatic capacitor, etc.
PBDE (Polybrominated diphenyl ethers)	0.1wt%	flame retardants of the resin, etc.

*0.1Wt% means that 1g is allowed for a product of 1kg.

- Notes for materials use
   SUM22L or SUM31L of the special steel include up to 0.35% of lead.
   Therefore, when using SUM22L or SUM31L 100g, it is big influence because a product contains 0.35g of lead.

- Notes for surface treatment use
   Previous galvanizing had Hexavalent chrome, but trivalent chrome is substituted by RoHS.
   When galvanizing, check hexavalent chrome-free.

第1章 4 線、文字および文章、尺度

1-4-1 線

一般的に、図面に使用する線は8種類です。
- ・太い実線
- ・太い破線
- ・太い一点鎖線
- ・細い二点鎖線
- ・細い実線
- ・細い破線
- ・細い一点鎖線
- ・極太の実線

線の種類は、適用に従い**表1-2**に示すように用いられます。

表1-2　線の種類と適用（抜粋）

用途名称	線の種類		適用
外形線	太い実線	———————	対象物の見える部分の形状を表すのに用いる。
寸法線	細い実線		寸法を記入するのに用いる。
寸法補助線			寸法を記入するために図形から引き出すのに用いる。
引き出し線			記述・記号などを示すために引き出すのに用いる。
回転断面線			図形内にその部分の切り口を90度回転して表すのに用いる。
中心線			図形に中心線を簡略して表すのに用いる。
かくれ線	太い破線または細い破線	— — — — —	対象物の見えない部分の形状を表すのに用いる。
中心線	細い一点鎖線	—・—・—・—	a) 図形の中心をあらわすのに用いる。 b) 中心が移動する中心軌跡を表すのに用いる。
ピッチ線			繰り返し図形のピッチを取る基準を表すのに用いる。
特定指定線	太い一点鎖線	—・—・—・—	特殊な加工を施す部分など特別な要求事項を適用すべき領域を表すのに用いる。
想像線	細い二点鎖線	—・・—・・—	a) 加工前または加工後の形状を表すのに用いる。 b) 工具、治具などの位置を参考に示すのに用いる。 c) 図示された断面の手前にある部分を表すのに用いる。
破断線	フリーハンドによる細い実線またはジグザグ線	〜〜〜	対象物の一部を破った境界、または一部を取り去った境界を表すのに用いる。
切断線	細い一点鎖線で、端部及び方向が変わる部分を太くしたもの	⌐_⌐	断面図を描く場合、その断面位置を対応する図に表すのに用いる。
ハッチング	細い実線を規則的に並べたもの	/////	図形の限定された特定の部分を他の部分と区別するのに用いる。例えば、断面図の切り口を示す。
特殊用途線	細い実線		外形線およびかくれ線の延長を表すのに用いる
	極太の実線	▬▬▬▬▬	圧延鋼板、ガラスなど薄肉部を単線図示するのに用いる。

Chapter 1 4 Lines, Characters and texts, Scale

1-4-1 Lines

Generally, the lines to use for a drawing are eight types.

- Thick continuous line
- Thick dashed line
- Thick long-dashed dotted line
- Thin long-dashed double-dotted line
- Thin continuous line
- Thin dashed line
- Thin long-dashed dotted line
- Extra thick continuous line

Lines are used as shown in **Table 1-2** depending on the application.

Table 1-2 Type of lines and applications (an extract)

Name of application	Type of lines		application
Visible outline	Thick continuous line	————	To be used for expressing the shape of visible part of object.
Dimension line	Thin continuous line		To be used for inscribing dimension.
Projection line			To be used for pointing to a dimension placed outside the view.
Leader line			To be used for pointing to a description, symbol, etc. indicated outside the view.
Outline of revolved section in line			To be used for expressing the section of the part by revolving 90 degree in the view.
Center line			To be used for expressing simply the center line of view.
Hidden outline	Thick dashed line or Thin dashed line	– – – –	To be used for expressing the shape of invisible part of object.
Center line	Thin long-dashed dotted line	— - — - —	a) To be used for expressing the center of view. b) To be used for expressing the center trace when the center has travelled.
Pitch line			To be used for indicating the reference to take pitch of view of repetitive features.
Special designed line	Thick dashed dotted line	— - — -	To be used for expressing the applicable area of special requirements where the part is to be given special processing.
Fictitious outline	Thin long-dashed double-dotted line	— -- — --	a) To be used for expressing the shape before or after processing. b) To be used for expressing the position of tool, jig, etc. for reference. c) To be used for expressing the viewer's side of the drawn section.
Break line	Thin freehand line or zigzag line	～～～	To be used for expressing the limit of partial or interrupted view and section.
Cutting-plane line	Thin dashed dotted line whose end part and direction changing part are made thick	⌐_·_⌐	To be used for expressing the cutting position on the corresponding drawing when drawing the sectional view.
Hatching	Thin continuous lines spaced regularly	/////	To be used for distinguishing the limited specific part of view from other parts. e.g. It indicates the section of sectional view.
Line for special use	Thin continuous line	————	To be used for expressing the extension of visible outline and hidden outline.
	Extra thick continuous line	▬▬▬▬	To be used for indicating clearly the single line drawing of thin wall part of rolled steel plate or a glass.

Chapter1 What are the fundamentals of drawings!

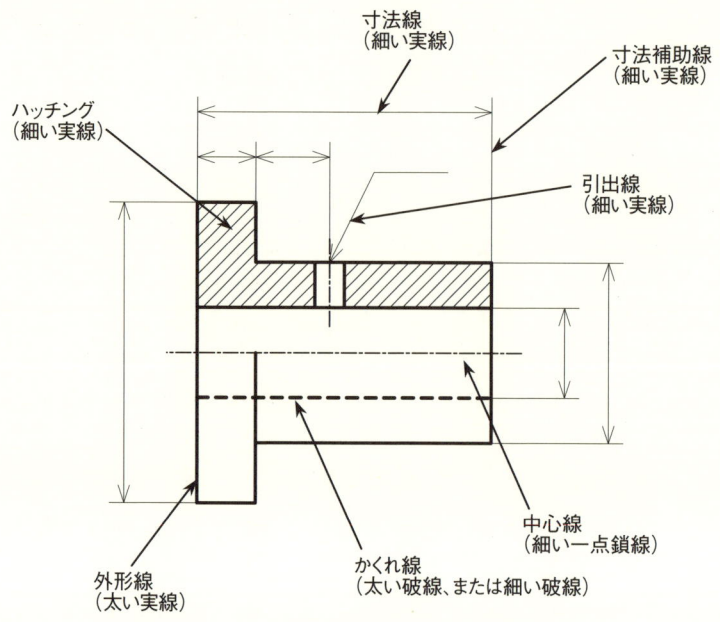

図1-7　代表的な線の適用と種類

2種類以上の線が同じ場所で重なる場合は、線の優先順に描きます。
①外形線
②かくれ線
③切断線
④中心線
⑤寸法補助線

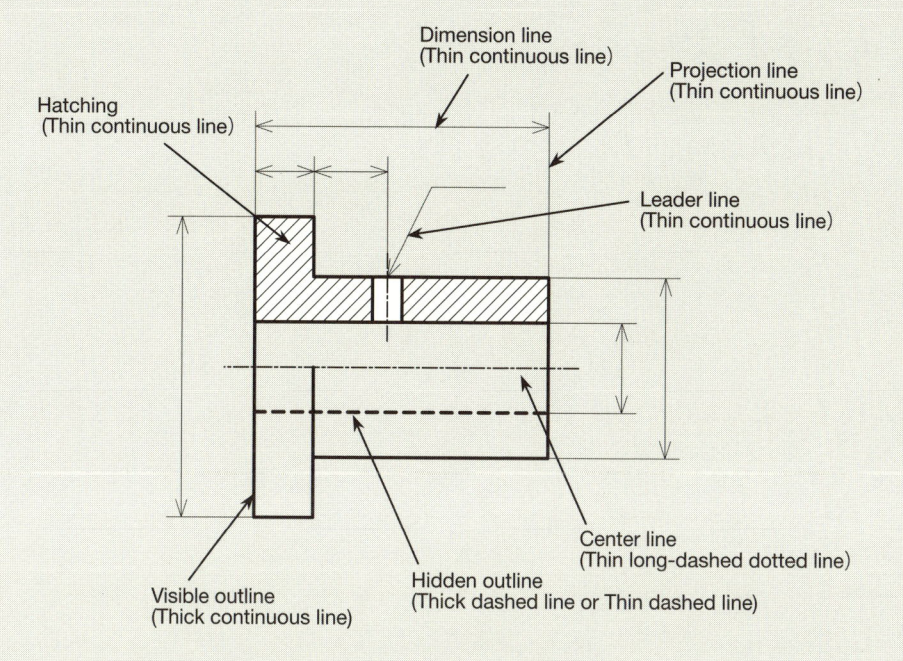

Figure 1-7 Typical application and types of lines

When lines more than two types are piled up at the same place, lines are drawn in priority.
1. Visible outline
2. Hidden outline
3. Cutting-plane line
4. Center line
5. Projection line

Chapter1   What are the fundamentals of drawings!

1-4-2 文字および文章

1) 文字の種類は次によります。
   - 漢字は常用漢字表を用いるのがよい。
   - 仮名は、ひらがなまたはカタカナのいずれかを用い、一連の図面において混用しない。ただし、外来語（例えば、ボルト、フランジなど）や注意を促す表記（例えば、塗装のダレなど）にカタカナを用いることは混用とみなしません。
2) 文字の高さはJISで規定されていますが、常識的な範囲内で読みやすい大きさであれば他の高さを使ってもかまいません。
3) 文章表現は次によります。
   - 文章は口語体で左横書きとします。
   - 図面注記は簡潔明瞭に書きます。

　海外のメーカーに図面を提出する場合は、和文と英文の注記を併記すべきと思います。なぜなら、日本の検査部門などの担当者が英文を理解できない場合があるからです。

　二か国語で表記する場合は、和文を上に、下に英文を記述します（**図1-8**）。

　投影図に示す英文は、大文字で表すとよいでしょう。

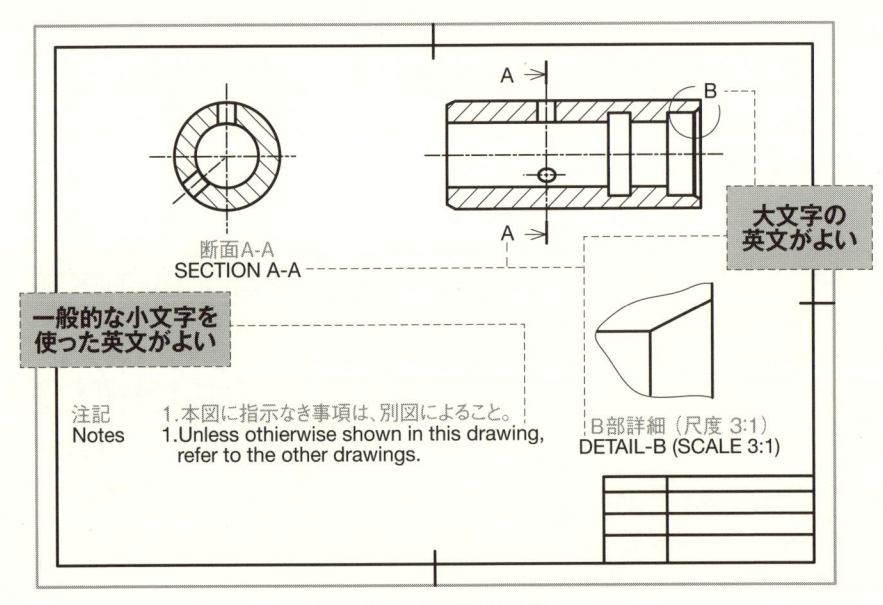

断面A-A
SECTION A-A

B部詳細（尺度 3:1）
DETAIL-B (SCALE 3:1)

注記　1.本図に指示なき事項は、別図によること。
Notes　1.Unless othierwise shown in this drawing, refer to the other drawings.

一般的な小文字を使った英文がよい

大文字の英文がよい

図1-8　2か国語の注記

1-4-2 Characters and texts

-1. Types of characters are as follows:
  - KANJI (Chinese characters) should preferably follow the List of KANJI for common use.
  - For KANA, either KATAKANA (Japanese angular syllabary) or HIRAGANA (Japanese cursive syllabary) is used. Do not use it mixing in the same series of drawings. However, the use of KATAKANA to foreign words (e.g. bolt, flange, etc.) and notes (e.g. sagging of paint, etc.) are not mixture.
-2. The height of characters is decided by JIS, but you can use the other height if it is easy to read within common-sense.
-3. The expression of texts is as follows:
  - The texts are written from left to right using colloquial style.
  - Notes in drawings should be simple and clear.

Write together notes of Japanese and English, when submitting drawings to an overseas maker. Because Japanese staff, such as an inspection section, cannot understand English sentence.

When writing bilingually, Japanese put above and English put below. Refer to Figure 1-8.

English texts shown in views are good to express a capital letter.

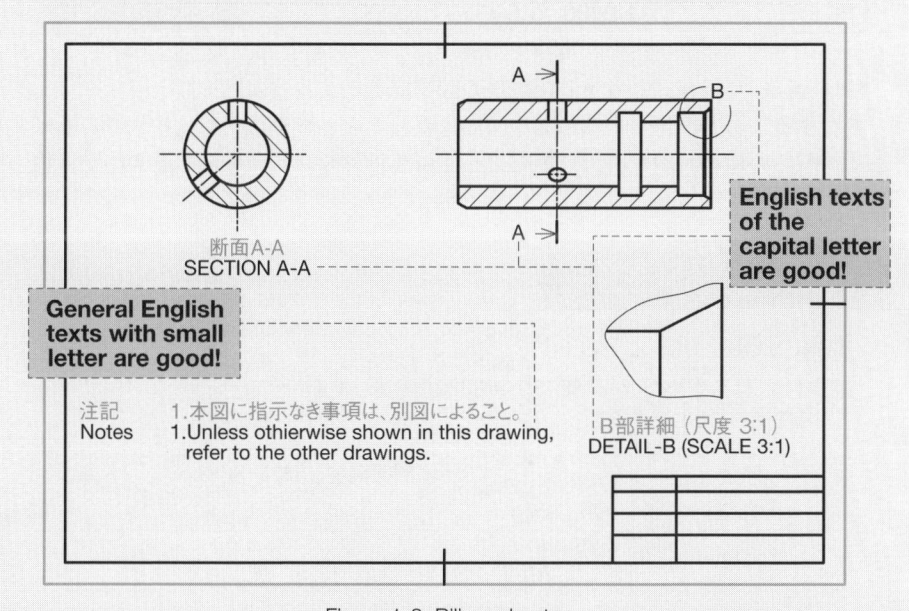

Figure 1-8 Bilingual notes

1-4-3　尺度

　図面に選ばれる尺度は、対象物の複雑さや表示の目的によります。実物大で描くことのできる用紙を選択すれば、部品の大きさをイメージしやすくなります。

　尺度は変更することができ、実物と同じ尺度を「現尺」、実物を拡大して描く尺度を「倍尺」、実物を縮小して描く尺度を「縮尺」と呼びます。

　尺度の指示は表題欄に記入します（**表1-3**）。
　尺度は、A：Bで表します。
　A：図面に描いた長さ　　　　　B：対象物の実際の長さ
　例）　現尺の場合　1：1　　倍尺の場合　X：1　　縮尺の場合　1：X

表1-3　推奨する尺度

分類	推奨尺度
倍尺	50:1　20:1　10:1　5:1　2:1
現尺	1:1
縮尺	1:2　1:5　1:10 1:20　1:50　1:100 1:200　1:500　1:1000 1:2000　1:5000　1:10000

> 推奨尺度は目安やから、必要に応じて中間の尺度を使ってもええで！

　倍尺で描いた場合、参考として現尺の図を描き加えるとよいでしょう。
　そうすることで、加工者は部品を図面に重ねて大きさや形状を比較することができます。この場合、現尺の図は簡略化して対象物の輪郭だけを示したものでも構いません。

φ(@°▽°@)　メモメモ

中間の尺度　JIS Z 8314

国際標準にはありませんが、JISでは中間の尺度があります。
これらを使うこともできます。

分類	中間の尺度（抜粋）
倍尺	$\sqrt{2}$:1　etc.
縮尺	$1:\sqrt{2}$ 1:1.5　1:2.5　1:3　1:4　1:6　1:15　1:25　etc.

1-4-3 Scale

The scale to be chosen for a drawing shall depend on the complexity of the object and the purpose of the expression. To choose the sheet that can draw in full scale is easy to understand of object's size.

Scale can be changed, same size as the object is called "full scale", expanded size of the object is called "enlargement scale", and reduced size of the object is called "reduction scale".

The scale shall be placed in the title block. Refer to Table 1-3.
The scale is expressed by A: B
A: Drawing length   B: Actual length of object
e.g.)   In case of full scale, 1 : 1   In case of enlargement scale, X : 1
        In case of reduction scale, 1 : X

Table 1-3 Recommended scales

Category	Recommended scale
Enlargement scales	50:1 20:1 10:1 5:1 2:1
Full scale	1:1
Reduction scales	1:2 1:5 1:10 1:20 1:50 1:100 1:200 1:500 1:1000 1:2000 1:5000 1:10000

Recommended scale is a guide line. You can use middle scale as necessity!

In case of enlargement scale, full scale figure may be added for reference in the drawing.

So, worker can put work-piece on the drawing and can compare size or shape. In this case, full scale can be simplified to express only the profile of the object.

Note ;-)

Middle scale JIS Z 8314

Although it is not in international standards, JIS has middle scales.
You can use these too.

Category	middle scale (an extract)
Enlargement scales	$\sqrt{2}$:1 etc.
Reduction scales	1:$\sqrt{2}$ 1:1.5 1:2.5 1:3 1:4 1:6 1:15 1:25 etc.

1-4-4　CAD製図の注意点

　CADを使って図面を描く場合、尺度設定を間違えると、現尺と倍尺（あるいは縮尺）の寸法数値が混在することがあります。尺度を変えるときには注意しましょう。

　尺度を変更した図面を用いてＤＲ（設計審査：Design Review）をする場合は、尺度による勘違いに注意しなければいけません。
　CADの組立図をプロジェクターで壁に映す場合、画面の大きさによって隙間の確保や強度の保証が難しくなります。
　そのため、DRの際には現尺の図面を準備して実際の大きさを把握しなければいけません。

φ(＠˚▽˚＠)　メモメモ

尺度変更による錯覚
　倍尺図面では弱いものでも強そうに見えたり、縮尺図面では強いものでも弱く見えたりしてしまうことがよくあります。
　このような勘違いは品質やコストに影響するので、日ごろから図面と現物を見比べて大きさを把握するセンスを磨きましょう。

1-4-4 Point of CAD drafting

When using CAD, if scale setup is mistaken, then the dimension value of a full scale and an enlarged/ reduction scale may be mixed. Be careful when you changing scale.

Be careful of the misunderstanding by a scale, when you do DR (Design Review) using the drawing which changed the scale.
When projecting the assembly drawing of CAD on a wall with a projector at DR, it becomes difficult to secure a gap or to guarantee strength by a screen size.
Therefore, in DR, the assembly drawing printed by the full scale must also be prepared, and an actual size must be checked.

Note ;-)

Illusion by scale change

We often see strongly as for a weak object in an enlarged scale, or weakly as for a strong object in a reduction scale.
Such an illusion influences quality and cost, let's develop a size sense by comparing the actual object with the drawing always.

用語集　glossary

本章で使用した単語です。他の用語も使うことができます。

The list of words used for this chapter is shown. You can use another words, too.

図面	drawing	用紙	sheet
A列	A series paper	輪郭線	border line
中心マーク	centering mark	格子参照方式	grid reference system
表題欄	title block	担当	in charge
検図	check	承認	approval
日付	date	材質記号	material sign
金属	metal	鉄	iron
鋼	steel	非鉄	nonferrous metal
炭素	carbon	樹脂	resin
表面処理	surface treatment	電気めっき	electroplating
亜鉛	zinc	環境	environment
六価クロム	hexavalent chromium	三価クロム	trivalent chrome
鉛	lead	実線	continuous line
破線	dashed line	一点鎖線	long-dashed line
二点鎖線	long-dashed double-dotted line	外形線	visible outline
かくれ線	hidden outline	切断線	cutting-plane line
中心線	center line	寸法補助線	projection line
文字	character	文章	text
漢字	Chinese character	かな	Japanese syllabary
ひらがな	Japanese cursive syllabary	カタカナ	Japanese angular syllabary
口語体	colloquial style	尺度	scale
倍尺	enlargement scale	現尺	full scale
縮尺	reduction scale	設計審査	design review
錯覚	illusion	品質	quality
コスト	cost		

第2章
投影法って、なんやねん!

Chapter2
What is the projection method!

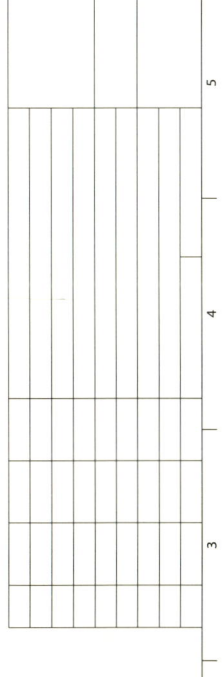

この章を学習すると、次の知識を得ることができます。
1）対象物の投影
2）第三角法
3）正面図の表し方と向き
4）わかりやすい投影図の表し方

After studying this chapter, you will be able to get knowledge as follows:
-1. Projection of the object
-2. Third angle projection
-3. Expression and direction of front view
-4. Expression of the simply views

第2章　1　対象物の投影

- 対象物の情報を最も明瞭に示す投影図を、正面図または主投影図とする。
- 投影図（断面図を含む）の数は必要かつ十分であるものとする。
- 可能な限り、隠れ線の使用は避ける。
- 不必要な細部の繰り返しを避ける。

投影図の名称を**図2-1**に表します。

正面図（主投影図）が選ばれると、関連する他の投影図の角度は90°と90°の倍数になります。

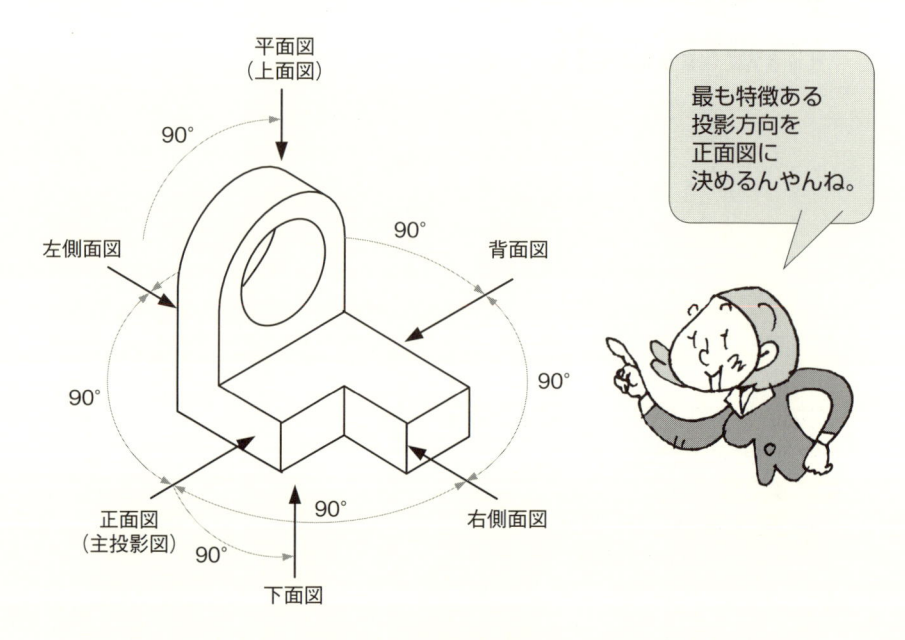

図2-1　投影図の名称

| Chapter2 | 1 | **Projection of the object** |

> - The view showing the most clearly information on the object should be front view or principal view.
> - The number of views (including sectional views) should be necessary and sufficient.
> - The need for hidden outlines as much as possible should be avoided.
> - The unnecessary repetition of details should be avoided.

The name of the views is shown in figure 2-1.

When the front view (principal view) is selected, the angle that related other views make is 90° and multiple of 90°.

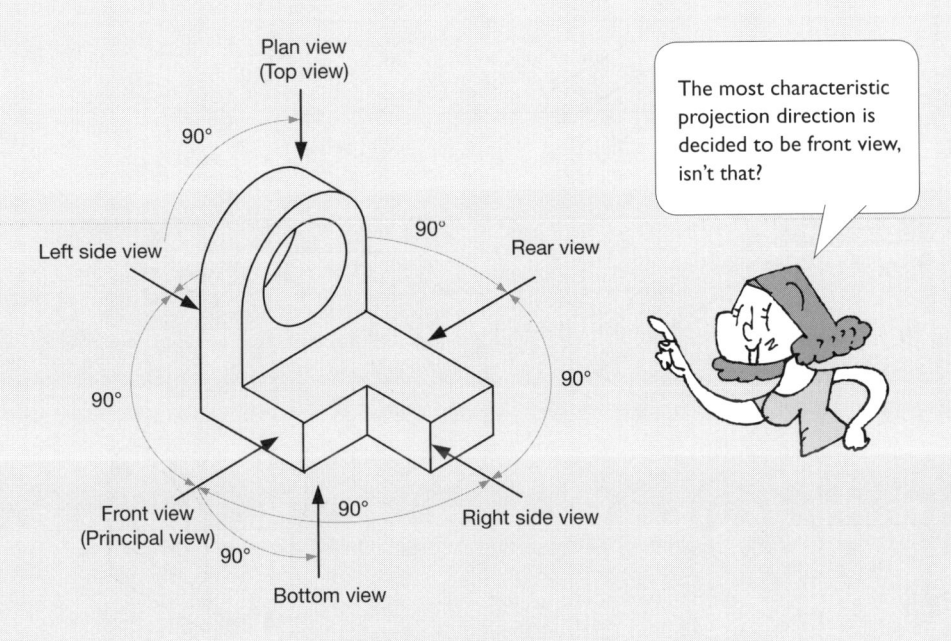

The most characteristic projection direction is decided to be front view, isn't that?

Figure 2-1 Name of views

Chapter2 What is the projection method!

対象物の端点や輪郭から垂直線または投影線を投影面に描きます。

正面図は前面に置いた面に投影されます。同様に、平面図は上に置いた水平面に投影され、右側面図は右に置いた側立面に投影されます（**図2-2**）。

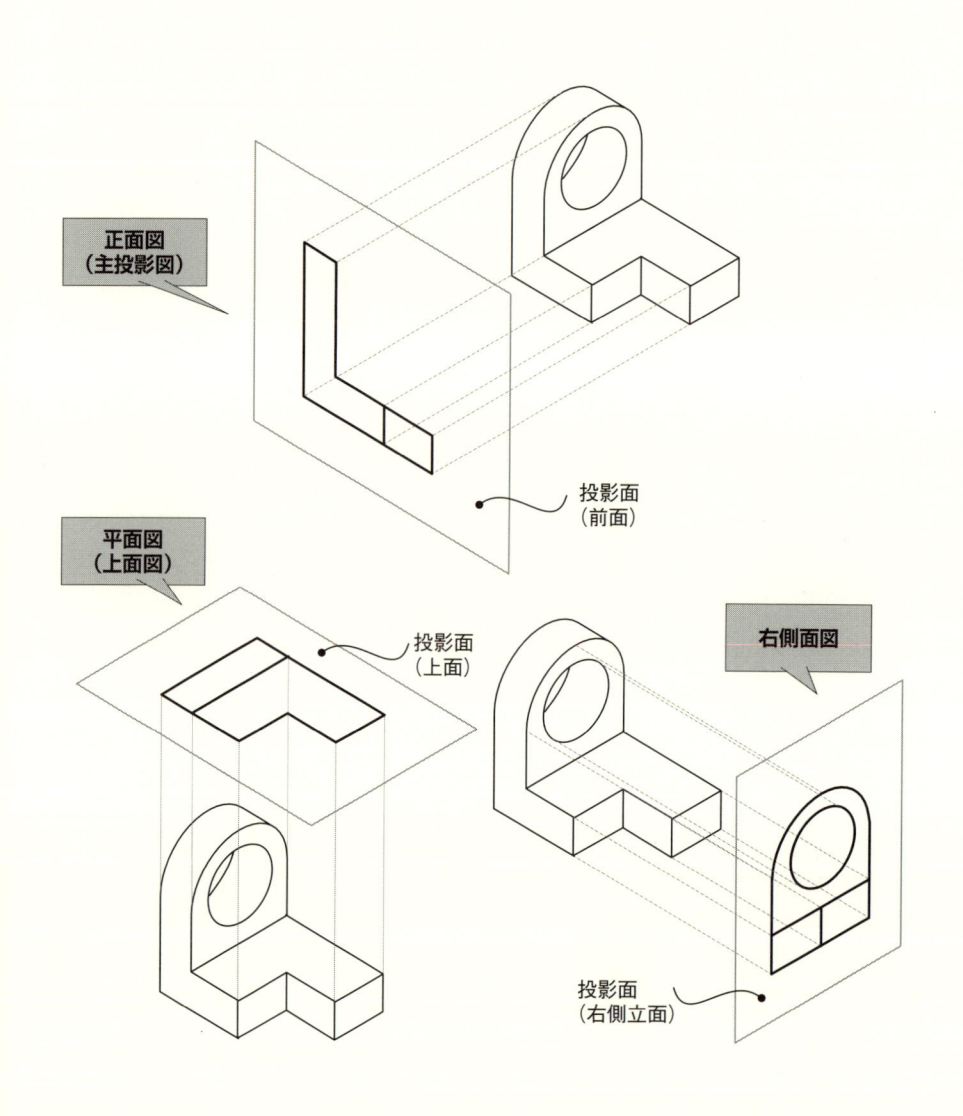

図2-2　投影図の描き方

Perpendicular lines or projection lines are drawn from all points on the edges or profiles of the object to the plane of projection.

The front view is projected onto the frontal plane. Similarly, the plan view is projected onto the upper horizontal plane. The right side view is projected onto the right profile plane. Refer to Figure 2-2.

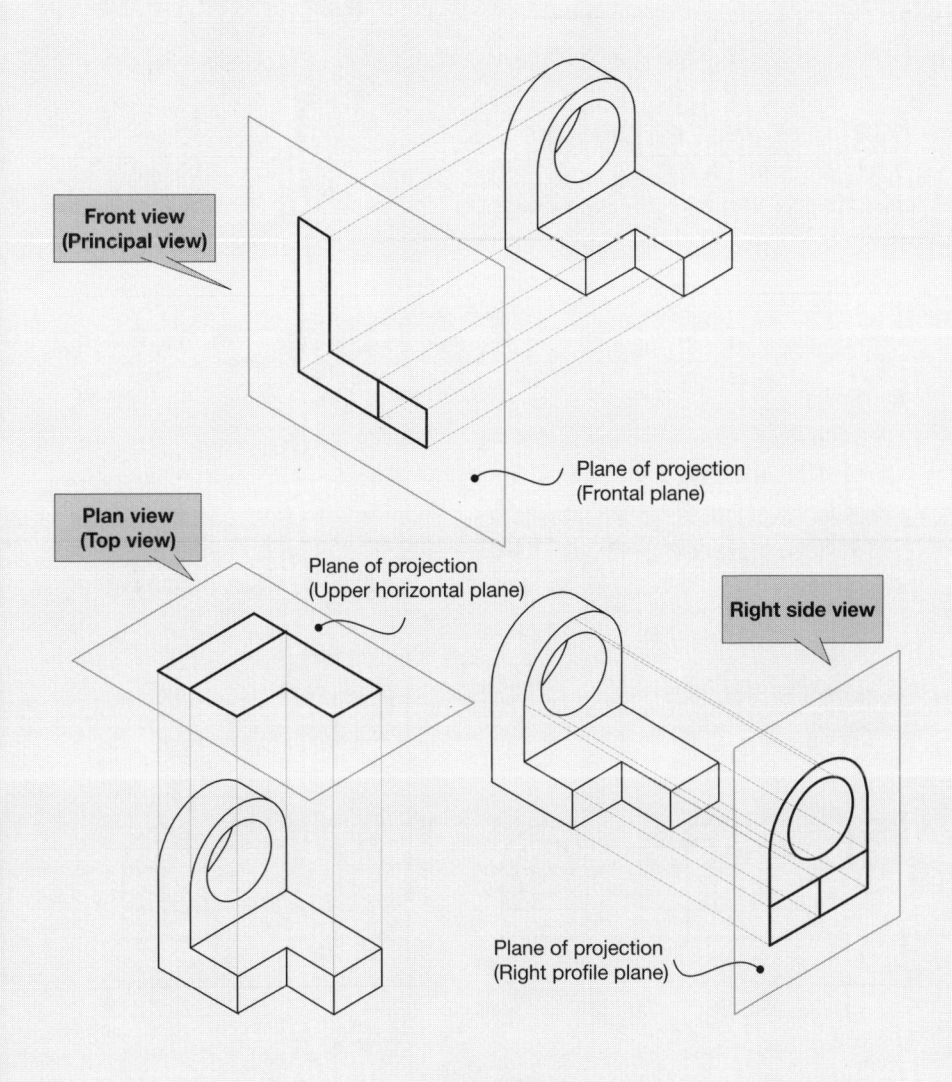

Figure 2-2 Drawing view

Chapter2 What is the projection method!

第2章 2 第三角法

　2次元の紙の上で3次元の対象物の投影図を示すために、同じ平面上に配置するよう、投影図を展開する必要があります。

　そこで、第三角法の考え方を紹介しましょう。

　投影物の主要な面に平行な面を当てると、ガラスの箱が形成されます。

　その箱を展開することで、対象物の6つの投影図ができあがります（**図2-3**）。

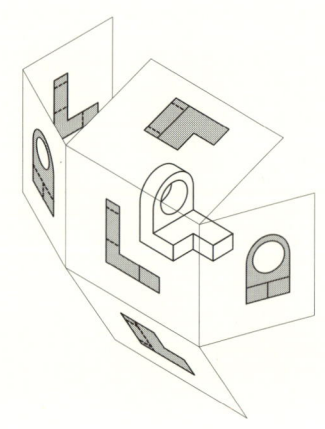

図2-3　ガラスの箱と第三角法

　第三角法は、正面図を基準とし、他の投影図は次のように配置します（**図2-4**）。
・平面図は、正面図の上側に配置する。
・下面図は、正面図の下側に配置する。
・左側面図は、正面図の左側に配置する。
・右側面図は、正面図の右側に配置する。
・背面図は、都合によって正面図の左側または右側に配置することができる。

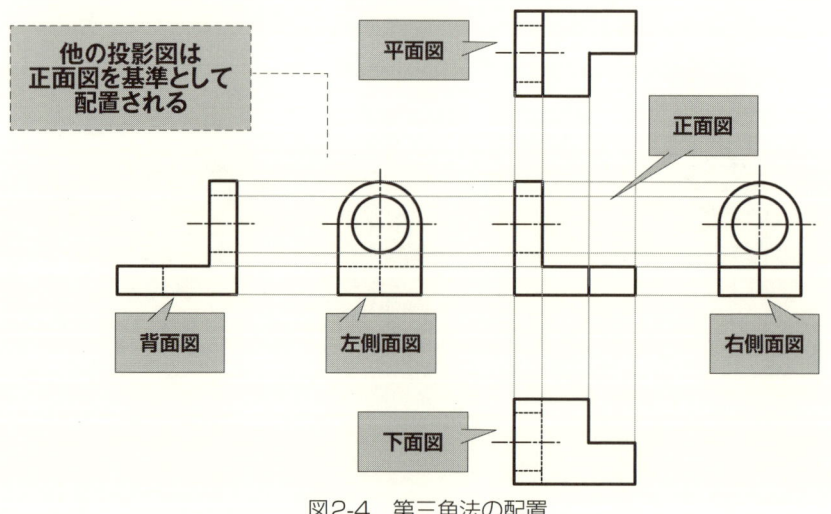

図2-4　第三角法の配置

Chapter2 2 Third angle projection

To express views of a 3D object on a 2D piece of paper, it is necessary to unfold the views such that they arrange in the same plane.

So, let me introduce the basic approach of the third angle projection.

Placing parallel planes to the principal planes of the object forms a glass box.

By unfolding grass box, six views of the object are produced. Refer to Figure 2-3.

Figure 2-3 Glass box and third angle projection

In third angle projection, other views are arranged on the basis of front view as follows: Refer to Figure 2-4.
- Plan view is put at the top of front view.
- Bottom view is put at the bottom of front view.
- Left side view is put at the left side of front view.
- Right side view is put at the right side of front view.
- Rear view can be put on the left or right side as required.

Other views are put on the basis of the front view.

Plan view

Front view

Rear view   Left side view   Right side view

Bottom view

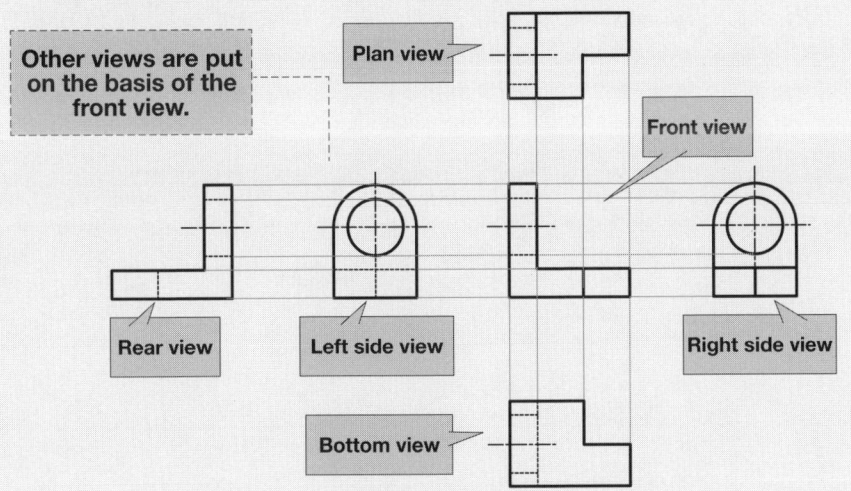

Figure 2-4 Arrangement of third angle projection view

Chapter2 What is the projection method!

選択した正面図が変わると、投影図の配置も変化します（**図2-5**）。
下図の配置は、図2-4とは異なりますが、第三角法として正しい図面です。

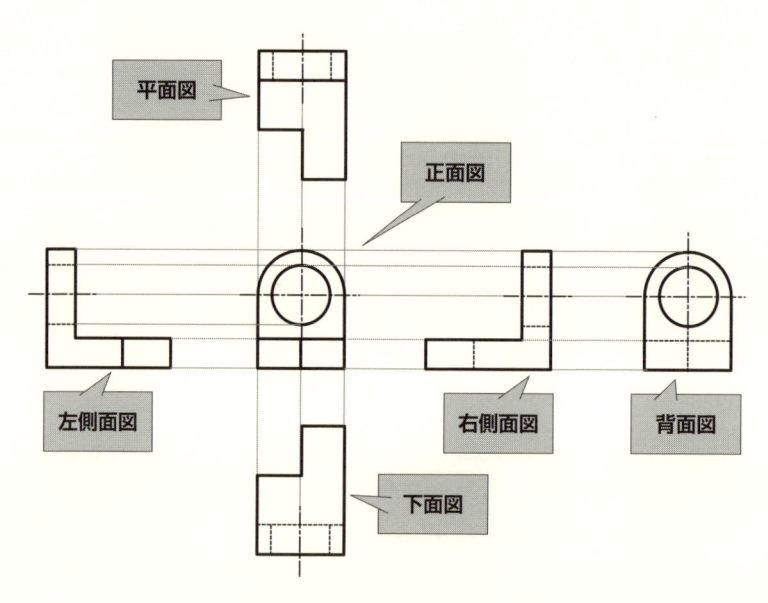

図2-5　正面図を変えたときの配置

When front view was changed, the arrangement of views also changes. Refer to Figure 2-5.

Although the following arrangement differs in Figure 2-4, this is correct drawing as third angle projection.

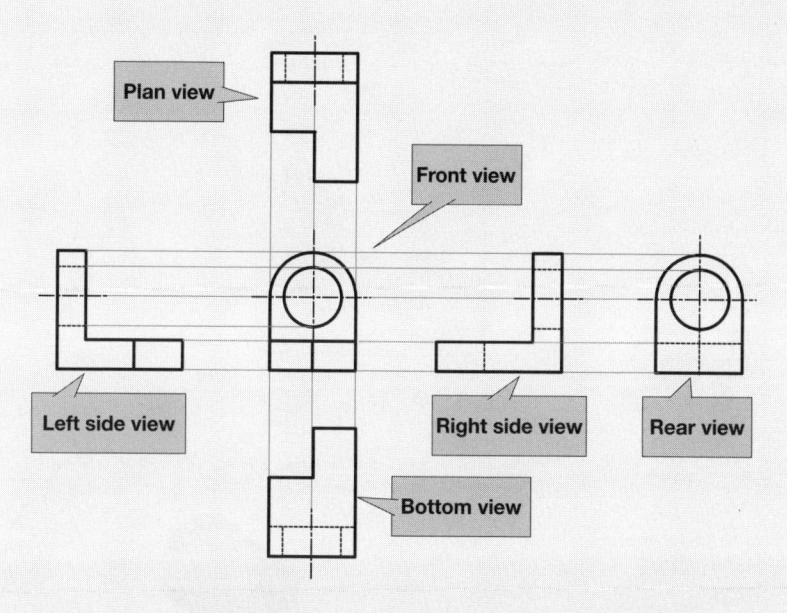

Figure 2-5 Arrangement of when the front view was changed

投影法の記号は表題欄の中またはその付近に示します（**図2-6**）。

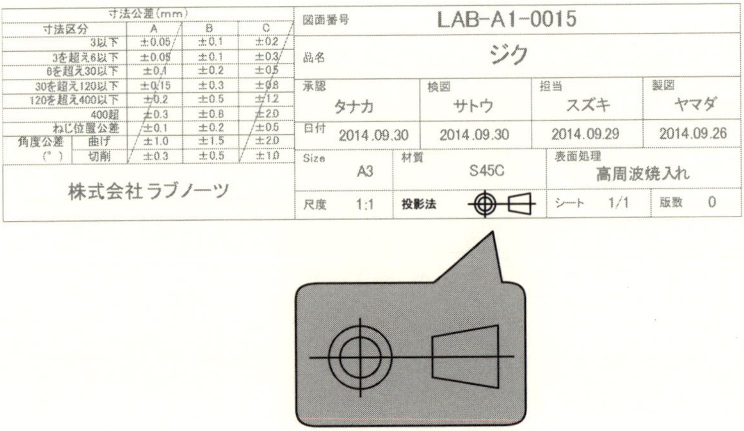

図2-6　第三角法の記号

> この記号は、円すいの正面と左側面の配置を表してるんがわかるやろ！

Symbol of projection method is indicated in the title block or nearby. Refer to figure 2-6.

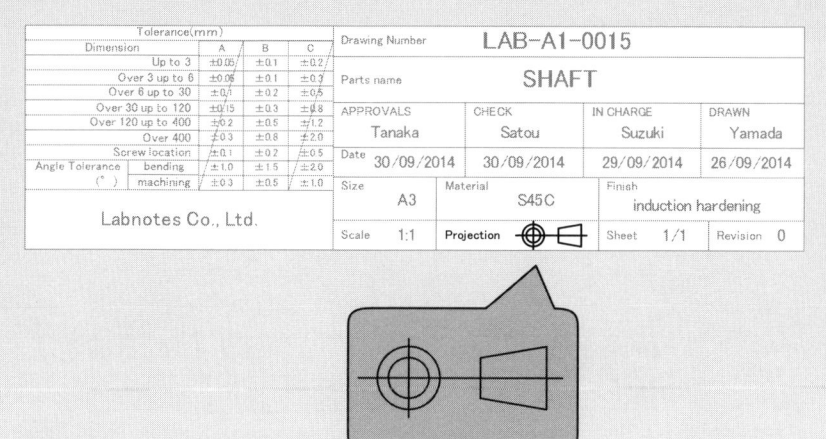

Tolerance(mm)				Drawing Number	LAB-A1-0015				
Dimension	A	B	C						
Up to 3	±0.05	±0.1	±0.2	Parts name	SHAFT				
Over 3 up to 6	±0.06	±0.1	±0.3						
Over 6 up to 30	±0.1	±0.2	±0.5						
Over 30 up to 120	±0.15	±0.3	±0.8	APPROVALS	CHECK		IN CHARGE		DRAWN
Over 120 up to 400	±0.2	±0.5	±1.2	Tanaka	Satou		Suzuki		Yamada
Over 400	±0.3	±0.8	±2.0						
Screw location	±0.1	±0.2	±0.5	Date 30/09/2014	30/09/2014		29/09/2014		26/09/2014
Angle Tolerance bending	±1.0	±1.5	±2.0						
(°) machining	±0.3	±0.5	±1.0	Size A3	Material S45C		Finish induction hardening		
Labnotes Co., Ltd.				Scale 1:1	Projection ⊕⊟		Sheet 1/1		Revision 0

Figure 2-6 Symbol of third angle projection

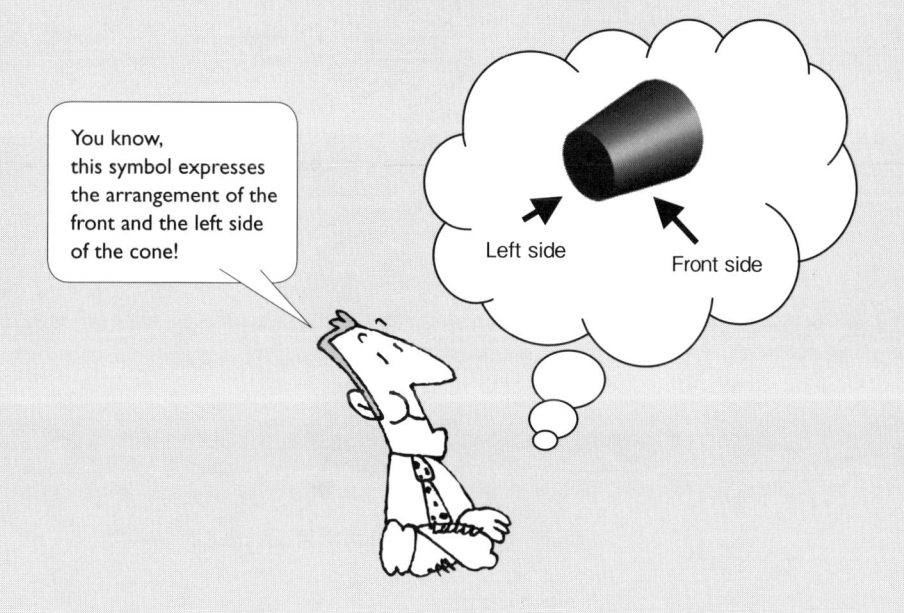

You know, this symbol expresses the arrangement of the front and the left side of the cone!

Left side

Front side

Chapter2 What is the projection method!

φ(@°▽°@) メモメモ

第一角法（参考情報）

第一角法は、正面図を基準とし、他の投影図は次のように配置します。
- 平面図は、正面図の下側に配置する。
- 下面図は、正面図の上側に配置する。
- 左側面図は、正面図の右側に配置する。
- 右側面図は、正面図の左側に配置する。
- 背面図は、都合によって正面図の左側または右側に配置することができる。

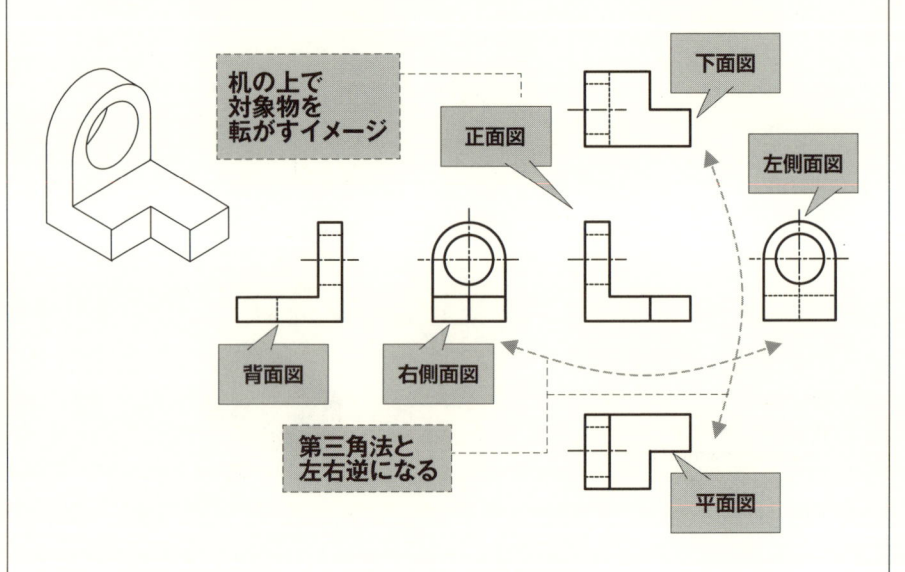

机の上で対象物を転がすイメージ

下面図 / 正面図 / 左側面図 / 背面図 / 右側面図 / 平面図

第三角法と左右逆になる

第一角法の記号を表題欄の中またはその付近に示します。

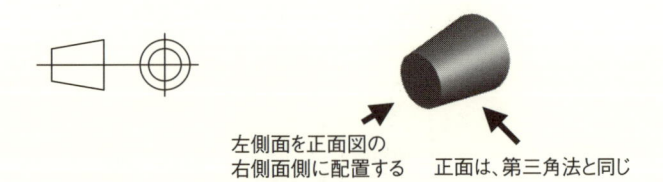

左側面を正面図の右側面側に配置する　　正面は、第三角法と同じ

次のように、国によって投影法が異なります。
- 第三角法…日本、アメリカ、カナダ、韓国など
- 第一角法…ドイツ、イギリス、フランス、中国など

Note ;-)

First angle projection (Reference information)

In the first angle projection, other views are arranged on the basis of front view as follows:
- Plan view is put at the bottom of front view.
- Bottom view is put at the top of front view.
- Left side view is put at the right side of front view.
- Right side view is put at the left side of front view.
- Rear view can be put on the left or right side as required.

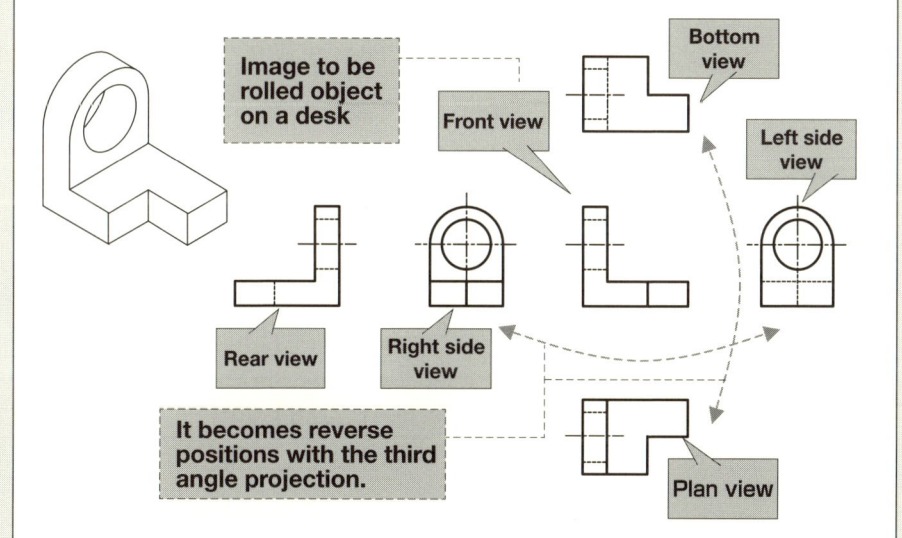

Image to be rolled object on a desk

Bottom view

Front view

Left side view

Rear view

Right side view

It becomes reverse positions with the third angle projection.

Plan view

Symbol of first angle projection is indicated in the title block or nearby.

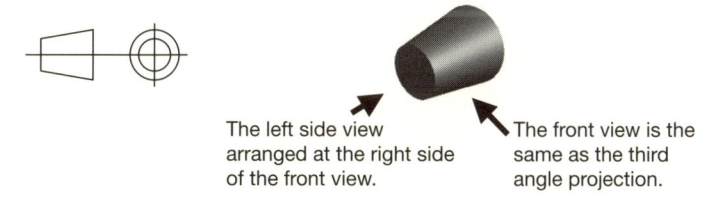

The left side view arranged at the right side of the front view.

The front view is the same as the third angle projection.

Projection method differs among countries as follows:
- Third angle projection --- Japan, United States, Canada, Korea, etc.
- First angle projection --- Germany, U.K., France, China, etc.

| 第2章 | 3 | 正面図の表し方と向き |

　正面図は、対象物の形状や機能を最も明瞭に表す面を描きます。なお、図面の目的に応じて、対象物を図示する向きは、次のいずれかによります。
・組立図など、主として機能を表す図面では、対象物を使用する向きで描く。
・部品図では、投影図は加工の向きに合わせて描く（**図2-7**、**図2-8**）。
・特別な理由がない場合には、対象物を横長に置いた向きで描く。

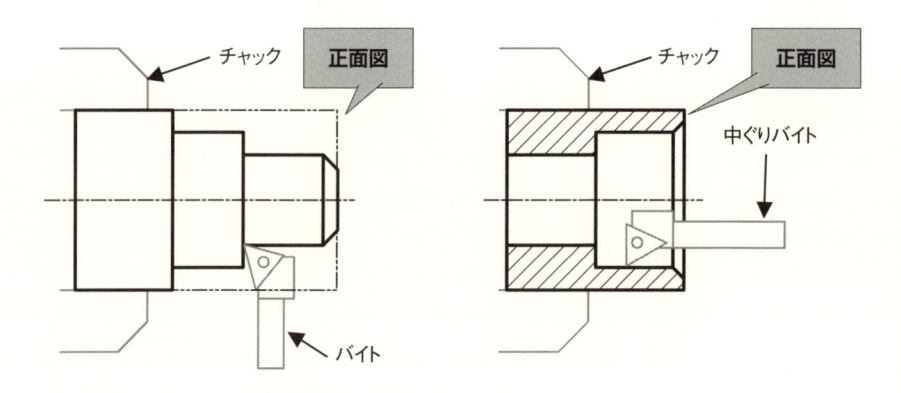

図2-7　旋盤加工の向き

φ(@°▽°@)　メモメモ

旋盤

　旋盤は、主軸に取り付けた材料を回転させ、バイトによって切削や穴あけ加工を行う工作機械です。旋盤は材料の左側を固定して、右側から切削します。
　したがって、投影図の向きは加工に合わせてあげることがマナーです。

Chapter 2   3   Expression and direction of front view

For front view, the surface indicating most clearly the shape and function of the object is drawn. Further, depending on the purpose of drawing, the object is drawn in the following directions:
- In the drawing expressing mainly the function such as assembly drawings, the object is drawn in the use direction.
- In work-piece drawings, views are drawn according to direction of processing. Refer to **Figures 2-7 and 2-8**.
- When there are no special reasons, the object is drawn in the direction of being placed horizontally.

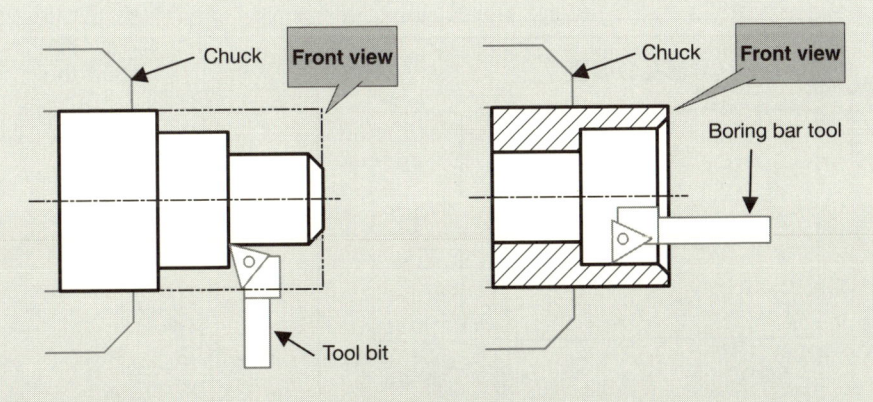

Figure 2-7 Direction of turning

Note ;-)

Lathe

The lathe is a machine tool which rotates the work-piece to cutting or drilling with tools. The lathe chucks the left side of materials and cuts it from the right side.

So, it is etiquette to set direction of the view with processing.

短い（小さい）部品の場合

エンドミルの軌跡

平面図

ドリル

エンドミル

正面図

長い（大きい）部品の場合

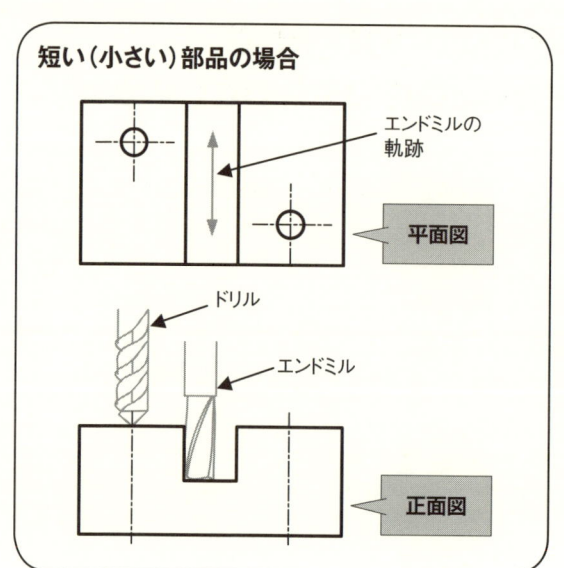

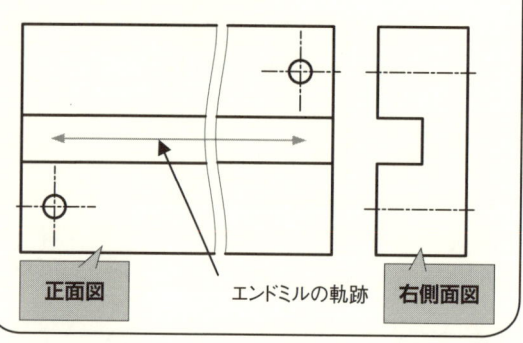

正面図　　エンドミルの軌跡　　右側面図

> フライス加工は上から削るから、正面図は上向きにすればいいんやね！

> そやけど、長い部品はフライス盤に横向きにセットするから、正面図は横向きにしたほうがええんや！

図2-8　フライス加工の向き

φ(@°▽°@)　メモメモ

立て形フライス盤

　立て形フライス盤は、部材から材料を除去するために垂直スピンドルを中心に回転するカッターを使います。
　平面加工や溝加工、穴あけ加工ができます。

When short/small work-piece

Trace of the endmill

Plan view

Drill

Endmill

Front view

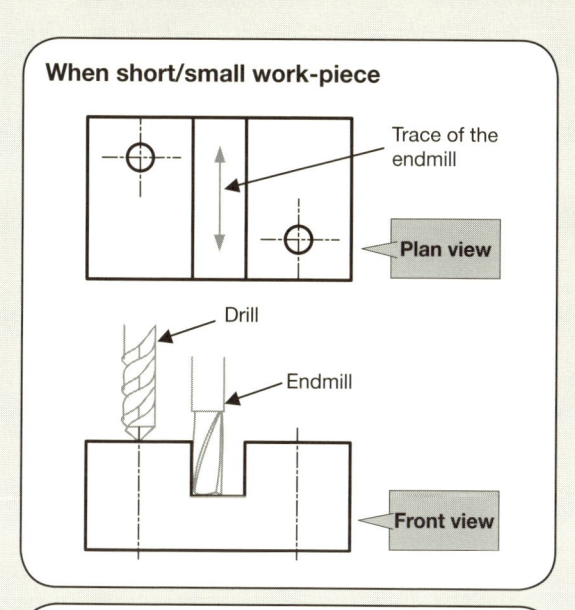

> Milling is cut from top. So, front view is turned upward, right?

When long/large work-piece

Front view

Trace of the endmill

Right side view

> But, long part is set sideways on the milling machine.
> So, front view is turned sideways.

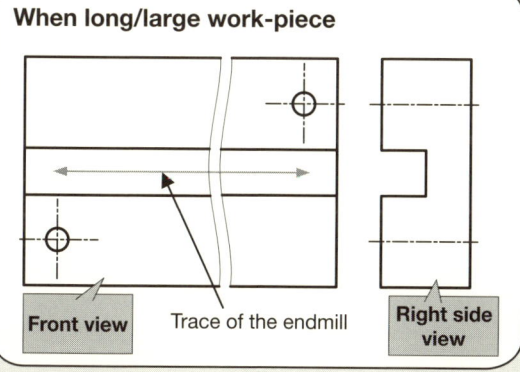

Figure 2-8 Direction of milling

Note ;-)

Vertical milling machine

The vertical milling machine uses cutter which rotates on the vertical spindle to remove material from a work-piece.

It can do flattening, grooving, and drilling.

Chapter2 What is the projection method!

一部に特定の形をもつ投影図は、なるべくその部分が図の上側に表れるように描きます。例えば、キー溝をもつ穴、表面に穴または溝をもつ管、切欠きをもつリングなど。

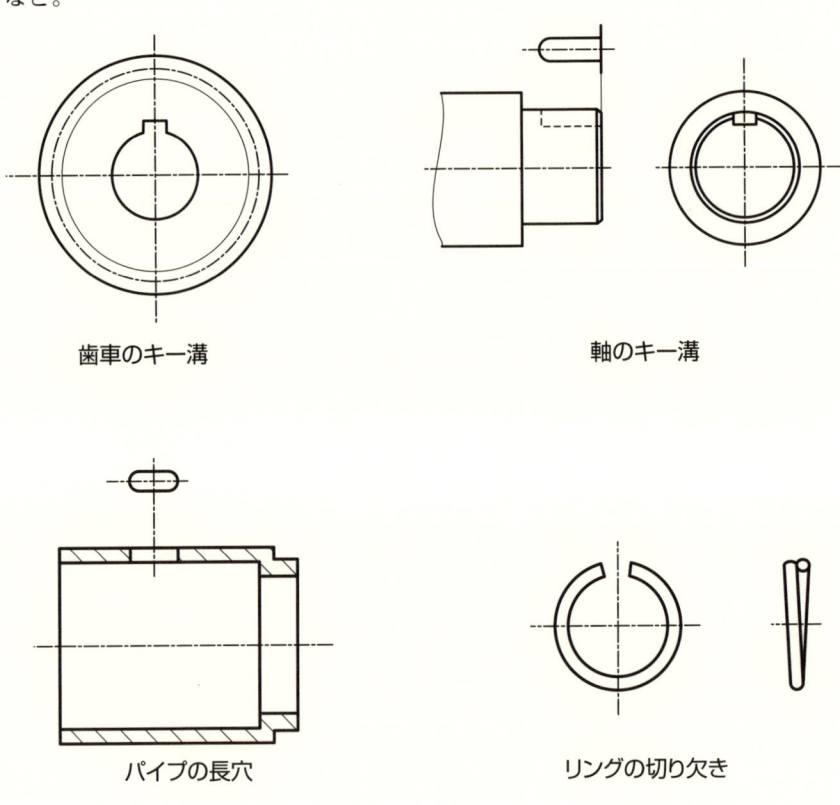

歯車のキー溝

軸のキー溝

パイプの長穴

リングの切り欠き

図2-9　上側に表れるように描く特定の形状

Views having a specific shaped part are drawn to appear the part is positioned on the upper side of view. For example, in case where the hole having keyway, tube having holes or groove in the surface, ring having cut splits etc.

Keyway of the gear

Keyway of the shaft

Slot of the pipe

Cut splits of ring

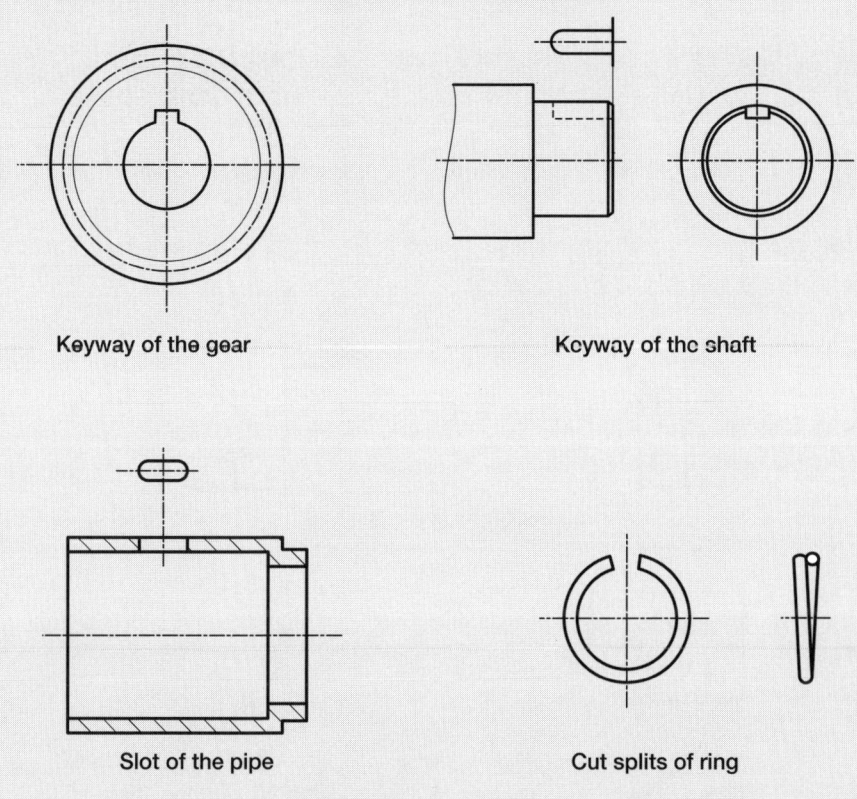

Figure 2-9 Specific shaped part to appear on the upper position

第2章　4　わかりやすい投影図の表し方

完全に対象物をあらわすために6面全ての投影図が必要とは限りません。それでは、図2-4の図形を完全に表すためにどのくらいの数の投影図が必要でしょうか？

投影図の数が少なすぎると、情報が少なくて形状を把握することが不可能となるのです（**図2-10**）。

投影図の数は、必要最低限に限られ、曖昧さなしで完全に対象物を詳細に描写するのに十分であるべきです。

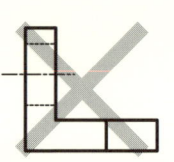

a) 情報が少なすぎる投影図

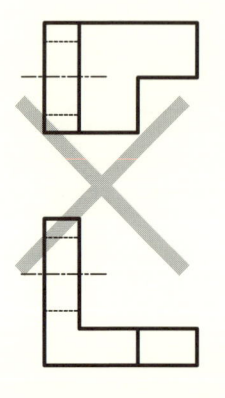

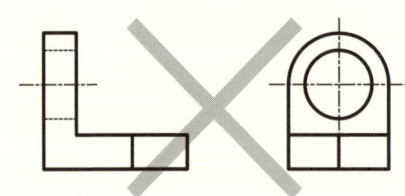

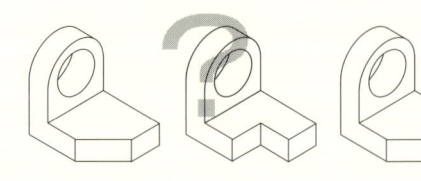

b) 形状を確定できない投影図

図2-10　不適切な投影図

| Chapter2 | 4 | **Expression of the simply views** |

To express the object completely, all the views of the 6 faces are not necessity. Well, how many views are necessary to express the object completely of Figure 2-4?

When there are too few views, it becomes impossible to understand the shape for little information. Refer to Figure 2-10.

The number of views shall be limited to the minimum necessary but shall be sufficient to completely express the object without unclearly.

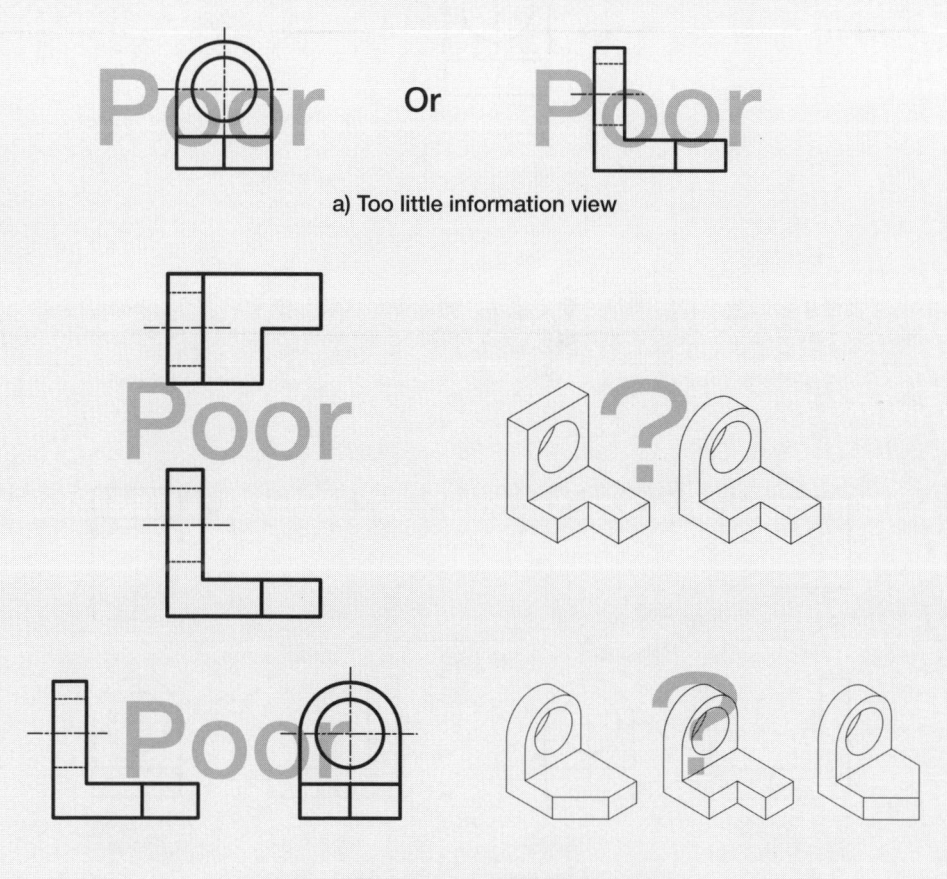

a) Too little information view

b) View which cannot determined a shape

Figure 2-10 Poor views

Chapter2 What is the projection method!

しかし、適切に投影図を組み合わせることで、この形状は**図2-11**のように2つの投影図で表すことができます。

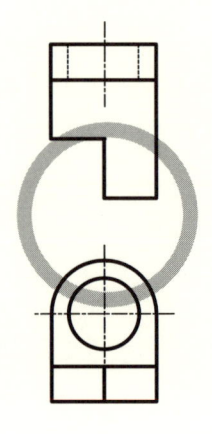

図2-11 最小限の組み合わせによる投影図

必要最小限の数の投影図が必要ですが、読み手が理解しやすいように投影図を増やしてあげるとよいでしょう（**図2-12**）。

または

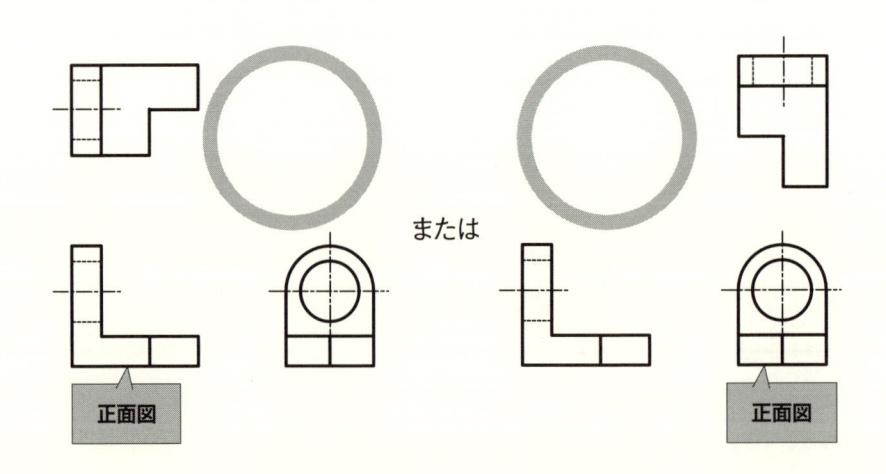

正面図　　　　　　　　　　　　　　　　正面図

図2-12 わかりやすい組み合わせの投影図

However, the combination of appropriate views can be expressed by two views as shown in the **Figure 2-11**.

Figure 2-11 Minimum combination views

The number of views of minimum necessary is needed. But, you may increase views to be easy to understand for readers. Refer to **Figure 2-12**.

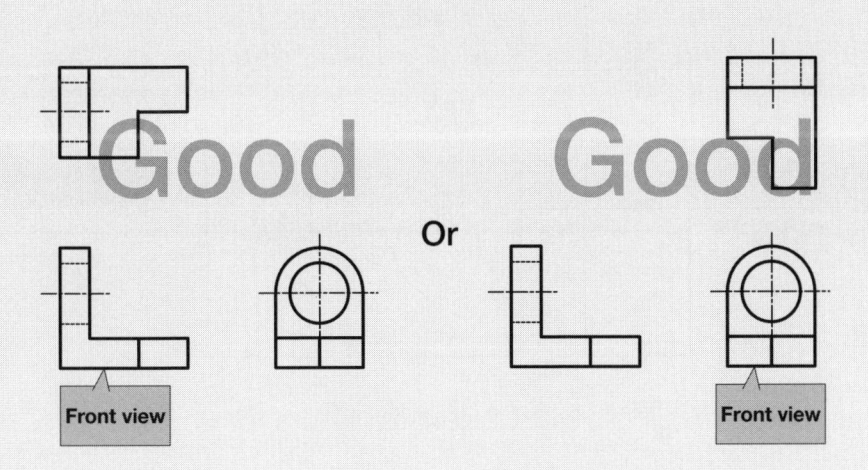

Front view

Or

Front view

Figure 2-12 Simply combinatorial views

また、互いに関連する投影図の配置は、なるべくかくれ線を用いずに表します。かくれ線の多い投影図を選択すると理解が難しくなります（**図2-13**）。

これは、実線の方が形状を理解しやすいためです。

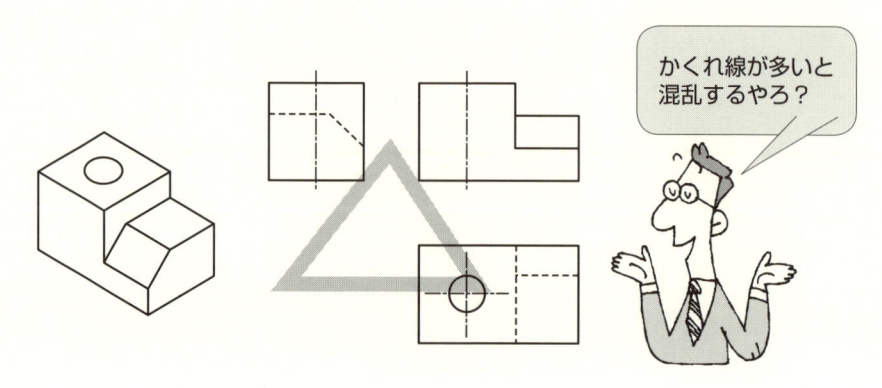

かくれ線が多いと混乱するやろ？

図2-13　かくれ線が多い投影図の選択

そのため、かくれ線は、理解を妨げないと判断した場合には、省略すべきでしょう（**図2-14**）。

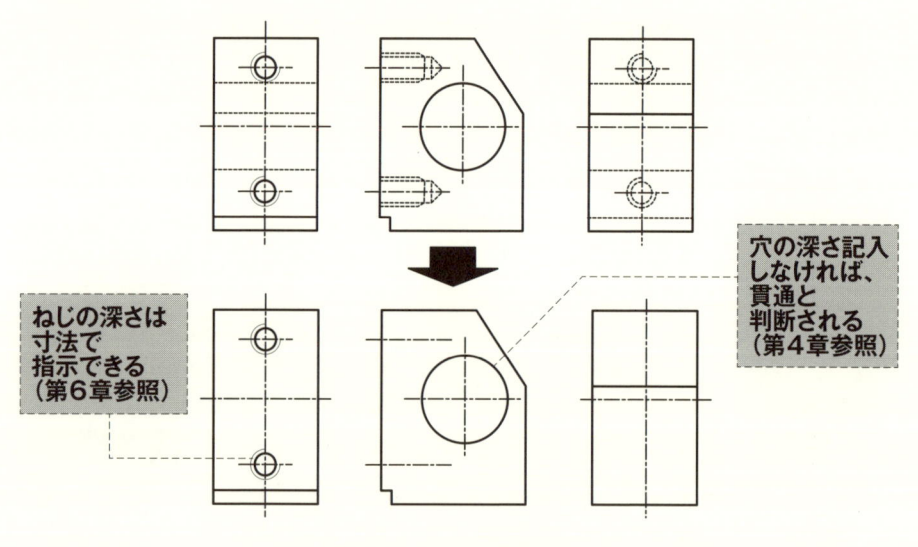

ねじの深さは寸法で指示できる（第6章参照）

穴の深さ記入しなければ、貫通と判断される（第4章参照）

図2-14　かくれ線の省略

Also, the arrangement of related views is expressed without using hidden outline as far as possible. When views with many hidden lines are chosen, it becomes difficult to understand the shape. Refer to Figure 2-13.

This is because visible outline is easier to understand a shape.

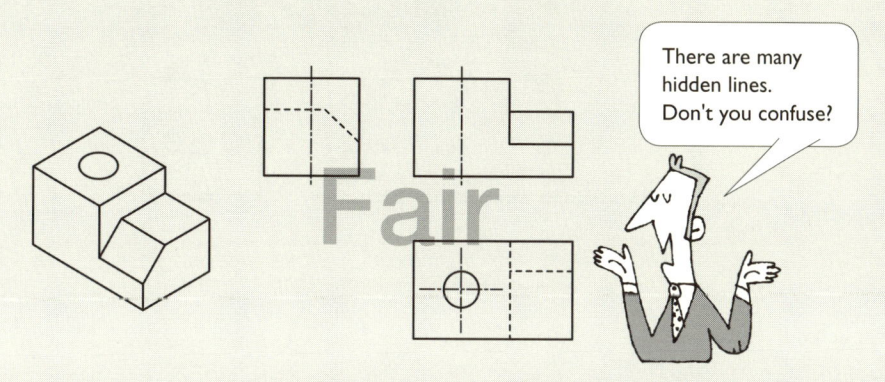

There are many hidden lines. Don't you confuse?

Figure 2-13 Selection of views with many hidden lines

That's why the hidden outlines should be omitted if the omission does not interfere with understanding. Refer to Figure 2-14.

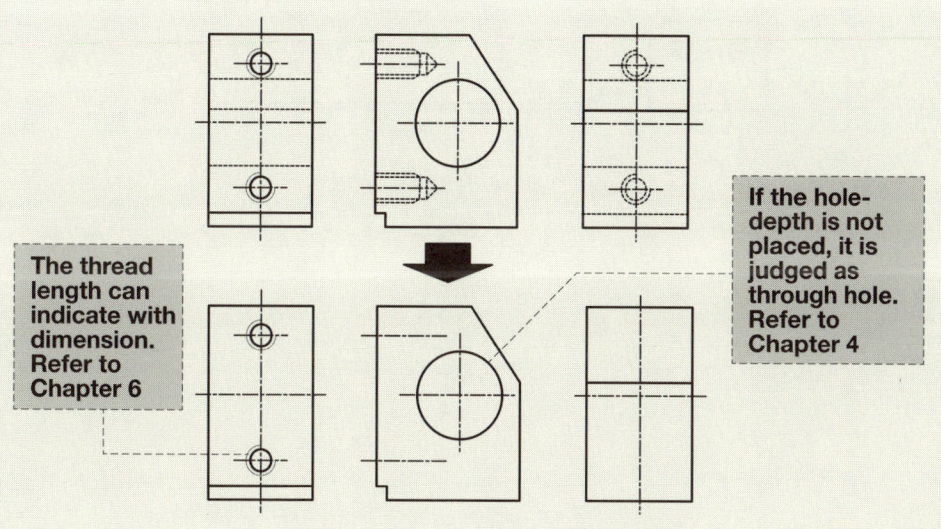

The thread length can indicate with dimension. Refer to Chapter 6

If the hole-depth is not placed, it is judged as through hole. Refer to Chapter 4

Figure 2-14 Omission of hidden outlines

Chapter2 What is the projection method!

ただし、かくれ線を用いないことで比較することが難しくなる場合は、かくれ線のある投影図を選択することができます（図2-15）。

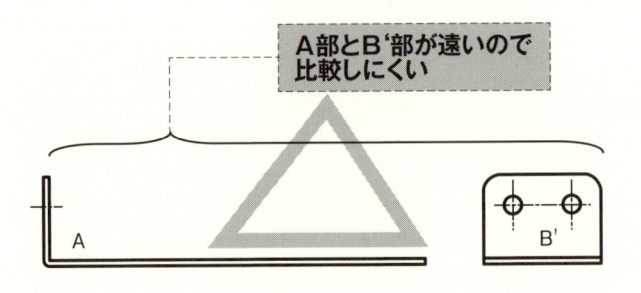

A部とB'部が遠いので
比較しにくい

a)かくれ線のない投影図の選択

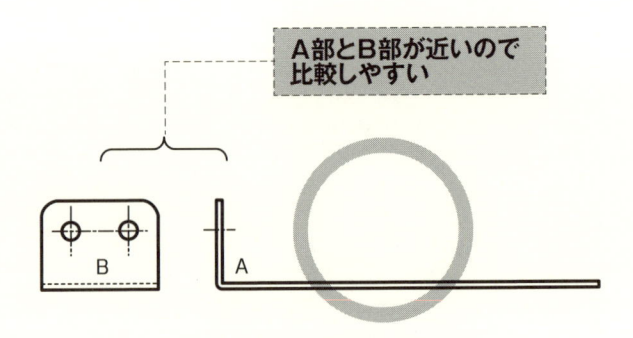

A部とB部が近いので
比較しやすい

b)かくれ線のある投影図の選択

図2-15 比較しやすい投影レイアウト

However, when it is difficult to compare without hidden outlines, you can select the view with hidden outline. Refer to Figure 2-15.

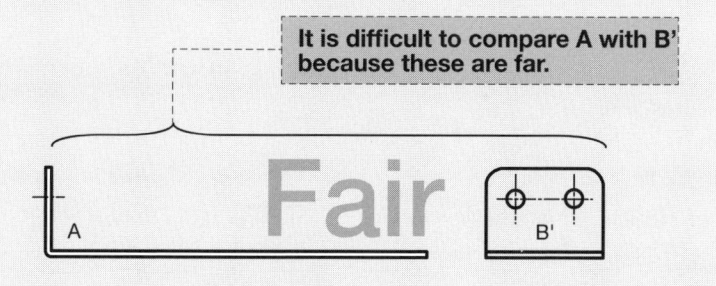

It is difficult to compare A with B' because these are far.

Fair

a) Selection of views without hidden outlines

It is easy to compare A with B because these are near.

Good

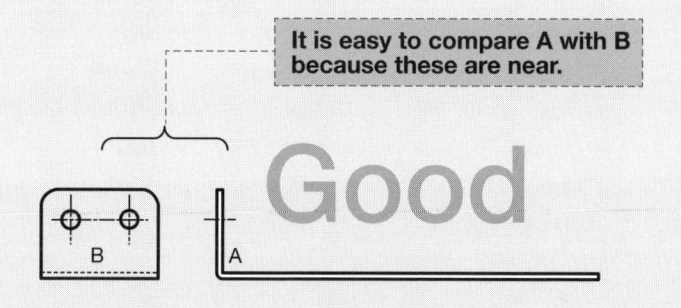

b) Selection of views with hidden outlines

Figure 2-15 Arrangement of views easy to compare

用語集　glossary

本章で使用した単語です。他の用語も使うことができます。
The list of words used for this chapter is shown. You can use another words, too.

対象物	object	投影図	view
主投影図	principal view	正面図	front view
右側面図	right side view	左側面図	left side view
平面図	plan view	下面図	bottom view
背面図	rear view	第三角法	third angle projection
第一角法	first angle projection	記号	symbol
旋盤	lathe	バイト（刃）	tool bit
チャック	chuck	フライス盤	milling machine
エンドミル	endmill	ドリル	drill
歯車	gear	キー溝	keyway
長穴	slot	溝	groove
管	pipe	切り欠き	split
リング	ring	省略	omit
ねじ深さ	thread length	穴深さ	hole depth

第3章
投影図のテクニックって、なんやねん!

Chapter3
What is the technique of the projection views!

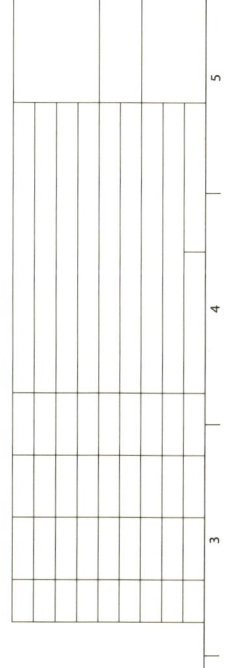

この章を学習すると、次の知識を得ることができます。
1) 対象物の一部のみを表す投影図
2) 断面図
3) 図形の省略
4) 特殊な図示

After studying this chapter, you will be able to get knowledge as follows:
-1. Views expressing only a part of the object
-2. Sectional view
-3. Omission of view
-4. Special expression of view

第3章　1　**対象物の一部のみを表す投影図**

> 形体の実際の描写が必要なのに正投影図上で表すことができない場合、形体は補助投影図で表現できます。
> 他に「部分投影図」、「矢示法」、「局部投影図」が補助投影図として表すことができます。

3-1-1　補助投影図

補助投影図は、傾斜面のように投影される面に展開した図のことです。（**図3-1**）。

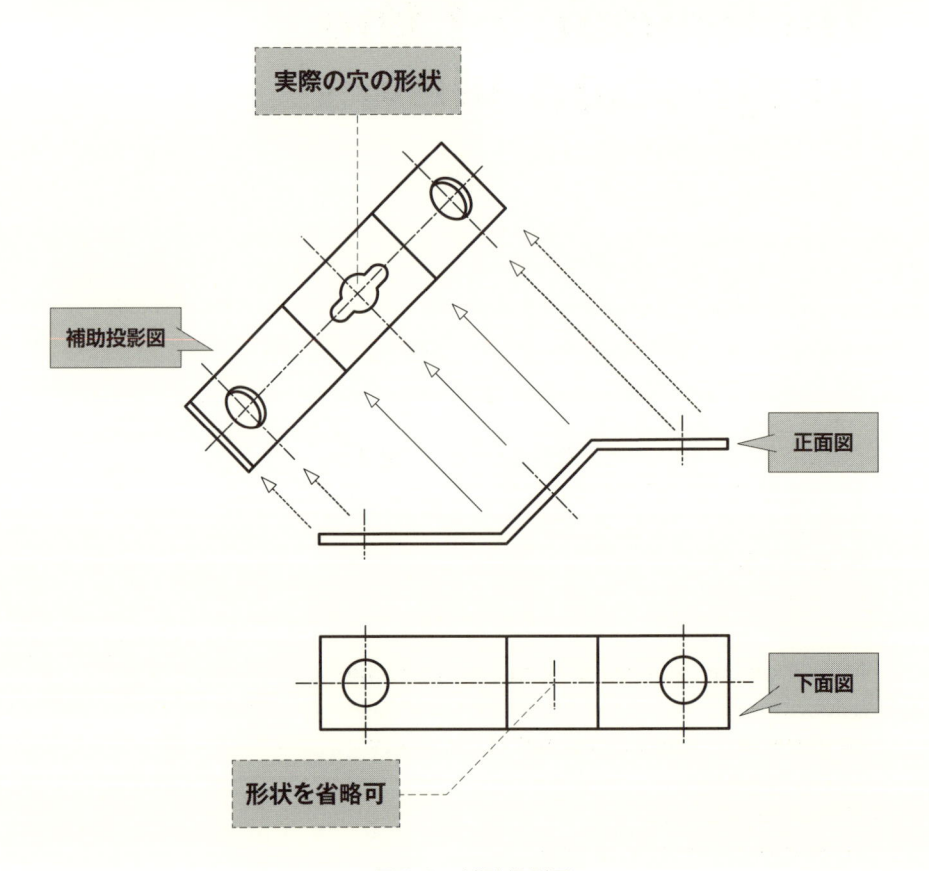

図3-1　補助投影図

| Chapter3 | 1 | **Views expressing only a part of the object** |

When actual representation of features is necessary, but cannot be expessed on the orthographic views, the features can be expressed in auxiliary views.

It can be shown as an auxiliary view by "Partial view" or "Reference arrow layout" or "Local view".

3-1-1 Auxiliary view

Like an inclined plane, auxiliary views are aligned with the views from which they are projected. Refer to Figure 3-1.

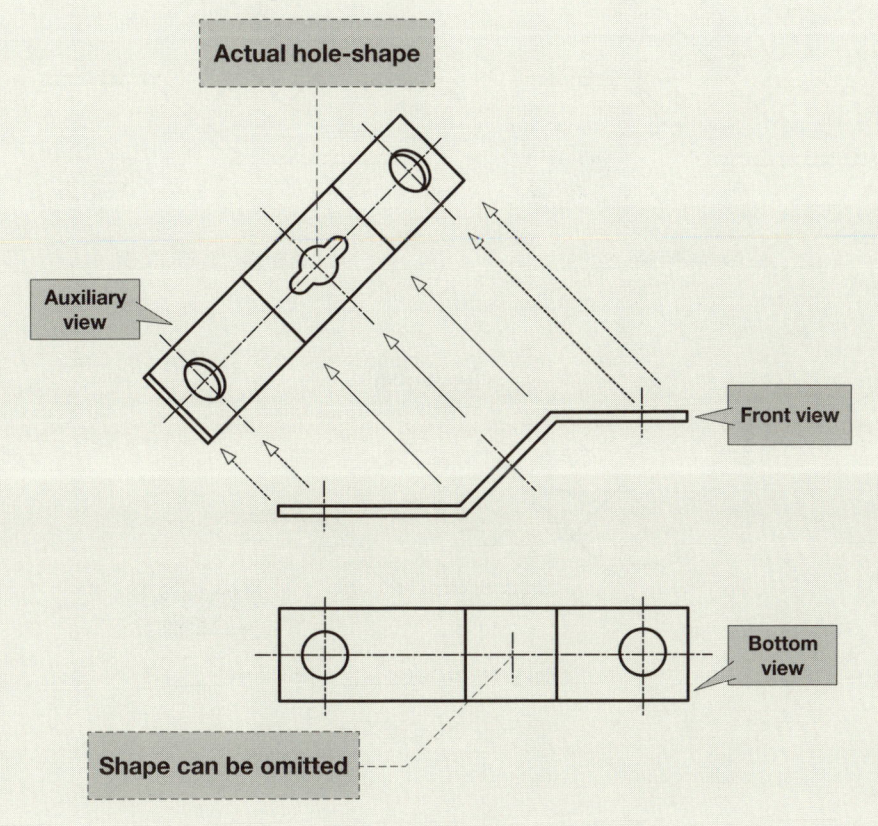

Figure 3-1 Aauxiliary view

Chapter3   What is the technique of the projection views!

a) 部分投影図

　対象物の必要な部分だけを示せば十分なとき、その必要な部分だけを部分投影図として表すことができます。

　この場合、省いた部分との境界は破断線を用いて示しますが、明確な場合は破断線を省略することも可能です。

　この投影を表すために、隣接する投影図との間に中心線や寸法補助線を結びます。

　紙面の関係などで、補助投影図を斜面に対向する位置に配置できない場合、折り曲げた中心線で結び投影関係を示すこともできます（**図3-2**）。

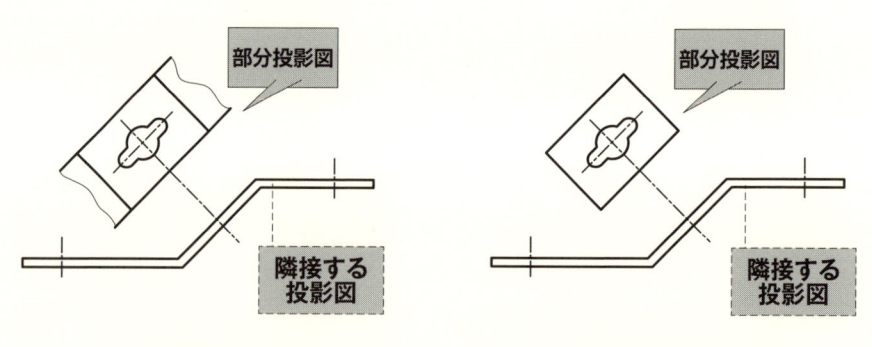

a) 破断線を用いた例　　　　　　　　b) 破断線を省略した例

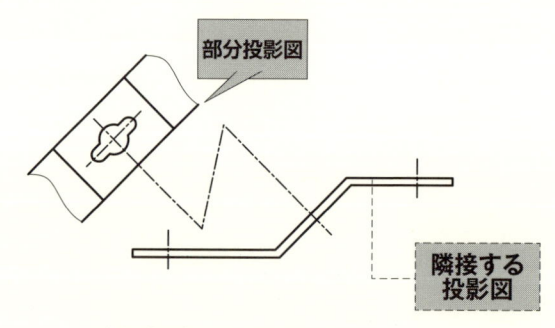

c) 折り曲げた中心線を結んだ例

図3-2　部分投影図

a) Partial view

When the indication of only the required part of the object is sufficient, the required part can be expressed as a partial view.

In this case, the boundary of the omitted part is indicated by break line. However, when it is clear enough, the break lines can be omitted.

Center line or projection line may continue between adjacent views to indicate this projection.

When the auxiliary view cannot be arranged at a position opposing the inclined plane due to the limitation of the sheet space or the like, the projection relation can be indicated by connecting with a folded center line. Refer to Figure 3-2.

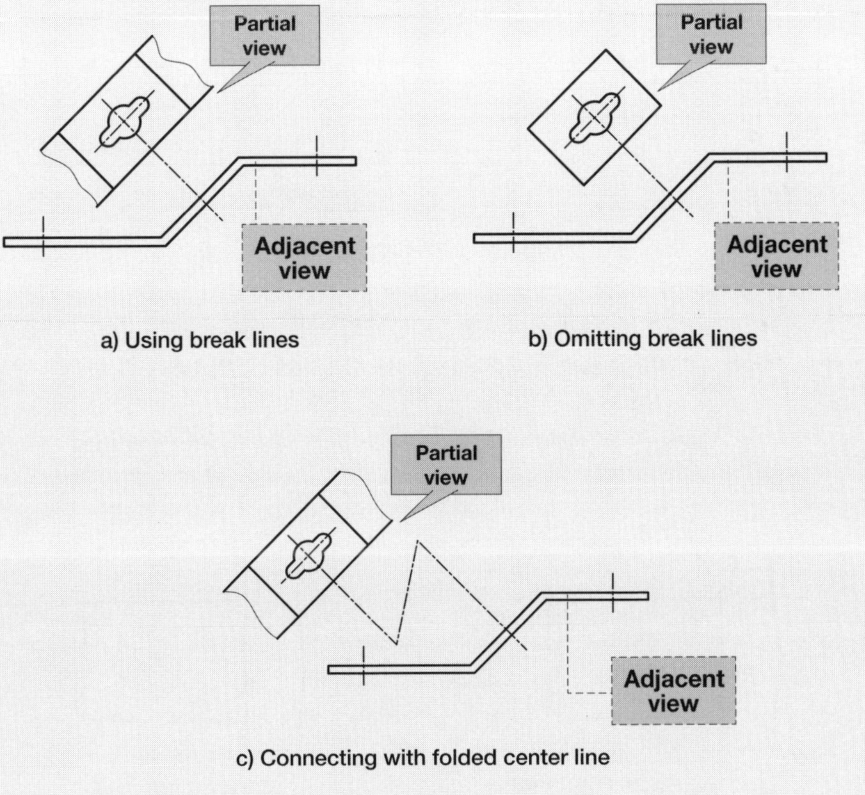

a) Using break lines

b) Omitting break lines

c) Connecting with folded center line

Figure 3-2 Partial view

b) 局部投影図

　穴やみぞなどの形状だけを示せば形状を理解できる場合は、詳細の不必要な繰り返しは避けるべきです（**図3-3**）。

　局部投影図は図3-3 b)のうちのどちらかを使用して、図面に記入することができます。

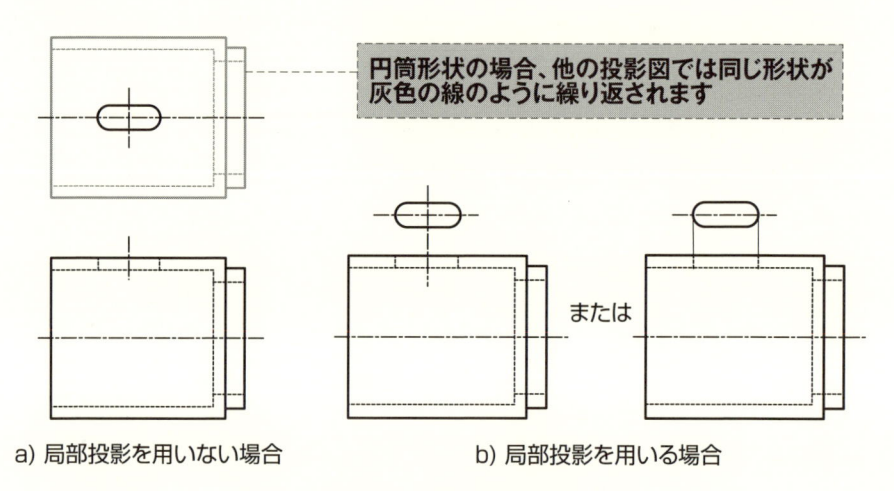

円筒形状の場合、他の投影図では同じ形状が灰色の線のように繰り返されます

a) 局部投影を用いない場合　　　　b) 局部投影を用いる場合

図3-3　局部投影図の例

　キー溝の表記には、局部投影図がよく用いられます（**図3-4**）。
　キーやそれらに対応するキー溝の寸法はJIS B 1301に記載されています。

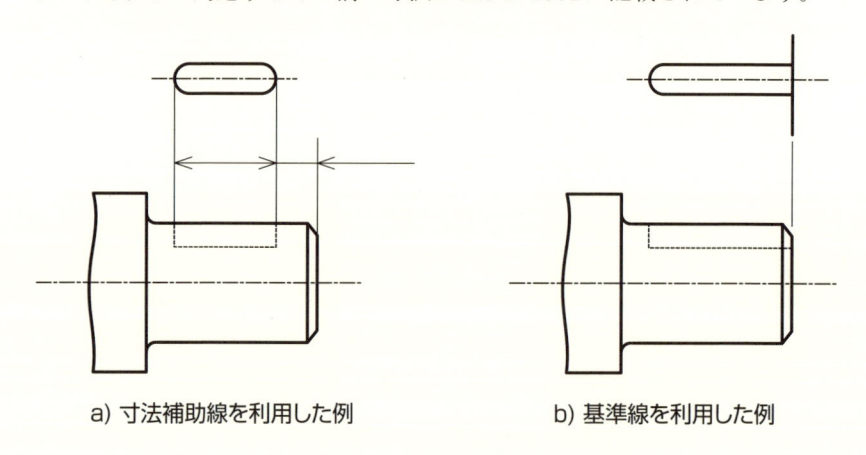

a) 寸法補助線を利用した例　　　　b) 基準線を利用した例

図3-4　キー溝の局部投影図の例

b) Local view

When shape can be understood if only shape of holes or grooves, etc. is shown, the unnecessary repetition of a detail shall be avoided. Refer to Figure 3-3.

Local view can be expressed on the drawing using either in Figure 3-3 b.

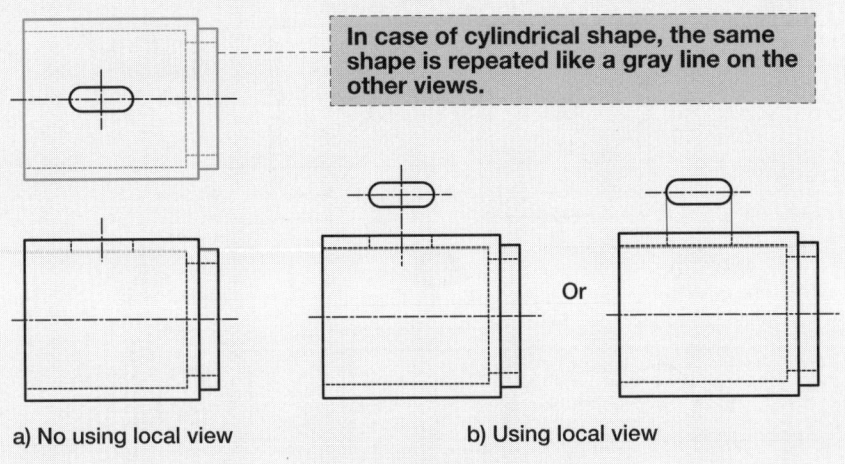

In case of cylindrical shape, the same shape is repeated like a gray line on the other views.

a) No using local view   b) Using local view

Figure 3-3 Example of local view

The local view is often used for the keyways. Refer to Figure 3-4.
The dimensions of key and their corresponding keyways are listed in JIS B 1301.

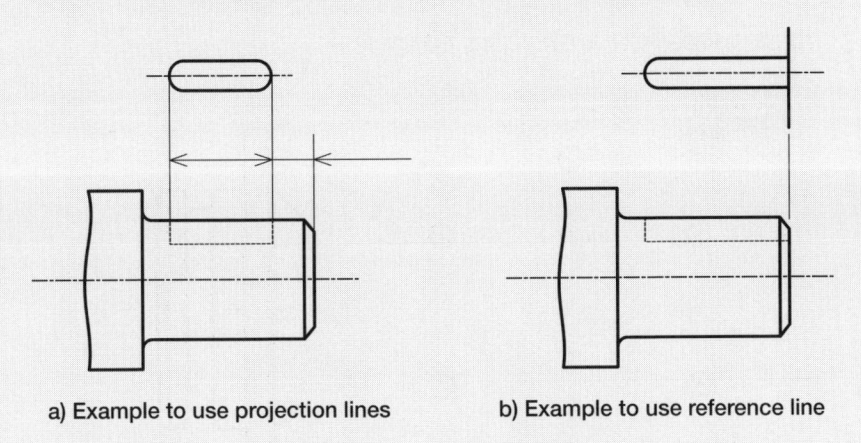

a) Example to use projection lines   b) Example to use reference line

Figure 3-4 Example of keyways local view

φ(@°▽°@)　メモメモ

補助投影図を選択するメリット

　投影図に外形線やかくれ線をすべて描くことで理解しにくくなる場合には、部分投影図に置き換えるべきです。

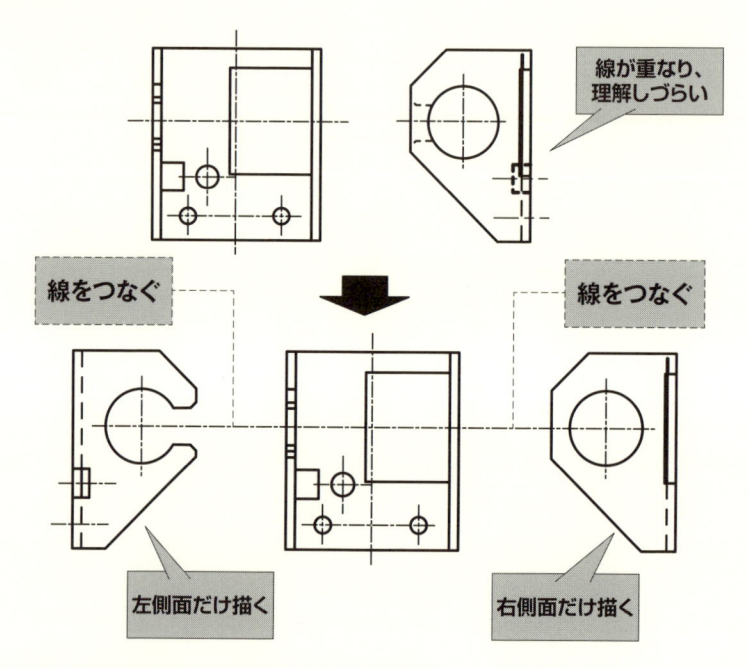

線が重なり、理解しづらい

線をつなぐ

線をつなぐ

左側面だけ描く

右側面だけ描く

　複数の穴も局部投影図として表すことができます。

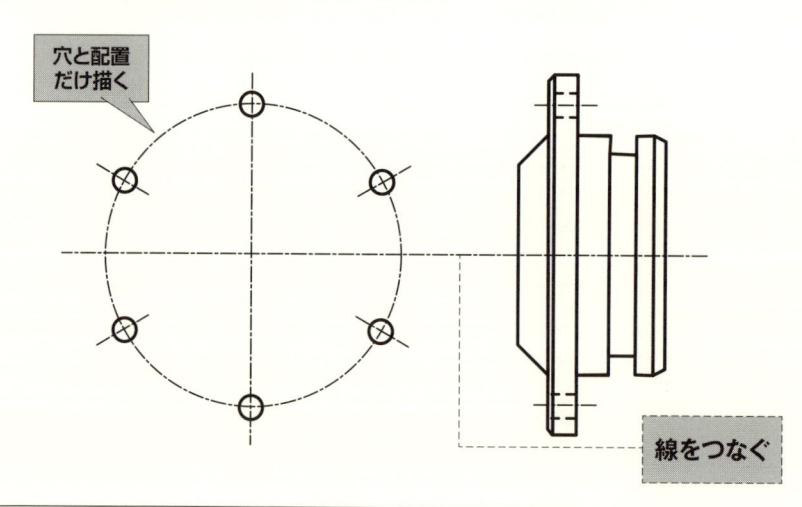

穴と配置だけ描く

線をつなぐ

Note ;-)

An advantage for choosing the auxiliary view

When drawing all visible outlines and hidden outlines in the view makes difficult to understand, partial view should be instead.

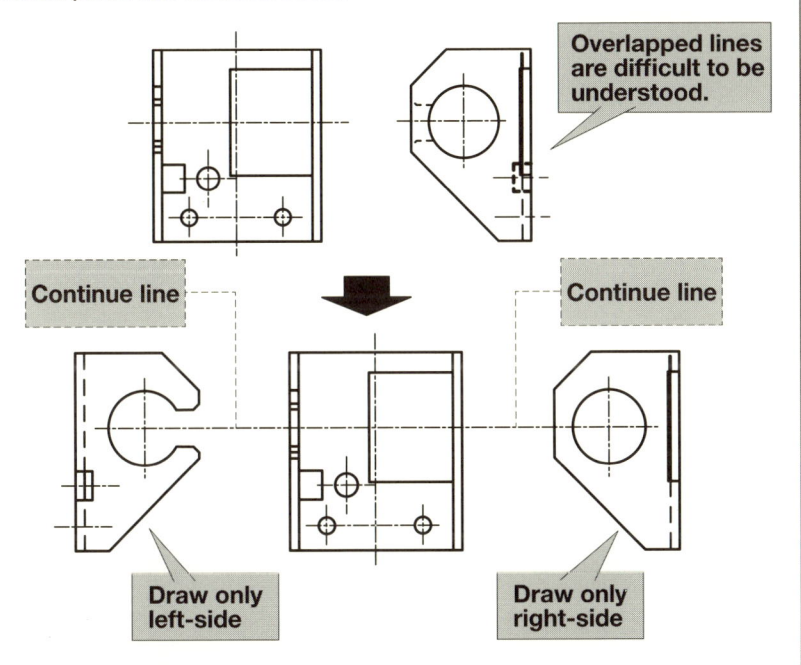

Overlapped lines are difficult to be understood.

Continue line

Continue line

Draw only left-side

Draw only right-side

Two or more holes can also be expressed as a local view.

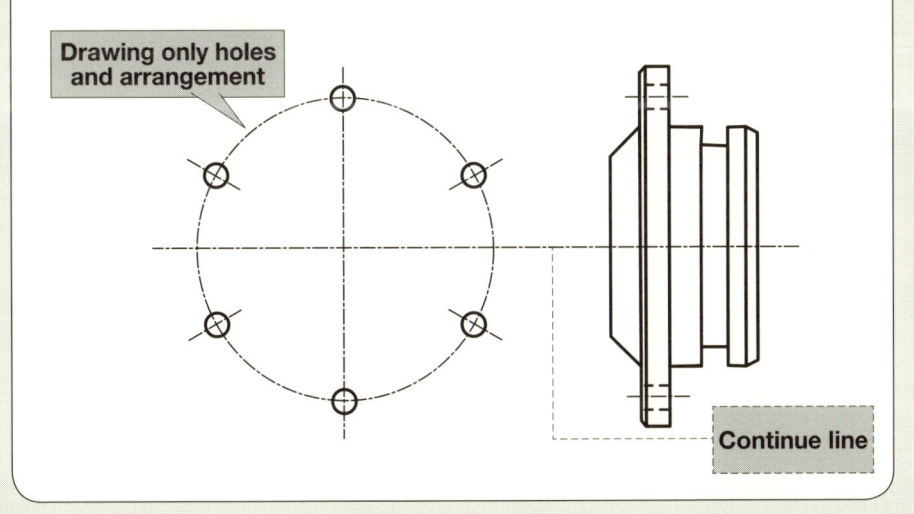

Drawing only holes and arrangement

Continue line

Chapter3 What is the technique of the projection views!

c) 矢示法

　紙面の制約により、第三角法に従わない配置を使う場合、矢の方向から見た投影図を任意の位置に配置することができます（**図3-5**）。

　この場合、矢印と大文字のアルファベットで指示します。この文字は、投影の向きに関係なくすべて上向きに書きます。

　アルファベットは、一般的にAから始めますが、他の箇所（断面図や部分拡大図）で、既にAを使用している場合は、繰り返さないよう次の文字（例えばB）を選択しましょう。

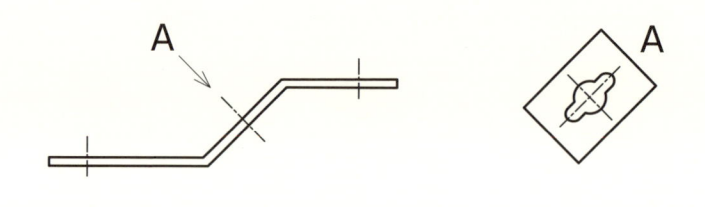

図3-5 矢示法

　大きな図面で投影図を離れた位置に配置する場合は、格子参照方式の記号を併記します（**図3-6**）。

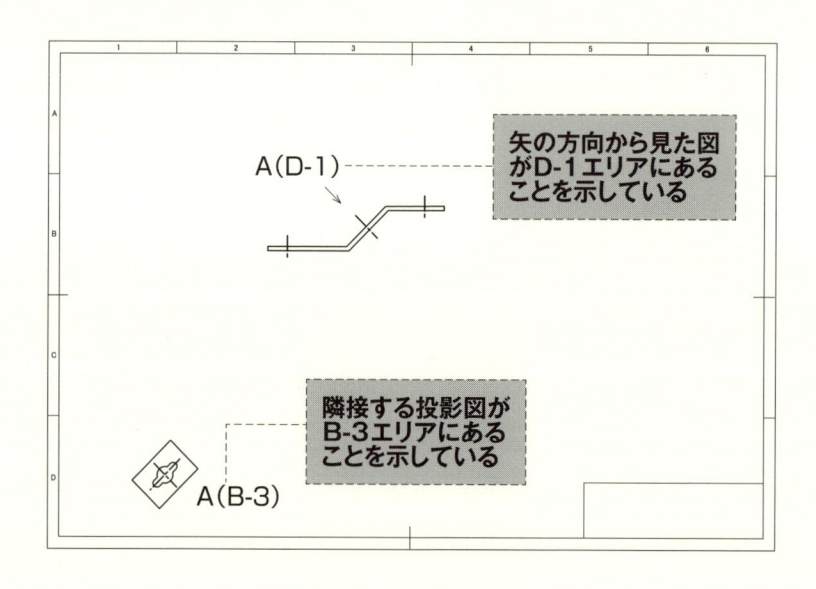

A(D-1) 矢の方向から見た図がD-1エリアにあることを示している

A(B-3) 隣接する投影図がB-3エリアにあることを示している

図3-6 格子参照方式を併記した矢示法

c) Reference arrow layout

In limitation of sheet space, when arrangement in accordance with third angle projection method is not used, views from arrow's direction can be arranged at any positions. Refer to Figure 3-5.

In this case, it is indicated by arrows and symbols of Latin alphabet capital letters. This letters are written upward irrespective of the projection direction.

Although Latin alphabet letters generally start with A, when letter-A is already used other parts (sectional views or enlarged view), you should choose next letters (e.g. B) so that it cannot repeat.

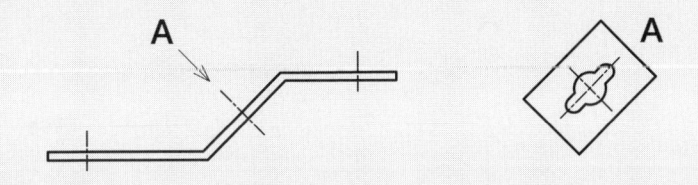

Figure 3-5 Reference arrow layouts

When the view is arranged in the separated place like a big sheet, the sign of a grid reference system is written together. Refer to Figure 3-6.

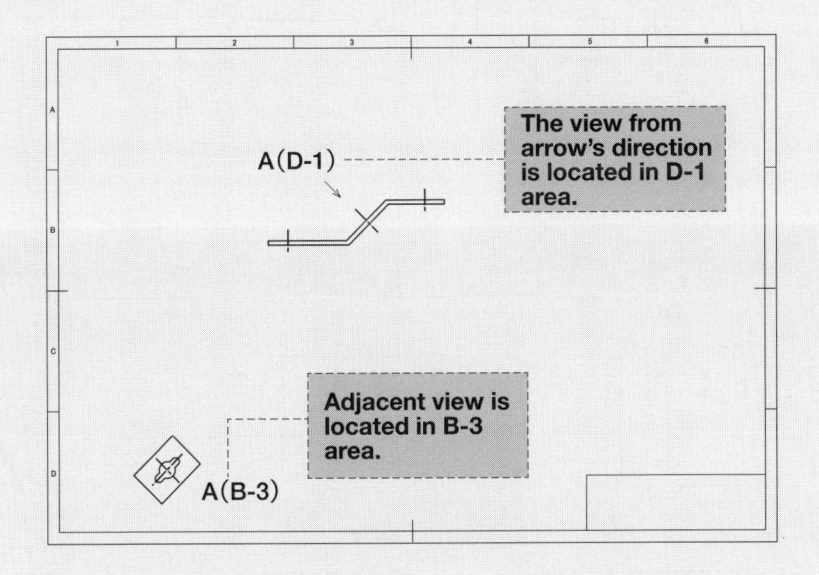

Figure 3-6 Reference arrow layouts with grid reference system sign

3-1-2　部分拡大図

　特定部分の図形が小さいために、その部分の詳細な図示や寸法記入ができない場合があります。主たる投影図において、選ばれた部分を細い実線で囲み、かつアルファベットの大文字で名前をつけます。また、部分拡大図を別の場所に描き、使用した名前と尺度の両方を記入します（**図3-7**　**図3-8**）。

　部分拡大図は図3-7のうちのどちらかを使用して、図面に記入することができます。

尺度1:1の図面の場合

A

A(2:1)　　A(2:1)

または

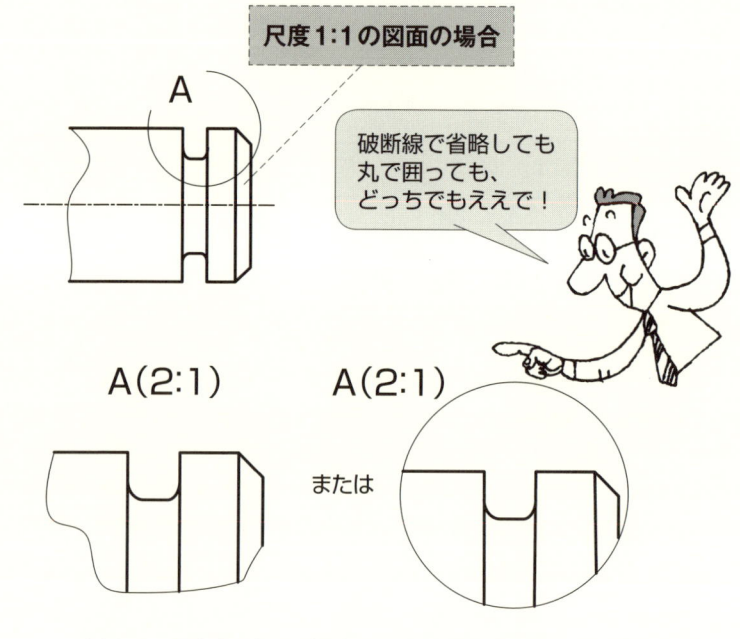

破断線で省略しても丸で囲っても、どっちでもええで！

図3-7　局部投影図　（尺度1:1の図面の場合）

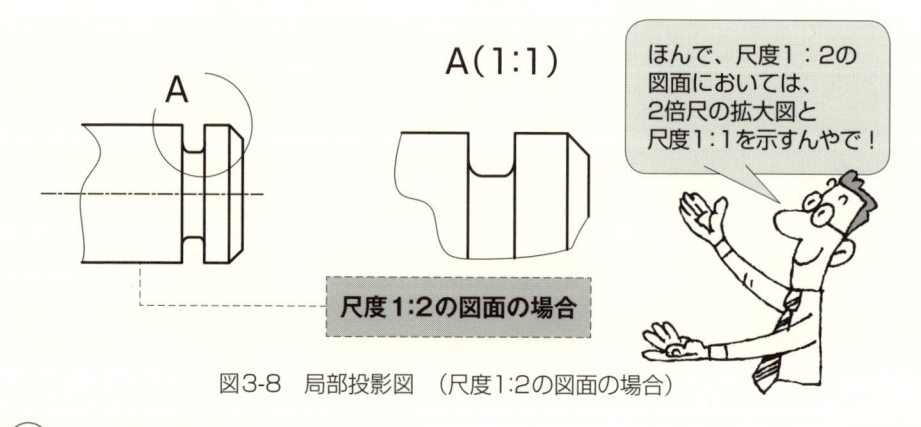

A

A(1:1)

尺度1:2の図面の場合

ほんで、尺度1：2の図面においては、2倍尺の拡大図と尺度1：1を示すんやで！

図3-8　局部投影図　（尺度1:2の図面の場合）

3-1-2 Enlarged view

Since the figure of a specific part is so small, details of the feature or dimensioning cannot be indicated. At main view, the selected part is framed by thin continuous line and having name of Latin alphabet capital letters.

And enlarged view is placed somewhere else, and have to be put both name and scale used. Refer to Figures 3-7 and 3-8.

Enlarged view can be expressed on the drawing using either in Figure 3-7.

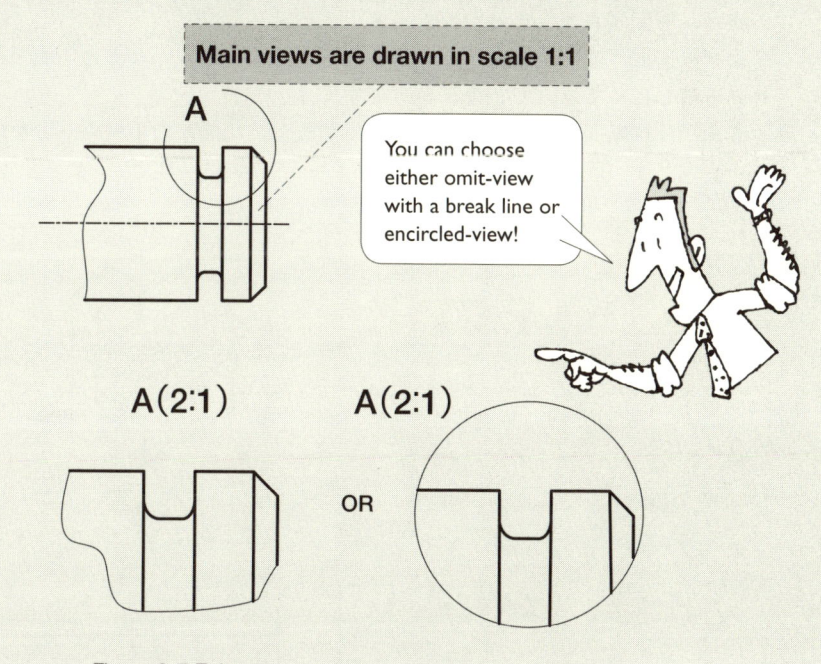

Main views are drawn in scale 1:1

A

You can choose either omit-view with a break line or encircled-view!

A(2:1)   A(2:1)

OR

Figure 3-7 Enlarged view (When the scale 1:1 drawing)

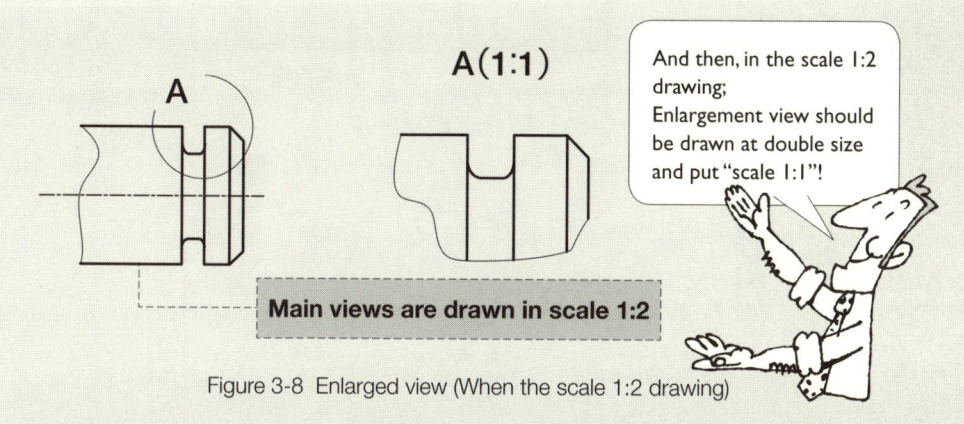

A   A(1:1)

And then, in the scale 1:2 drawing;
Enlargement view should be drawn at double size and put "scale 1:1"!

Main views are drawn in scale 1:2

Figure 3-8 Enlarged view (When the scale 1:2 drawing)

Chapter3 What is the technique of the projection views!

| 第3章 | 2 | 断面図 |

3-2-1 断面図のルール

断面図に関する決まりごとを最初に示します。

a) 隠れた部分をわかりやすくするために、かくれ線の記入は避けるべきです。そのため、断面図を使うことができます。

　なぜ断面にするのかというと、かくれ線（破線）が多くなると図面が複雑になり、理解しにくくなってしまうからです。

　断面図は切断面を用いて対象物を仮に切断し、切断前の手前の部分を取り除いた投影図です（**図3-9**）。

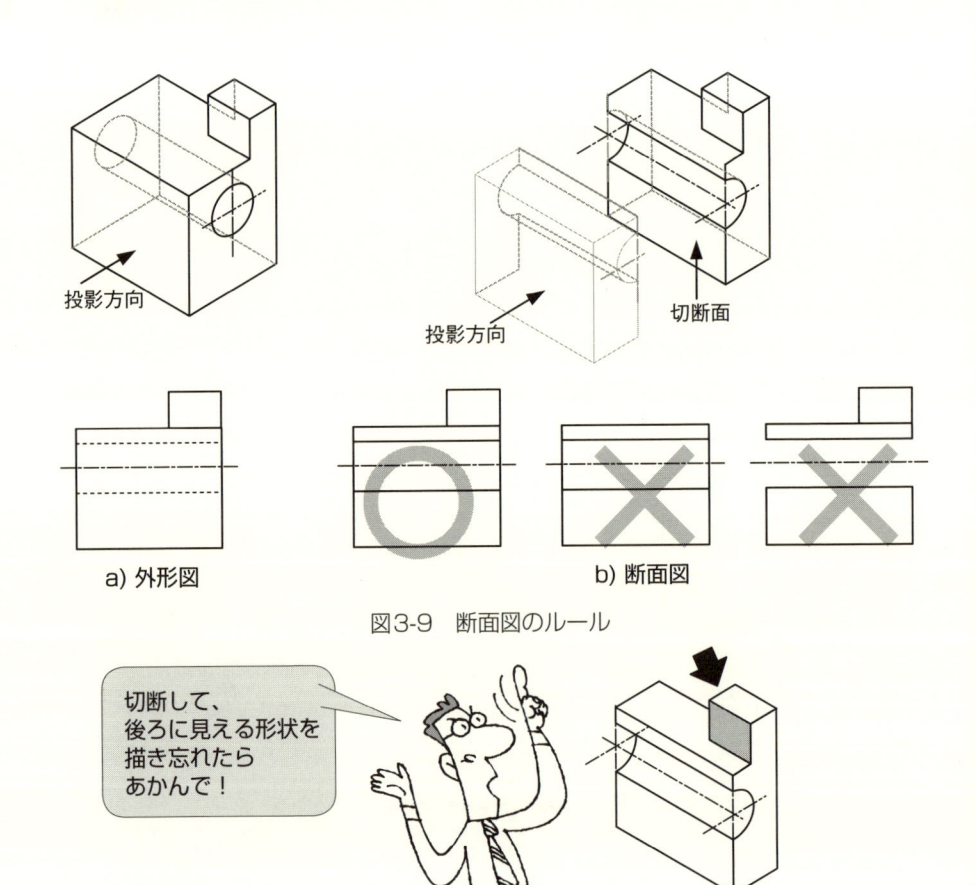

a) 外形図　　　　　b) 断面図

図3-9　断面図のルール

切断して、後ろに見える形状を描き忘れたらあかんで！

第3章　投影図のテクニックって、なんやねん！

Chapter 3 — 2 | Sectional view

3-2-1 Rule of sectional view

The rule about a sectional view is shown first.

a) In order to indicate the hidden part with clarity the need for hidden outlines shall be avoided. So, the sectional view can be used.

About necessity of the sectional view, when the hidden outline (dashed line) increases, the view will become complicated and it will become difficult to understand the shape.

Sectional view is a view which cut the object using the cutting plane and removed the object of front side before cutting. Refer to Figure 3-9.

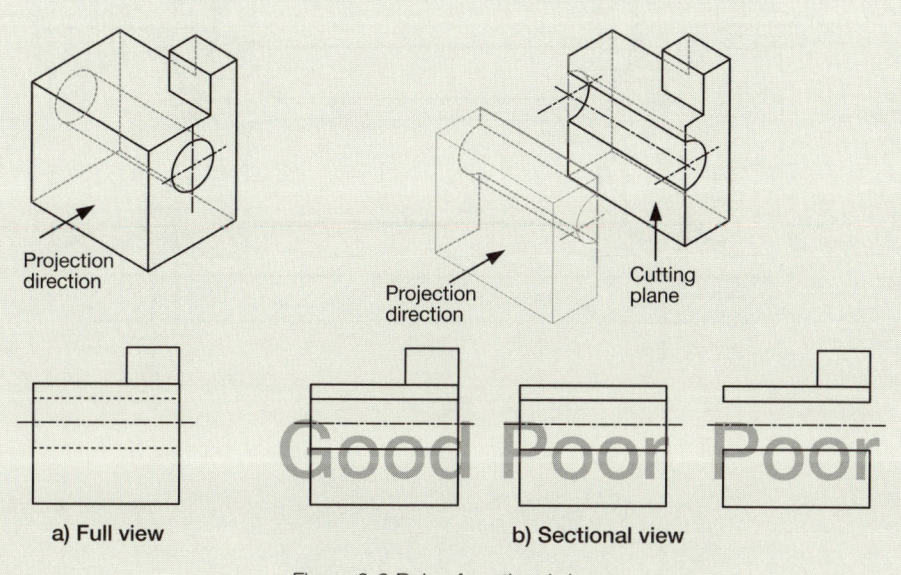

a) Full view   b) Sectional view

Figure 3-9 Rule of sectional view

Don't forget to draw the shape which is visible to backward after cutting!

Chapter3 What is the technique of the projection views!

b) しかし、切断したために理解を妨げるもの（例1参照）、または切断しても意味のないもの（例2参照）は長手方向に切断しません（**図3-10**）。

　　例1)　リブ、アーム、歯車の歯
　　例2)　軸、ピン、ボルト、ナット、座金、小ねじ、リベット、キー、鋼球、円筒ころ

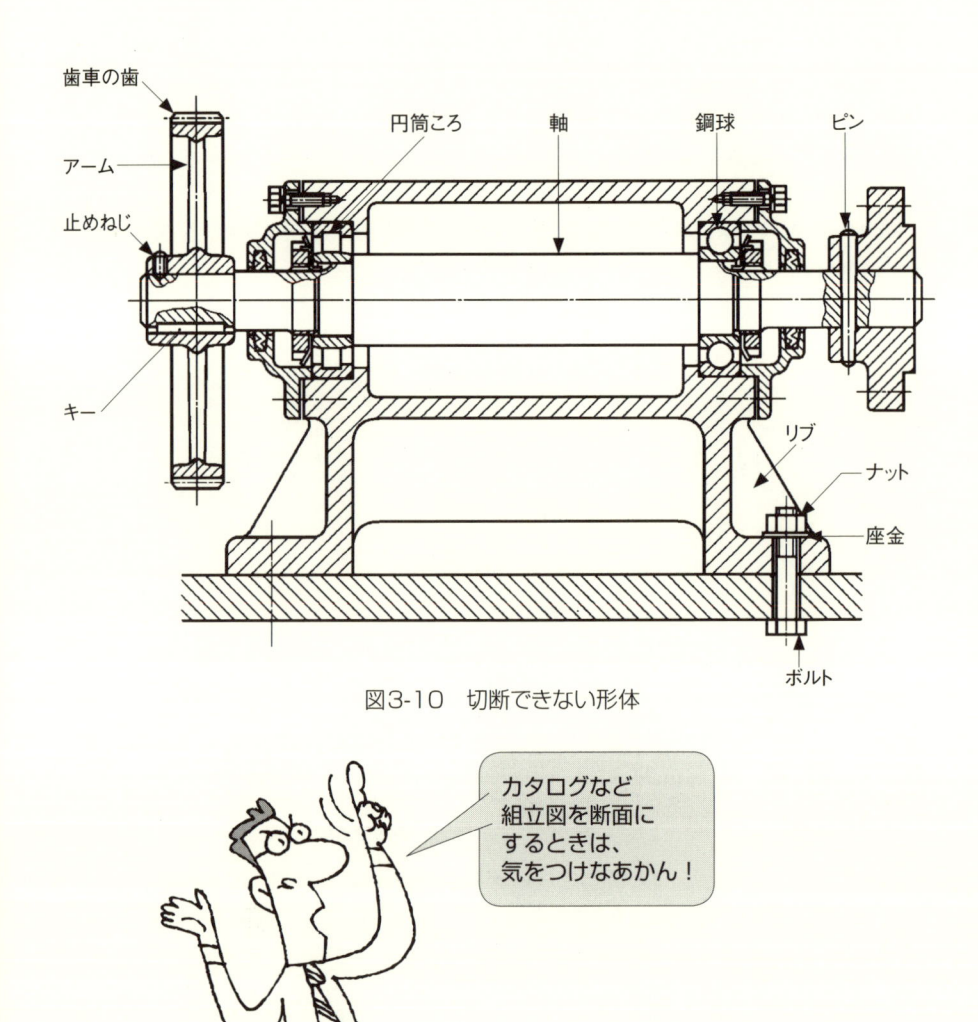

図3-10　切断できない形体

カタログなど組立図を断面にするときは、気をつけなあかん！

b) However, by cutting, the objects (Refer to e.g.1) which are hard to understand, or the nonsensical objects (Refer to e.g.2) shall not be cut in the longitudinal sections. Refer to Figure 3-10.
  e.g.1) Rib, Arm, Teeth of gear
  e.g.2) Shaft, Pin, Bolt, Nut, Washer, Machine screw, Rivet, Key, Steel ball, Cylindrical roller

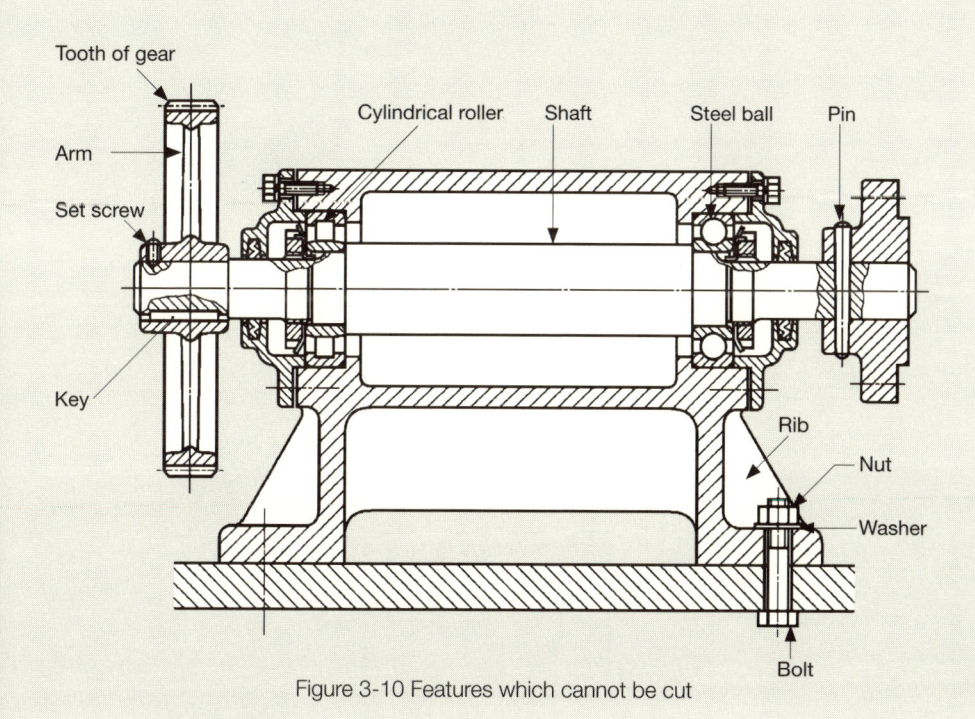

Figure 3-10 Features which cannot be cut

Take care, when you make assembly drawings, such as a catalog, into a section!

Chapter3 What is the technique of the projection views!

c) 切断面の位置を指示する場合は、細い一点鎖線を用いて指示します。また、両端および切断方向の変わる角は太い線を描きます。

　投影方向を示す場合には、細い一点鎖線の両端に投影方向を示す矢印を描きます。さらに切断面を指示する場合には、矢印によって投影方向を示し、大文字のアルファベットを矢印の端に記入します。また、切断面の識別記号（例えばA-A）は、断面図の真下、または真上に示します（**図3-11**）。

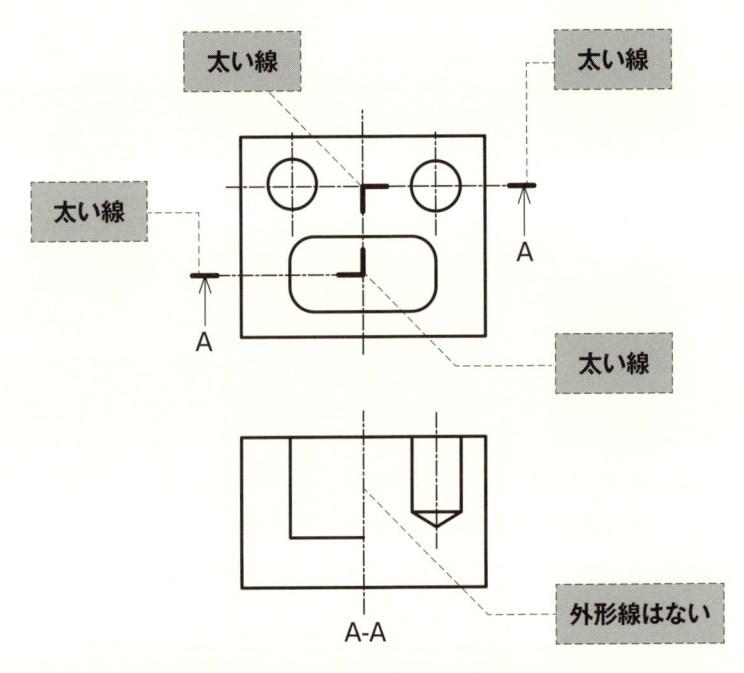

図3-11　2つの切断面と切断線

中心線を利用して、切断方向を変えるとええんやね！

c) When the position of cutting plane will be indicated, a thin long-dashed dotted line is used. And both end and corner where the cutting direction changes are placed thick line.

When the projection direction will be indicated, arrows indicating the projection direction are placed at both ends of the long-dashed dotted line. In addition, when the cutting plane will be indicated, the projection direction is indicated by arrows, and symbols of Latin alphabet capital letters are indicated at end of the arrows. And, the ID symbol of the cutting plane (e.g., A-A) is indicated below or above the sectional view. Refer to Figure 3-11.

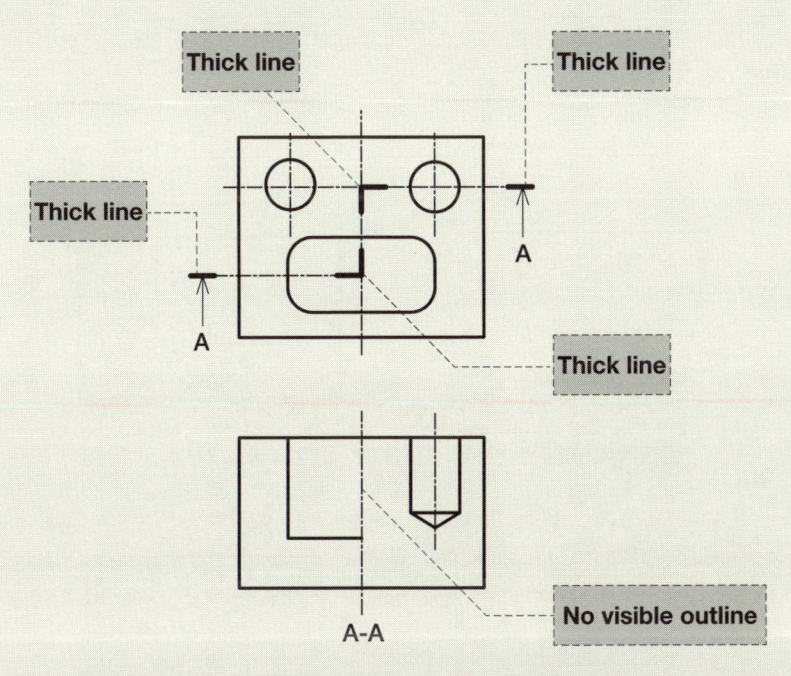

Figure 3-11 Cut by two planes and cutting-plane line

It is good to change the cutting direction using a center line, isn't it!

Chapter3 What is the technique of the projection views!

d) 断面の切り口を示すために、ハッチングは次のように使用します。
- ハッチングは細い実線で、主たる中心線に対して45°にするのがよい。
- 断面図に材料などを表示するため、特殊なハッチングを記入してもよい。この場合、その意味を図面内にはっきりと指示するか、該当規格を引用して示す（図3-12 a）。
- 同じ切断面上に現れる同一部品の平行な切り口には同一のハッチングを施す。ただし、切断面を区別する必要がある場合には、ハッチングをずらしてもよい（図3-12 b）。
- 隣接する切り口のハッチングは、線の向き、または角度を変えるか、その間隔を変えて区別する（図3-12 c）。
- ハッチング領域に文字や記号を記入しなければいけない場合は、ハッチングを中断します（図3-12 d）。

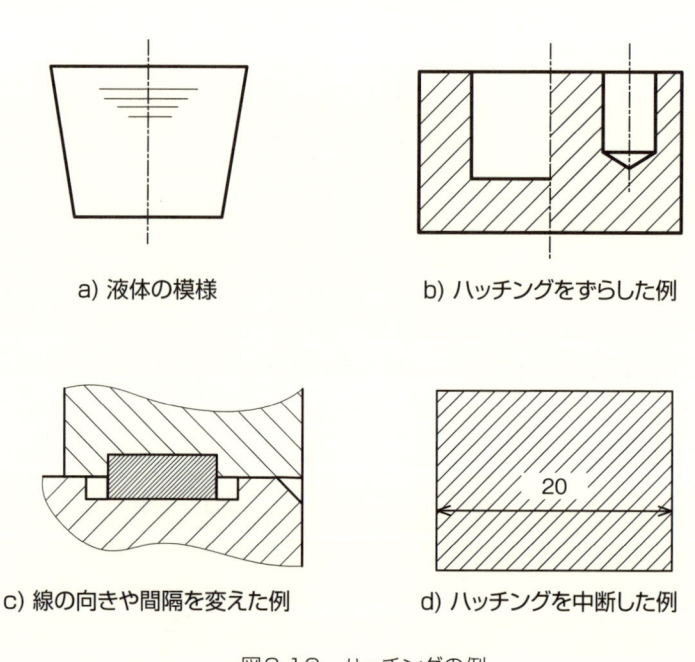

a) 液体の模様

b) ハッチングをずらした例

c) 線の向きや間隔を変えた例

d) ハッチングを中断した例

図3-12　ハッチングの例

d) In order to show the section, hatching is used as follows:
- Thin continuous lines are used as hatching, and are given at an angle of 45° to the main center lines.
- In order to indicate materials and the like on the sectional view, special hatching can be placed. In this case, designate clearly the meaning of the hatching in the drawing or indicate by citing the corresponding standard.
Refer to Figure 3-12 a.
- When sections of the same work-piece in parallel cutting planes are shown side by side, same hatching is used. However, hatching can be offset when difference between the cutting planes is required. Refer to Figure 3-12 b.
- The hatching of adjacent sections is separated by changing the direction of line or the angle or changing its interval. Refer to Figure 3-12 c.
- Hatching is interrupted when texts or symbols must be placed in hatched area. Refer to Figure 3-12 d.

a) Liquid pattern

b) Offset hatching

c) Hatching obtained by changing line direction and interval

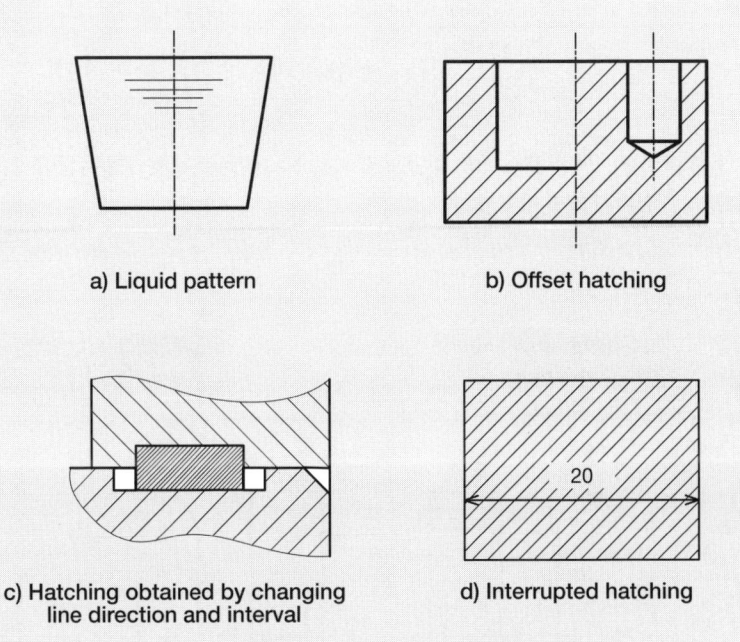

d) Interrupted hatching

Figure 3-12 Example of hatching

3-2-2 全断面図

　全断面図は、投影対象物の基本的な形状をもっとも明確に表すように描かなければいけません。
　対称形状の場合は、切断面を表す切断線は記入しません（**図3-13**）。

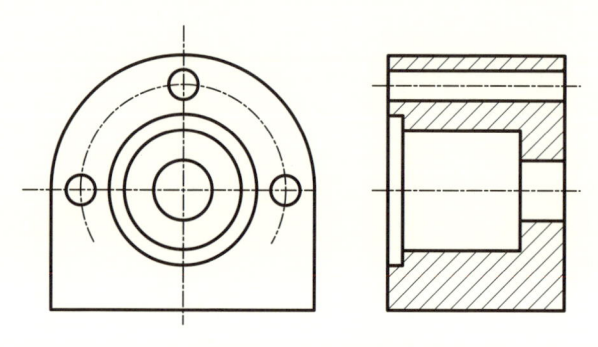

図3-13　全断面図の例

　必要がある場合は、特定部分の形をよく表すように切断面を選びます。この場合、切断線によって切断位置を示します。
　誤解を防ぐため、断面図は切断面の近辺に配置するのがよいでしょう（**図3-14**）。

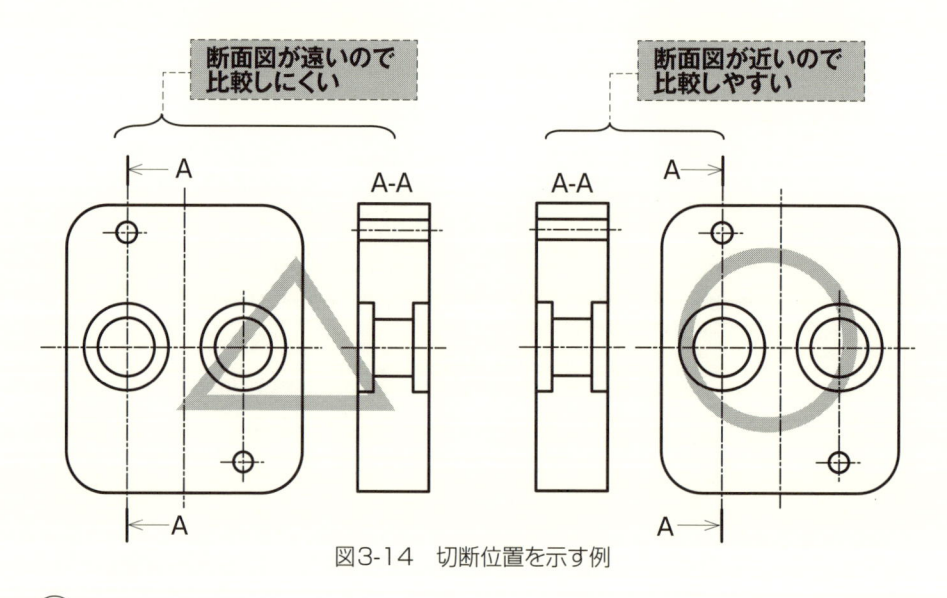

図3-14　切断位置を示す例

3-2-2 Full sectional view

The full sectional view has to be drawn to express most clearly the fundamental shape of the object.

In case of the symmetrical shape, not place of the cutting-plane line. Refer to Figure 3-13.

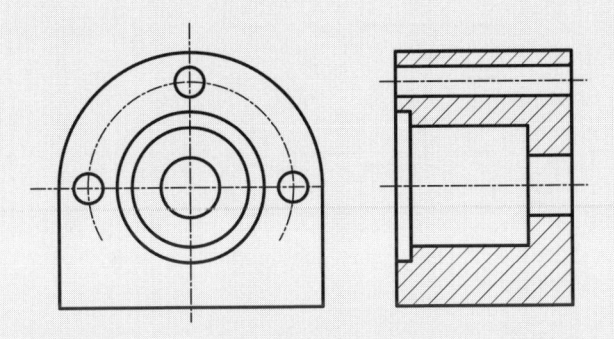

Figure 3-13 Full sectional view

If necessary, select the cutting plane to express clearly the shape of a specific part. In this case, indicate the cutting position with the cutting-plane line.

For prevention of misunderstanding, the sectional view will be good to arrange near a cutting plane. Refer to Figure 3-14.

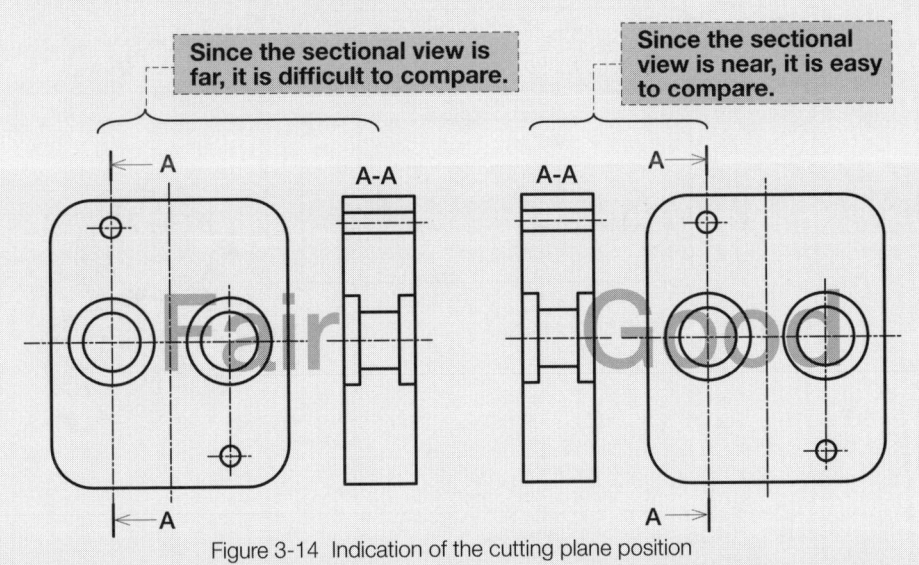

Since the sectional view is far, it is difficult to compare.

Since the sectional view is near, it is easy to compare.

Figure 3-14 Indication of the cutting plane position

Chapter3 What is the technique of the projection views!

3-2-3　片側断面図

対称形状の対象物は、外形図の半分と全断面図の半分を組み合わせて表すことができます（**図3-15**）。

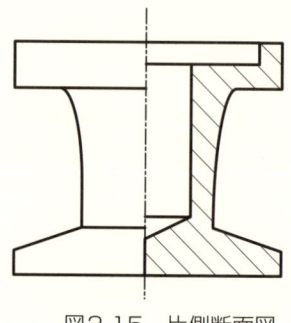

図3-15　片側断面図

3-2-4　部分断面図

外形図の一部を切り欠くことで必要とする部分の一部だけを断面で表すことができます。この場合は境界を破断線で示します（**図3-16**）。

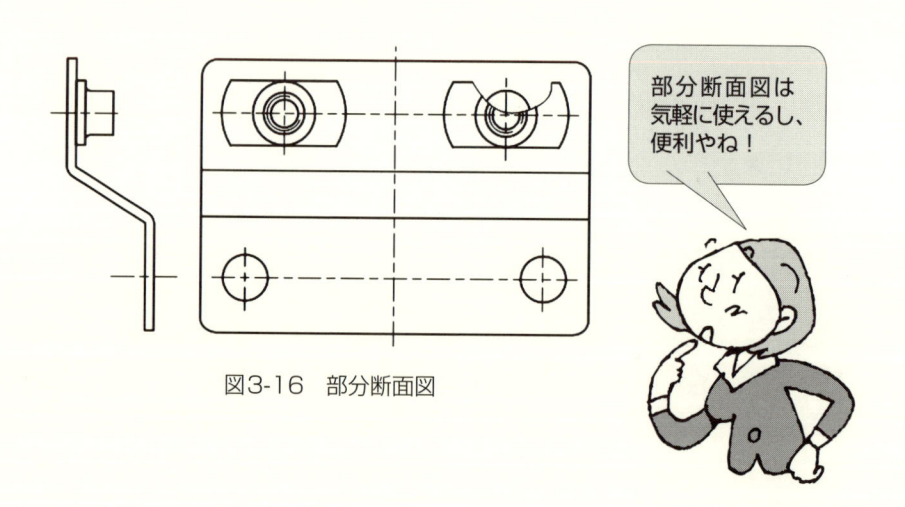

図3-16　部分断面図

部分断面図は気軽に使えるし、便利やね！

3-2-3 Half sectional view

The object of symmetrical shape can be expressed by combination of half of full view and half of full sectional view. Refer to Figure 3-15.

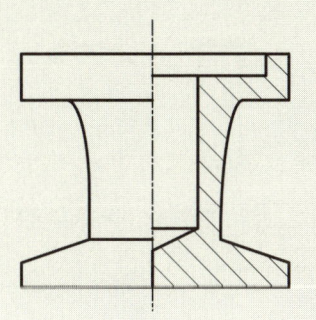

Figure 3-15 Half sectional view

3-2-4 Local(Detail) sectional view

By cutting a part of full view, only the required area can be expressed as the sectional view. In this case, indicate the boundary by a break line. Refer to Figure 3-16.

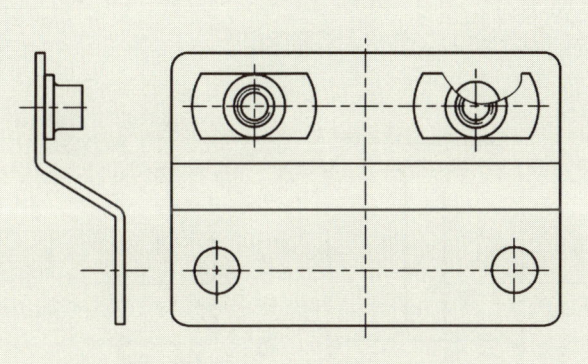

Figure 3-16 Local (Detail) sectional view

> Local sectional view can be used easily.
> So, it is convenient, isn't it!

3-2-5 回転図示断面図

　ハンドルや車輪などのアーム、リム、リブ、フック、軸、構造物の部材の切り口は、90度回転させて表すことができます（図3-17）。

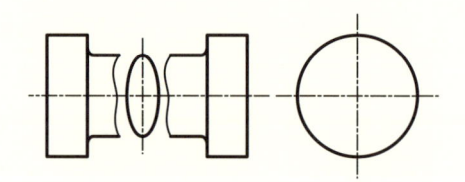

a）投影図を破断して描いた回転図断面図

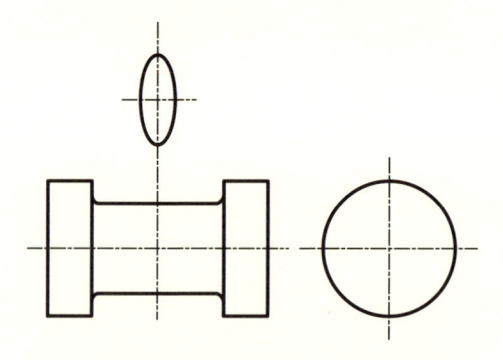

b）切断線の延長線上に描いた回転図断面図

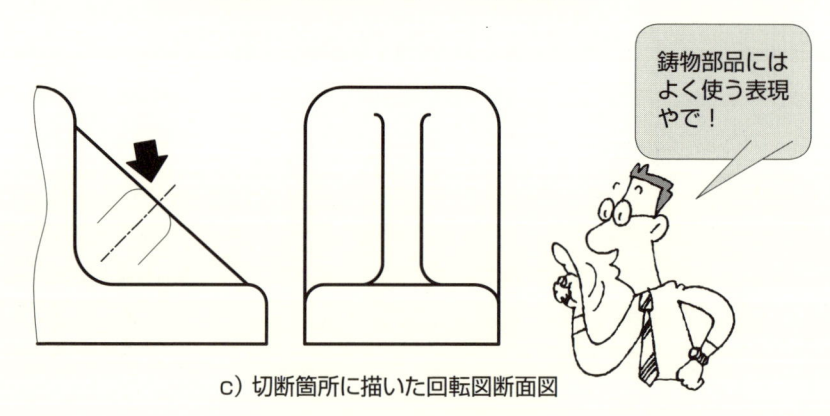

c）切断箇所に描いた回転図断面図

図3-17 回転図示断面図

鋳物部品には
よく使う表現
やで！

3-2-5 Revolved sectional view

The section of arm of handles or wheels, rib, hook, shaft, member of structures can be expressed by revolving 90°. Refer to Figure 3-17.

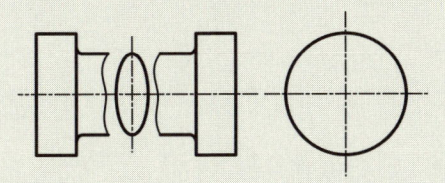

a) Revolved sectional view drawn by broken view

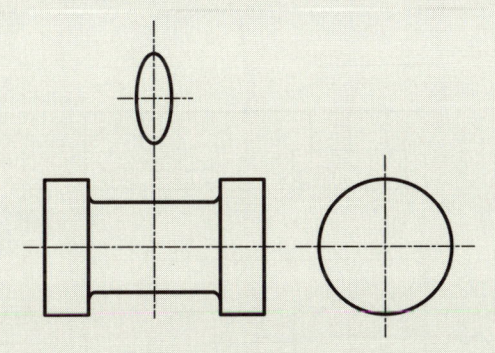

b) Revolved sectional view drawn on the extended line of a cutting-plane line

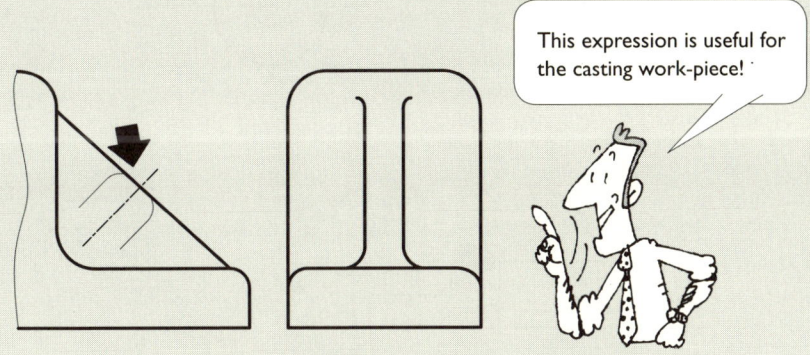

This expression is useful for the casting work-piece!

c) Revolved sectional view is drawn in the cutting position

Figure 3-17 Revolved sectional view

3-2-6 組み合わせによる断面図

　2つ以上の断面図を組み合わせて示す投影図は、次のようにします。
この場合、必要に応じて断面を見る方向を示す矢印と大文字のアルファベットをつけます。

a) オフセット断面図

　　断面図は、2つ以上の平行平面で切断した断面図を組み合わせて示すことができます。
　　この場合、切断線によって切断した位置を示し、オフセットによる断面図であることを示すために、2つの切断線を任意の位置でつなぎます（図3-18）。

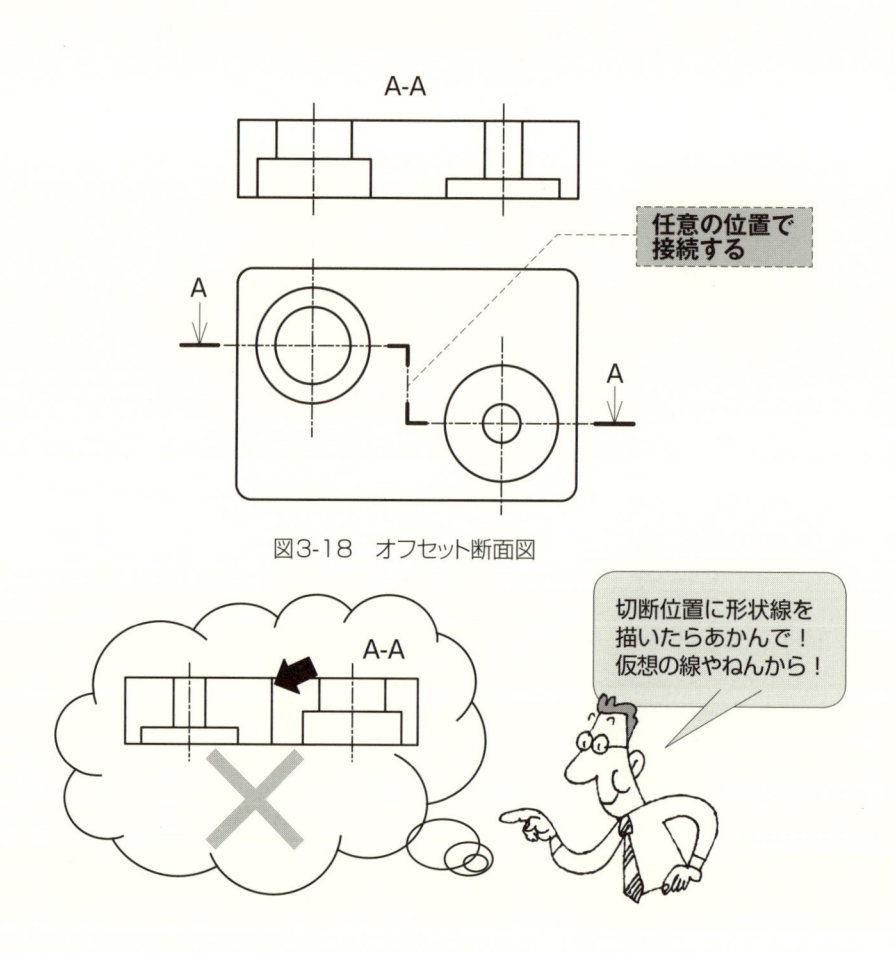

図3-18　オフセット断面図

切断位置に形状線を描いたらあかんで！
仮想の線やねんから！

3-2-6 Sectional view by combination

The view to be indicated by combination of over two sectional views is as follows:
In this case, put the arrow marks and Latin alphabet capital letters indicating the direction of viewing the sectional surface, as required.

a) Offset sectional view

The sectional view can be indicated by combining sectional view cut by over two parallel planes.

In this case, indicate cutting position by the cutting-plane line. Connect the two cutting-plane lines at an arbitrary position in order to indicate that it is the sectional view by offset. Refer to Figure 3-18.

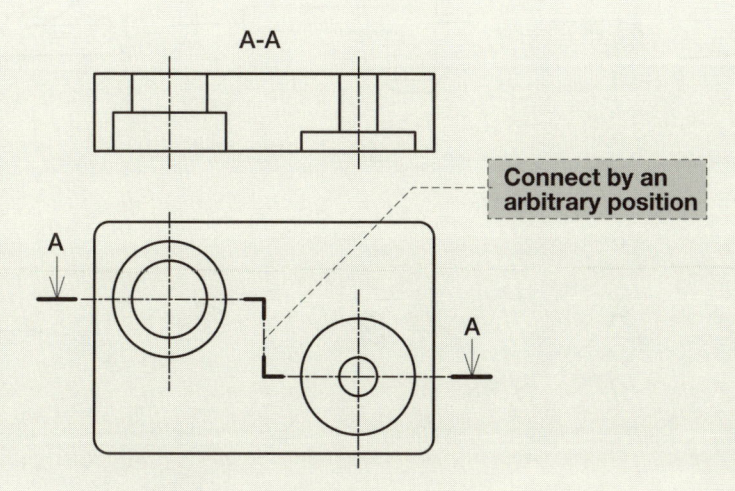

Figure 3-18 Example of offset sectional view

Don't draw outlines on the cutting position. It's a virtual line!

Chapter3 What is the technique of the projection views!

b) **整列した断面図**

　対称形状またはこれに近い形の対象物の場合には、中心線を境界にして、その片側を投影面に平行に切断し、他の側を投影面とある角度をもって切断することができます。この場合、切断面は角度のある方向から見た図を描きます。（図3-19）。

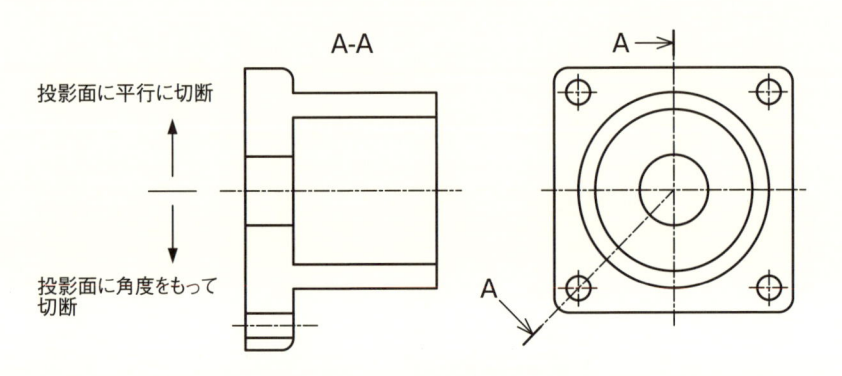

図3-19　整列した断面図

b) Aligned sectional view

In case of the object of symmetric shape or of any shape close to it, one side of the object can be cut in parallel to the projection plane by centerline as the border, and the other side at angle relating to the projection plane. In this case, the cutting plane draws seen from the direction of an angle. Refer to **Figure 3-19**.

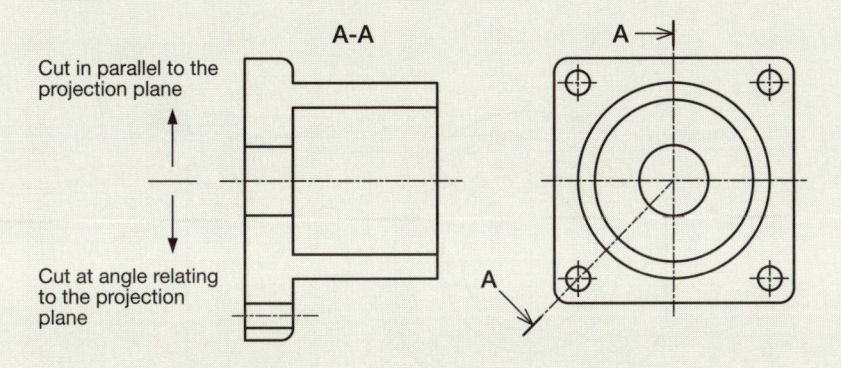

Figure 3-19 Aligned sectional view

Good   Poor

I have to wright letters upward.
Letters don't incline, right?

c) ピッチ円上の穴の断面
　ピッチ円上に配置する穴やねじは、側面図（断面を含む）においては、次のように表します（**図3-20**）。
・中心線はピッチ円が作る円筒の上端と下端の位置に描く。
・投影の向きに関係なく、その片側だけに1個の穴を示し、他の穴は省略する。

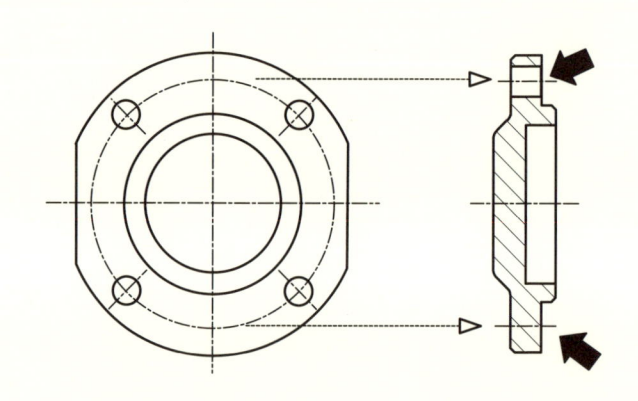

図3-20　側面図におけるピッチ円上の穴の表し方

はっきり言って、
このルールを知ってる
設計者は少ないねん‥

トホホ…

c) The section of the holes arranged on a pitch circle

For the holes or screw threads arranged on the pitch circle, in the side view (including sectional view) is expressed as follows: Refer to Figure 3-20.
- Center lines are drawn on the position of top and bottom of a circular cylinder made by pitch circle.
- Irrespective of the projection direction, one hole on just one side is indicated, and other holes are omitted.

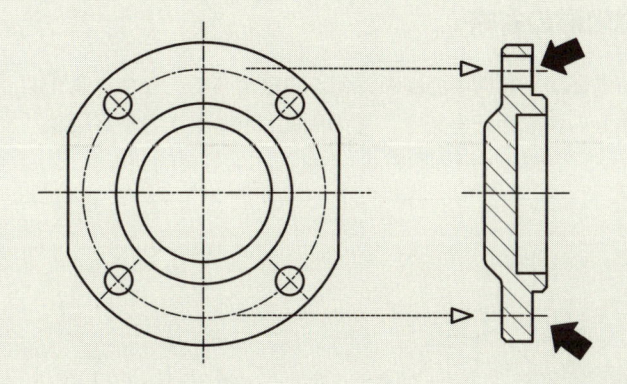

Figure 3-20 Expression of the hole on the pitch circle in side view

Honestly, many engineers don't know this rule…

第3章　3　図形の省略

多くの機械部品は、対称形状であるか、多数の穴があります。
このような場合に作図効率と紙面の節約のために、投影図を省略することができます。

3-3-1　対称図形の省略

図形が対象形の場合は、対称中心線の片側を省略することができます。
対称中心線の片側の図形だけを描き、その両端に短い2本の平行細線「＝」をつけます（**図3-21**）。この2本の平行細線は、対称図示記号と呼びます。

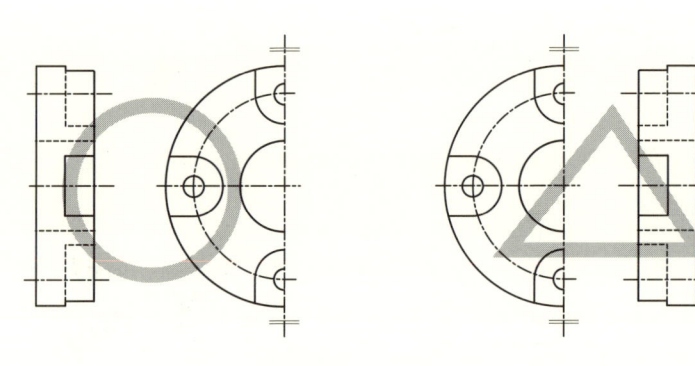

a) 外形図と並べる場合

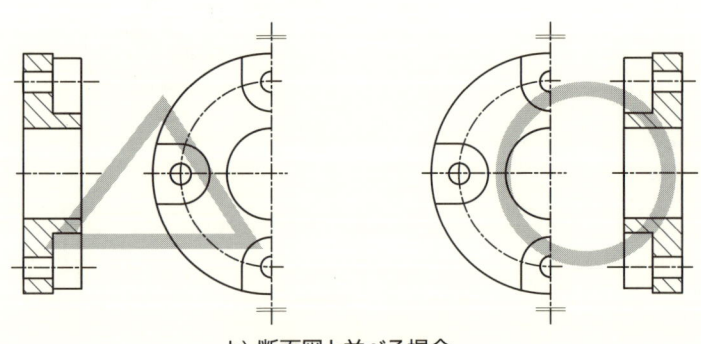

b) 断面図と並べる場合

図3-21　対称図形の省略

| Chapter3 | 3 | # Omission of view |

Many mechanical work-pieces are symmetrical-shaped, or have many holes.
In this case, views can be omitted for drawing efficiency and saving the space.

3-3-1 Omission of symmetrical shape

In the case of symmetrical shape, one side of the center line of symmetry can be omitted.

Only one side view of symmetrical center line is drawn, and two short parallel thin lines"=" are put at both end of it. Refer to Figure 3-21.

These two short parallel thin lines are called symmetry symbol ("TAISHOU ZUJI KIGOU" in Japanese).

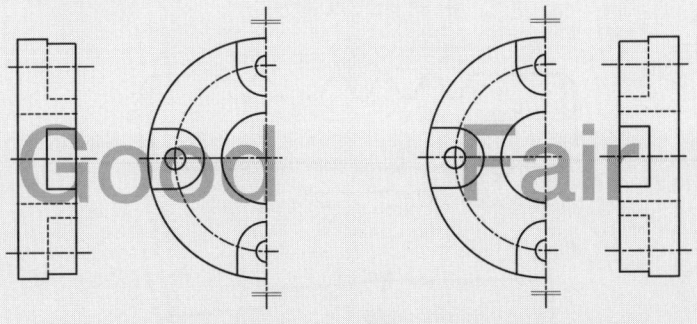

a) In case of placing with an outline view

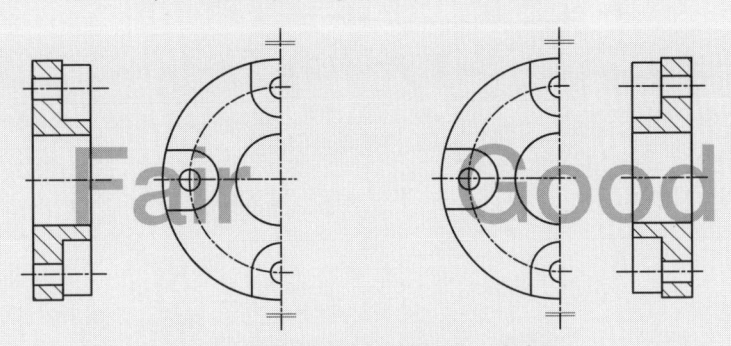

b) In case of placing with a sectional view

Figure 3-21 Omission of symmetrical shape

3-3-2 中間部分の省略

　次に示す形状の場合、投影図は、紙面を節約するために中間部分を切り取って、その主たる部分だけを近づけて図示することができます（**図3-22**）。
・同一断面形の部分
　　例）　軸、棒、管、形鋼
・同じ形が規則正しく並んでいる部分
　　例）　ラック、工作機の送りねじ
・長いテーパなどの部分
　　例) 長いテーパ軸

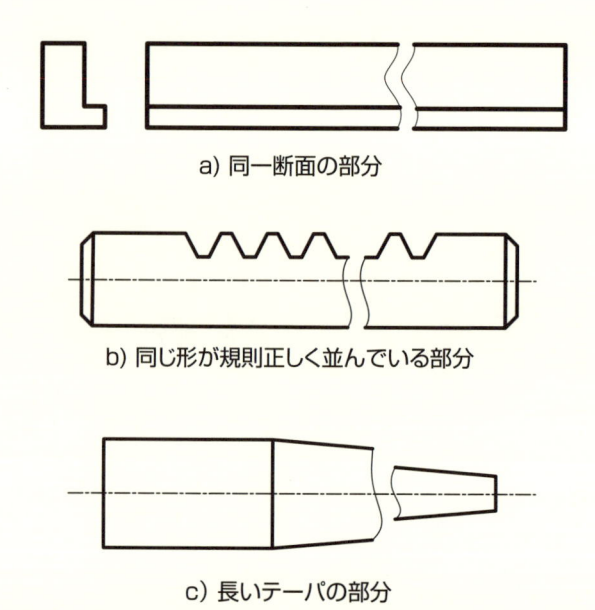

a) 同一断面の部分

b) 同じ形が規則正しく並んでいる部分

c) 長いテーパの部分

図3-22 中間部分の省略

3-3-2 Omission of intermediate parts

In case of the shape shown below, the view can be indicated by omitting the intermediate parts and arranged only the prior parts closing to them, so that space can be saved. Refer to Figure 3-22.
- The parts of same sectional shape.
    e.g.) Shaft, Bar, Tube, Steel section
- The parts which same shape are lined up in order.
    e.g.) Rack, Master screw of machine tool
- The parts of long taper and like.
    e.g.) Long taper shaft

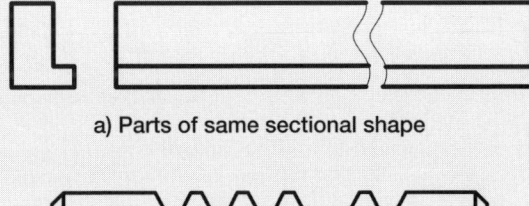

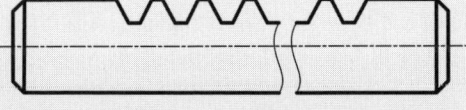

a) Parts of same sectional shape

b) Parts which same shape are lined up in order.

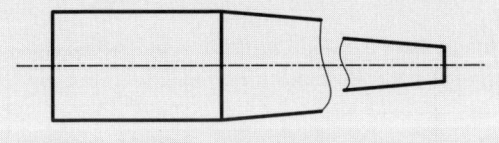

c) Parts of long taper

Figure 3-22 Omission of intermediate parts

第3章 4 特殊な図示

3-4-1 2つの面の交わり部の表示

a) 2つの面が交わる場合、交線を太い実線で表します（**図3-23**）。

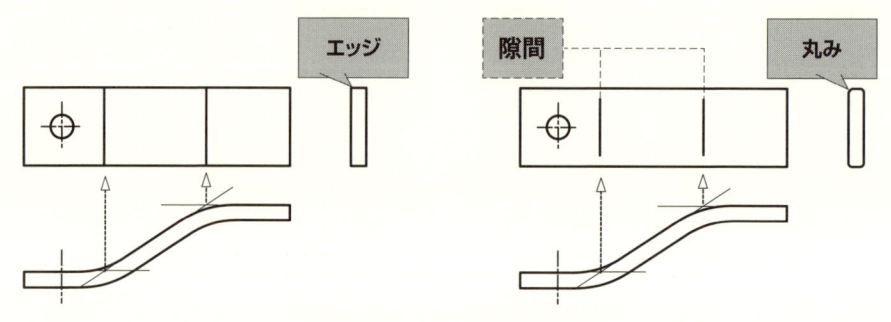

図3-23 交わり部の例

b）曲面同士または曲面と平面が交わる部分の線を相貫線といいます。
　相関線は、直線で表す場合と、正しい投影に近似させた曲線で表す場合があります（**図3-24**）。

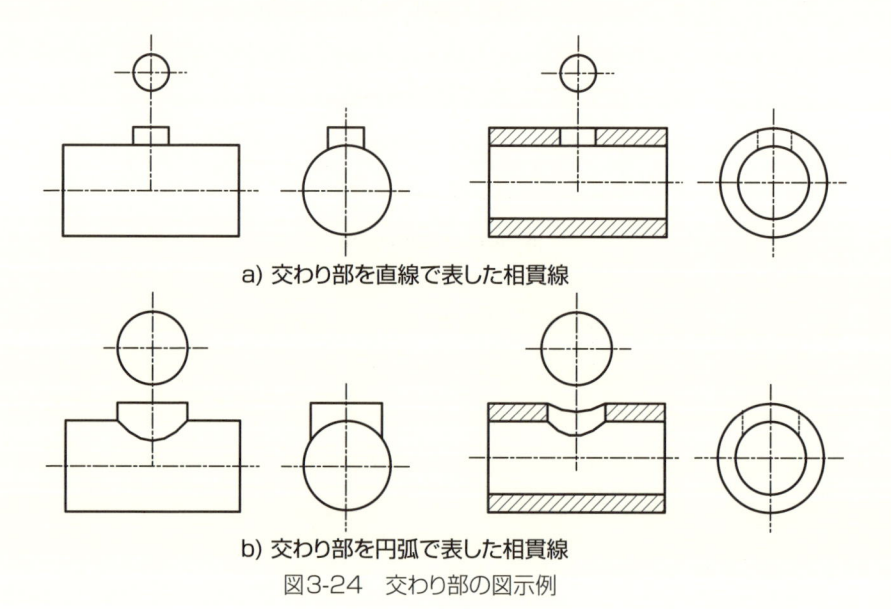

a) 交わり部を直線で表した相貫線

b) 交わり部を円弧で表した相貫線

図3-24 交わり部の図示例

| Chapter3 | 4 | **Special expression of view** |

3-4-1 Indication of two crossing surfaces

a) When two surfaces are crossed, line where surfaces intersect is drawn by thick continuous line. Refer to Figure 3-23.

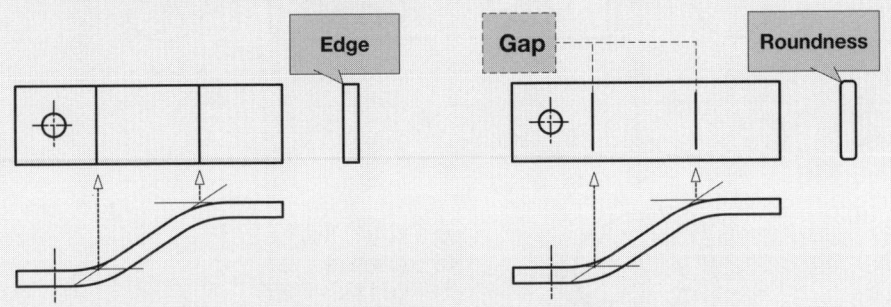

Figure 3-23 Examples of intersecting part

b) The crossing line of two curved surfaces or curved surface and flat surface is called intersection curve ("SOUKAN-SEN" in Japanese).

Intersection curve may be drawn by straight lines or by curved line similar to correct projection. Refer to Figure 3-24.

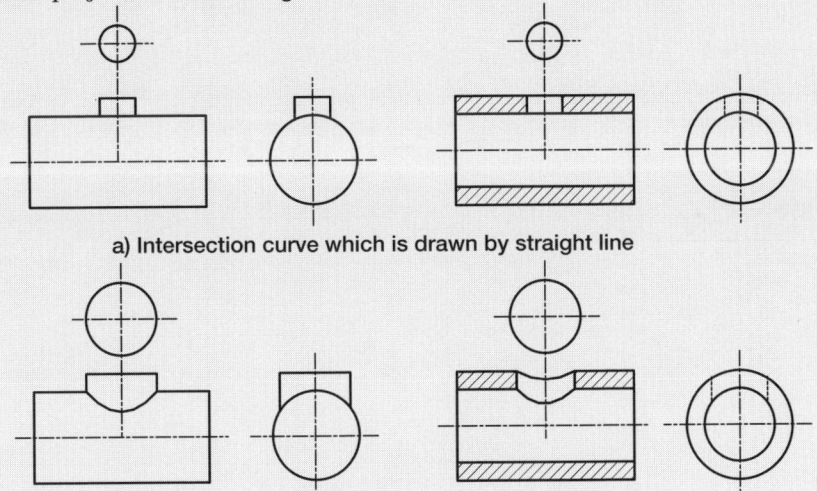

a) Intersection curve which is drawn by straight line

b) Intersection curve which is drawn by curved line

Figure 3-24 Indication of two crossing surface

3-4-2 曲面の中の平面の表し方

一部分だけが平面である場合には、細い実線で対角線を記入します（**図3-25**）。

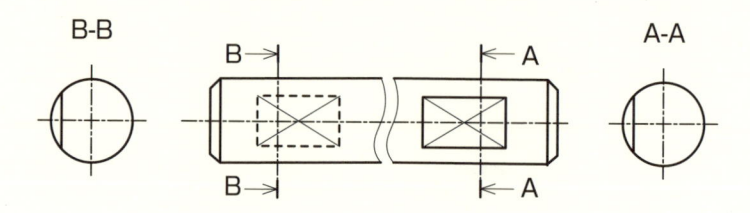

図3-25　曲面の中の平面部分の表し方

> かくれ線で平面をあらわす場合でも、対角線は細い実線で描くんやね！

3-4-2　Indications of flat surface in curved surface

When only a part is flat surface, thin continuous diagonal lines are drawn. Refer to Figure 3-25.

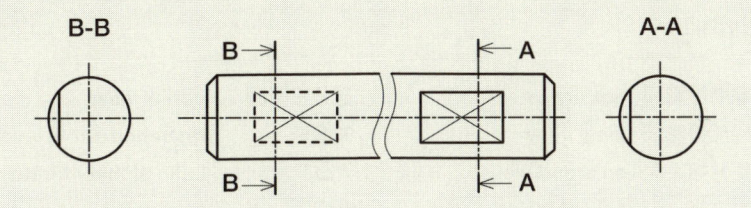

Figure 3-25　Indications of flat surface in curved surface

> Diagonal lines are drawn as the thin continuous line even hidden shape, aren't they?

用語集　glossary

本章で使用した単語です。他の用語も使うことができます。
The list of words used for this chapter is shown. You can use another words, too.

補助投影図	auxiliary view	部分投影図	partial view
局部投影図	local view	矢示法	reference arrow layout
部分拡大図	enlarged view	アルファベット	Latin alphabet letter
大文字	capital letter	断面図	sectional view
切断面	cutting plane	外形図	full view
切断線	cutting plane line	ハッチング	hatching
全断面図	full sectional view	対称形状	symmetrical shape
片側断面図	half sectional view	部分断面図	broken-out sectional view
破断線	break line	回転図示断面図	revolved sectional view
鋳物部品	casting work-piece	オフセット断面図	offset sectional view
任意の位置	arbitrary position	整列した断面図	aligned sectional view
ピッチ円	pitch circle	図形の省略	omission of view
平行細線	parallel thin lines	対称図示記号	symmetry symbol
相貫線	intersection curve	対角線	diagonal line

第4章
寸法記入のルールって、なんやねん!

Chapter4
What is the rule of dimensioning!

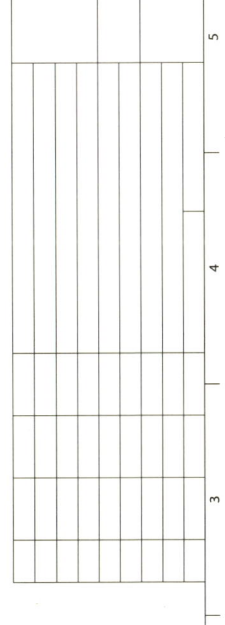

この章を学習すると、次の知識を得ることができます。
1) 寸法記入の基本
2) 寸法補助記号

After studying this chapter, you will be able to get knowledge as follows:
-1. Fundamentals of dimensioning
-2. General symbols for dimensioning

第4章　1　寸法記入の基本

> 　サイズ形体は、直線距離で定義され、2つの平行平面、円筒面あるいは球面からなるものです。
> 　サイズ形体は、また、角度の大きさでもそれぞれ定義され、互いに角度を持つ2つの平面（くさび形）、または円すい表面からなるものです。
> 　形体とは、点や線、面といった部品の特別な部分のことです。

寸法数値の単位は以下の通りです。
　　大きさ／長さ／位置…ミリメートル(mm)
　　角度…度（°）　分（'）　　秒（"）
寸法の単位をインチで表す必要がある場合は、注記を指定します。
　例）注記：特に明記しない限り、すべての寸法はインチで表されています。

　一般的に、寸法は寸法補助線を用いて寸法線を記入し、この上側に寸法数値を指示します。寸法線は指示する長さに平行に、または角度を測定する方向に引き、寸法線の両端には端末記号をつけます（**図4-1**）。

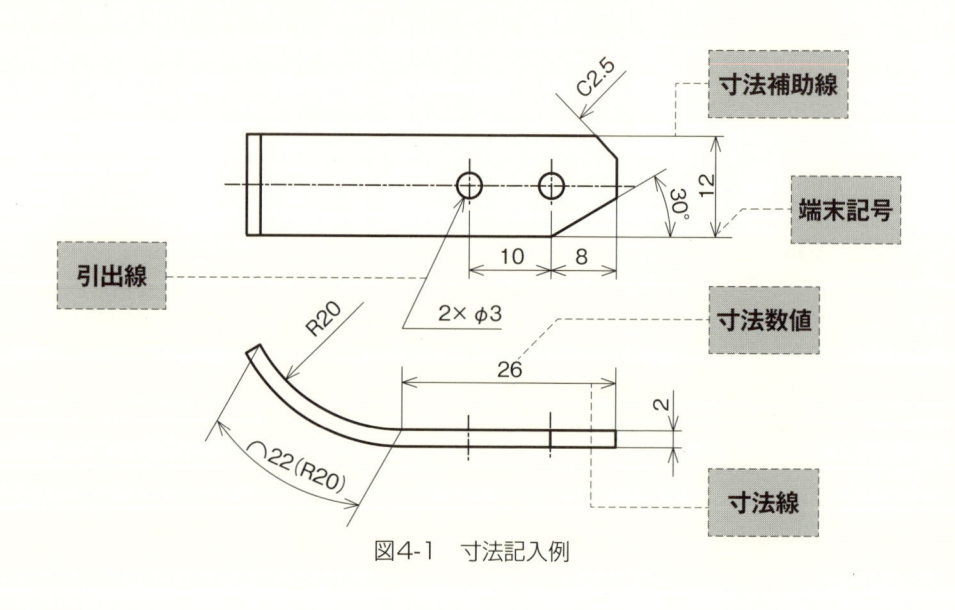

図4-1　寸法記入例

Chapter4	1	**Fundamentals of dimensioning**

> A feature-of-size consists of two parallel plane surfaces, a cylindrical surface or a spherical surface, in defined with a linear distance.
> A feature-of-size might also consist of two plane surfaces at an angle to each other (a wedge) or a conical surface, in defined with an angular size.
> A feature is a specific part of the work-pieces, such as a point, a line or a surface.

The unit of dimension is as follows,
    Size / Length/ Position ... in millimeter (mm)
    Angle ... in degrees (°), minutes ('), and seconds ('')
When the unit of dimensions is required to be expressed in inches, clarify the units used with a note on the drawing.
    e.g.) NOTE: Unless otherwise specified, all dimensions are in inches.

As a rule, the dimension line is placed by using projection lines and dimension value shall be indicated thereon. The dimension line is drawn in parallel to the direction of length to be indicated or of angle to be measured. At both ends of the dimension line, the end symbols are put. Refer to Figure 4-1.

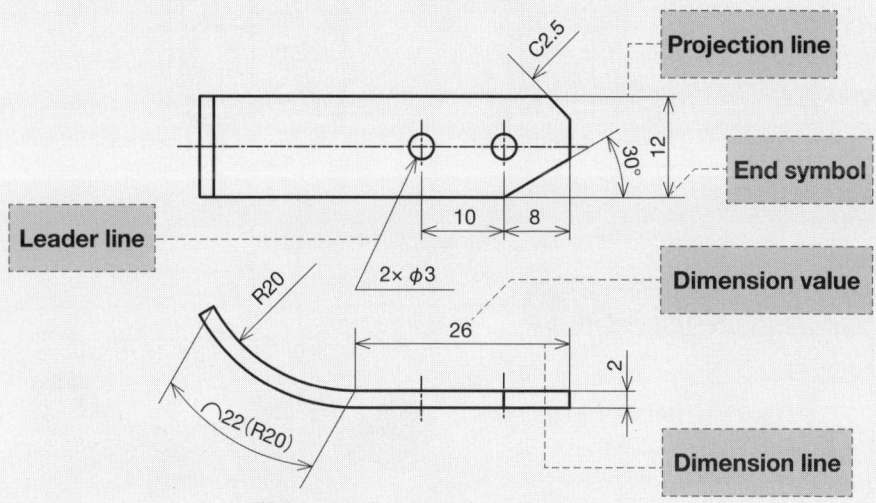

Figure 4-1 Example of dimensioning

図面には、特に明示しない限り、対象物の仕上がり寸法を示します。

　※鋳造部品では、鋳放し図、前加工図、最終機械加工図などがあります。それぞれ鋳放し寸法および前加工寸法、最終仕上がり寸法が指示されます（**図4-2**）。

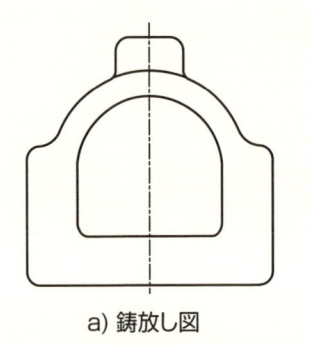

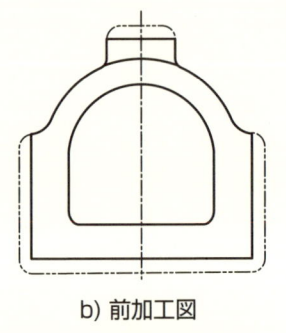

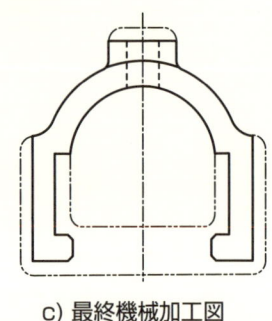

a) 鋳放し図　　　　b) 前加工図　　　　c) 最終機械加工図

図4-2　鋳物部品の図面

　寸法補助線を引き出すと図が紛らわしくなるときは、寸法補助線を使う必要はありません（**図4-3**）。

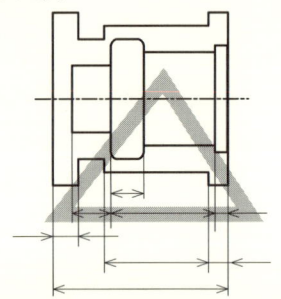

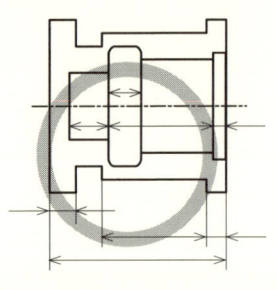

図4-3　寸法補助線を使わない寸法

φ(@°▽°@)　メモメモ

ブランク

　ブランクとは、鋳物や板金など、機械加工直前の部品のことです。
　写真は、切削加工前のナックルアームの鋳放し品です。

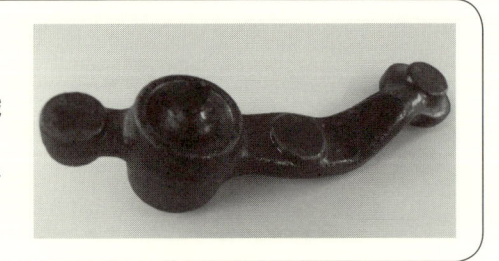

Unless otherwise indicated, the dimensions mean the finished dimensions of the object.

* In casting work-pieces, there are the as-cast drawings, pre-machining drawings, final machining drawings etc. As-cast dimensions and pre-machining dimensions and finished dimensions are indicated in each drawing. Refer to Figure 4-2.

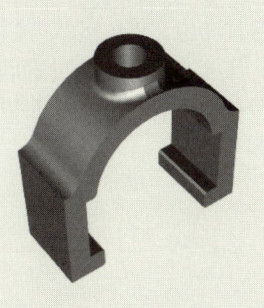

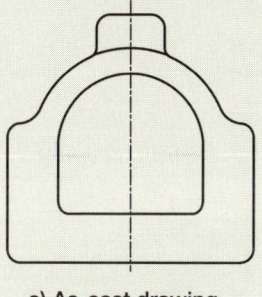

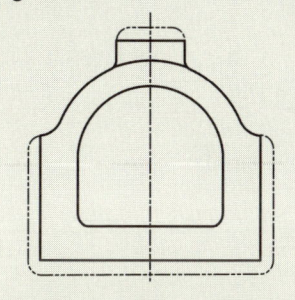

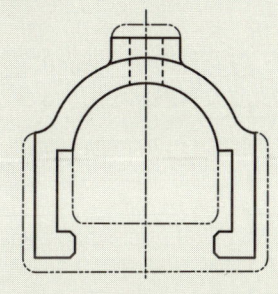

a) As-cast drawing   b) Pre-machining drawing   c) Final machining drawing

Figure 4-2 Drawings of casting work-piece

When the drawing becomes confusing by placing the projection lines, there is no need to use projection lines. Refer to Figure 4-3.

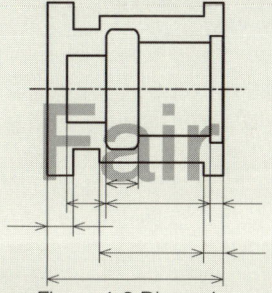

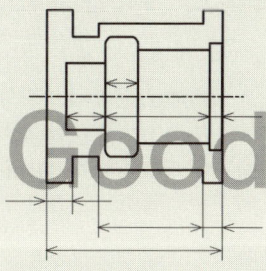

Figure 4-3 Dimensions which not to use projection lines

Note ;-)

Blank

Blank means work-piece before machining, such as a casting or sheet metal.

Photo is an as-cast knuckle arm before the machining.

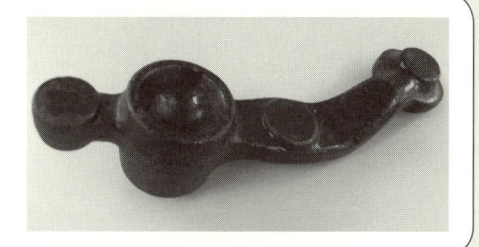

寸法補助線と投影図の間をわずかに離してもかまいませんが、1枚の図面の中、あるいは複数に分割される一連の図面のすべてで統一します（図4-4）。

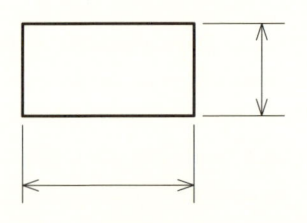

図4-4　隙間を設けた寸法補助線

　寸法補助線の間隔が狭くて矢印を記入する余地がない場合は、矢印の代わりに黒丸または斜線を用います（図4-5）。

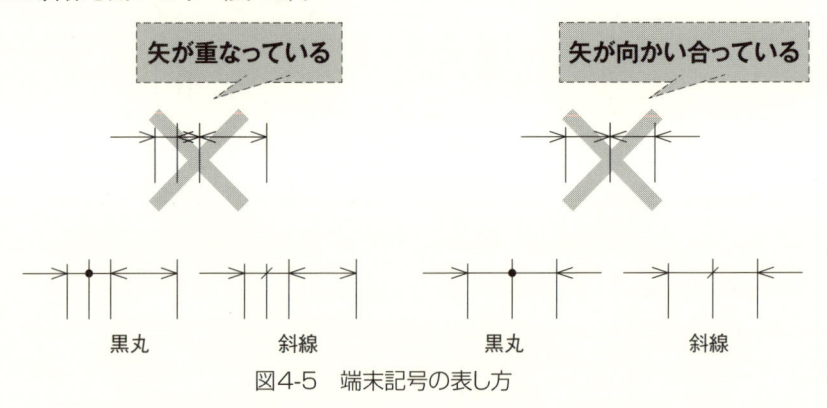

矢が重なっている　　　矢が向かい合っている

黒丸　　斜線　　黒丸　　斜線

図4-5　端末記号の表し方

　狭いスペースへの寸法記入は、部分拡大図を用いるか、引出線を寸法線から斜め方向に引き出し、その上に寸法数値を記入します。この場合、引出線の引き出す側の端には端末記号をつけません（図4-6）。

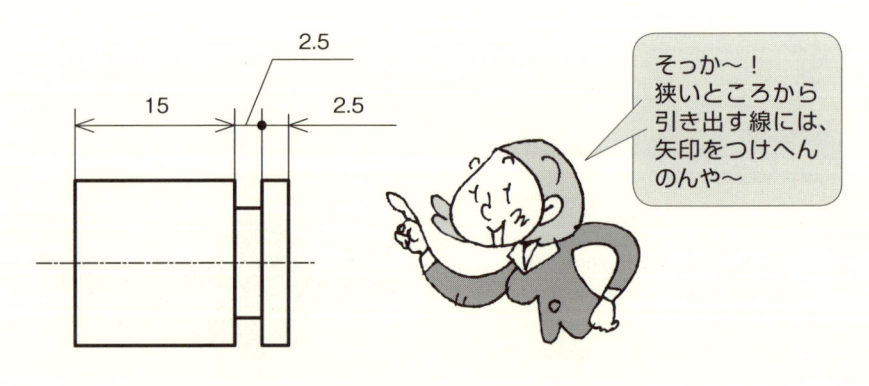

図4-6　引出線を用いた寸法数値の記入

そっか〜！狭いところから引き出す線には、矢印をつけへんのんや〜

第4章　寸法記入のルールって、なんやねん！

The projection lines can be placed with a little gap from the view, but they should be unifying in single-sheet drawing or through all sheets in multi-sheet drawings. Refer to **Figure 4-4**.

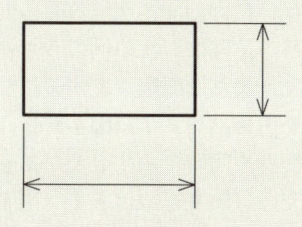

Figure 4-4 Little gap between view and projection lines

When the interval of projection lines is too narrow to place the arrow, black dots or slash can be used instead. Refer to **Figure 4-5**.

Arrows overlapped

Arrows face to face

Poor

Poor

Black dot   Slash   Black dot   Slash

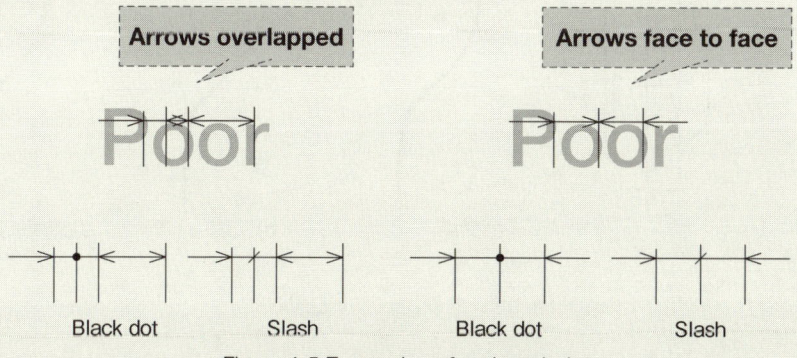

Figure 4-5 Expression of end symbols

For the dimensioning in a narrow space, enlarged view can be used or draw the leader line in oblique direction from the dimension line and place the dimension value above it. In this case, end symbol shall not be attached to the end of the drawer side of leader line. Refer to **Figure 4-6**.

That's it! For the dimensioning in a narrow space, the end symbol shall not be attached!

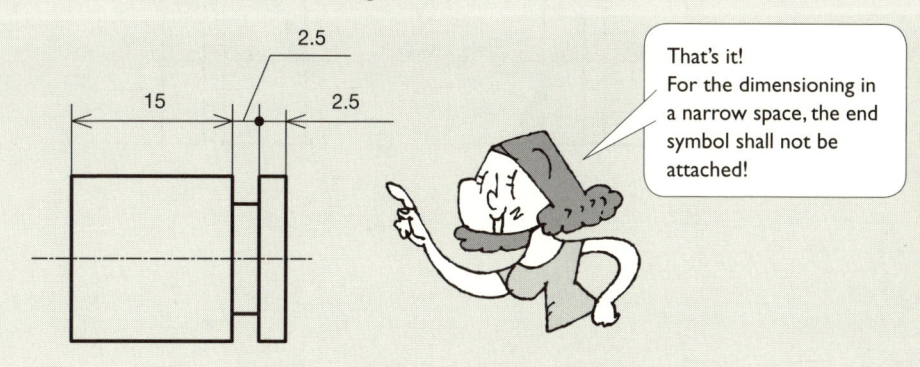

Figure 4-6 Dimensioning using leader lines

互いに傾斜する2つの面の間に丸みまたは面取りがされているとき、2つの面の交わる位置は、次の例のうち一つを使用して図面に表すことができます。
　丸みまたは面取りをする以前の形状線を細い実線で表し、その交点から寸法補助線を引き出します（**図4-7** a）。
　交点を明確に示す必要がある場合は、それぞれの線を互いに交差させるか、または交点に黒丸をつけます（図4-7 b c）。

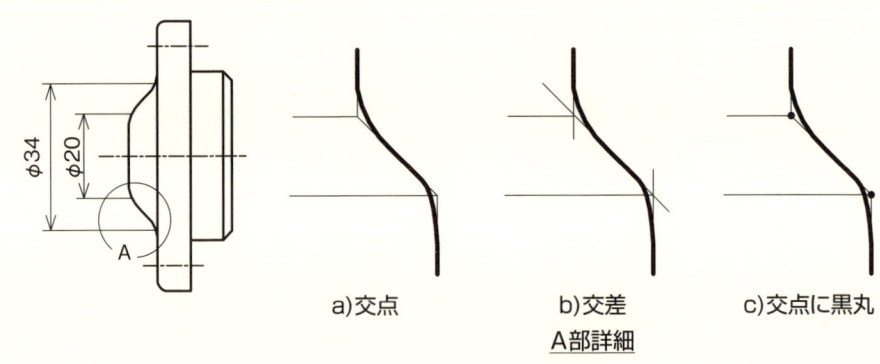

a)交点
A部詳細

b)交差

c)交点に黒丸

図4-7　丸みからの寸法補助線

　寸法補助線が重なって誤解を生じやすい場合、角度をつけた寸法補助線を使うことができます（**図4-8**）。

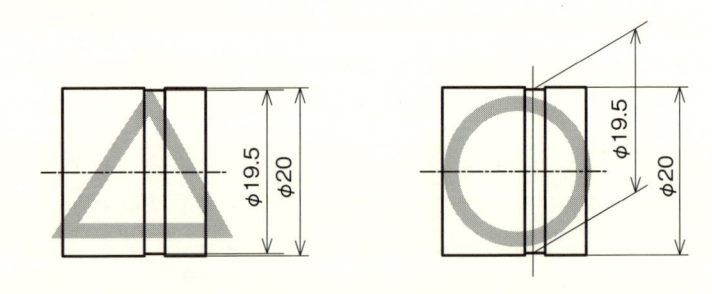

図4-8　角度をつけた寸法補助線

When rounding or chamfering is applied between two surfaces crossing with angles, the crossing point of two surfaces can be expressed on the drawing using one of following examples.

Profile lines before rounding or chamfering is expressed by thin continuous lines, and the projection lines are drawn from the crossing point of these. Refer to Figure 4-7 a.

When it is required to indicate the crossing point clearly, the lines are drawn to intersection each other or black dots are attached to the crossing point. Refer to Figure 4-7 b and c.

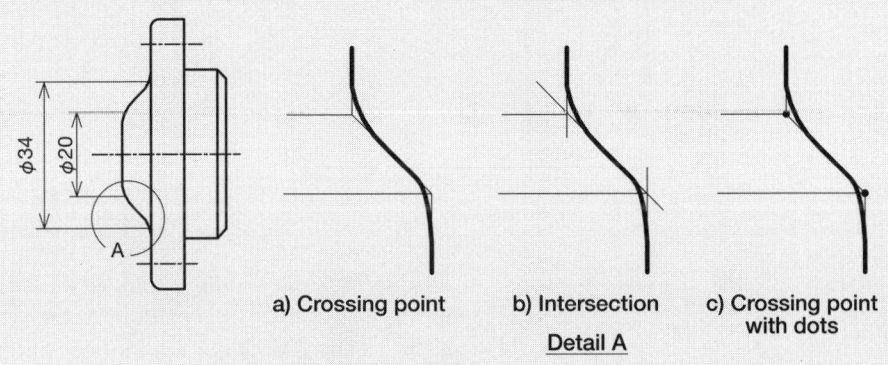

a) Crossing point   b) Intersection   c) Crossing point with dots

Detail A

Figure 4-7 Projection lines drawn from rounding

When projection lines overlap each other and it is easy to misunderstand, angled projection lines can be used. Refer to Figure 4-8.

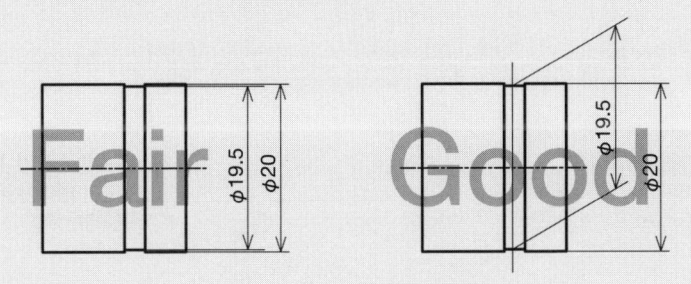

Figure 4-8 Angled projection lines

対称形状で中心線の片側だけを示した場合、寸法線はその中心線を越えるまで延長します。この場合、延長した寸法線の端には端末記号をつけません（**図4-9 a**）。
　ただし、誤解や混乱がなければ、寸法線は中心線を越えなくてもかまいません（図4-9 b）。

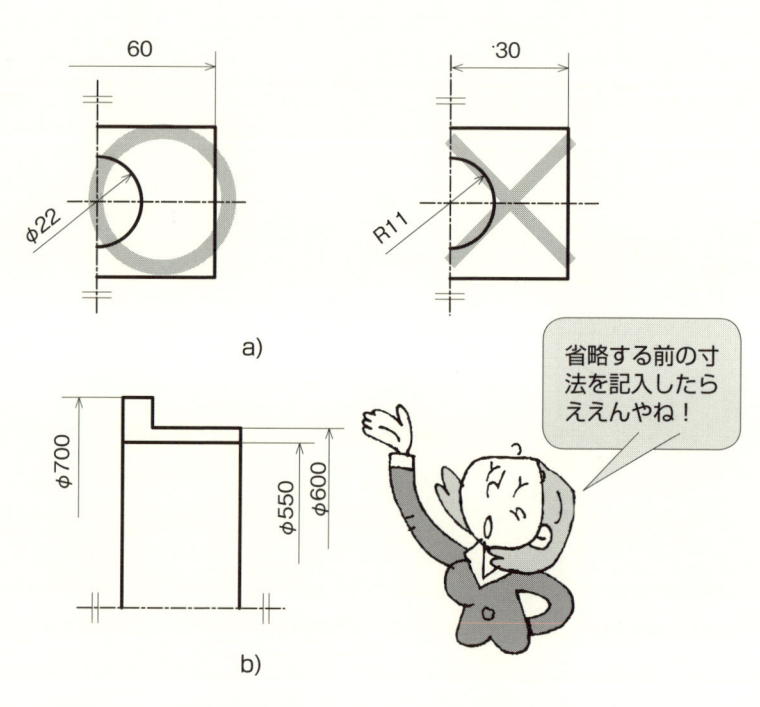

図4-9　半分省略した対称図形の寸法線の表し方

省略する前の寸法を記入したらええんやね！

寸法計測の決まりごとってあるん？

よー覚えとけよ！寸法計測の原則は2点間計測や！

When only one side of the center line is indicated in the symmetrical shape, dimension line is extended until exceeding center line

In this case, do not attach the end symbol on the end of extended dimension line. Refer to Figure 4-9 a.

However, when there is no misunderstanding or confusion, the dimension line does not have to exceed the center line. Refer to Figure 4-9 b.

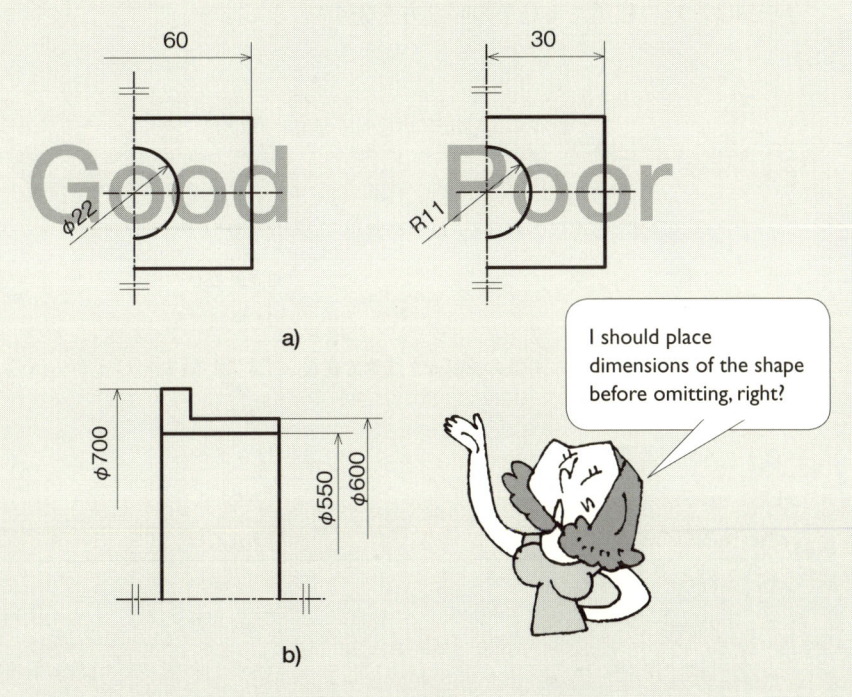

a)

b)

Figure 4-9 Expression of dimension line in half view of symmetrical shape

I should place dimensions of the shape before omitting, right?

Is there any rule of size measurements?

Keep mind!
Principle of size measurements is two-point measurements!

Chapter4 What is the rule of dimensioning!

第4章 2 寸法補助記号

寸法補助記号の種類

　寸法補助記号は、その要求を明確にするために、寸法数値と共に使用します。寸法補助記号の種類およびその呼び方を**表4-1**に示します。

表4-1　寸法補助記号の種類と呼び方

記号	意味	呼び方
φ	180°を超える円弧の直径または円の直径	「まる」または「ふぁい」
R	半径	「あーる」
CR	コントロール半径	「しーあーる」
Sφ	180°を超える球の円弧の直径または球の直径	「えすまる」または「えすふぁい」
SR	球の半径	「えすあーる」
C	45°の面取り	「しー」
□	正方形の辺	「かく」
⌒	円弧の長さ	「えんこ」
t	厚さ	「てぃー」
▽	穴深さ	「あなふかさ」
⊔	（浅い）ざぐり	「ざぐり」‥黒皮を少し削り取るもの
	深ざぐり	「ふかざぐり」
∨	皿ざぐり	「さらざぐり」

　その他の記号を**表4-2**に示します。

表4-2　その他記号の種類と呼び方

記号	意味	呼び方
◁	こう配	「こうばい」
▷	テーパ	「てーぱ」

Chapter 4 — 2. General symbols for dimensioning

Type of general symbols for dimensioning

General symbols for dimensioning are used with dimension value to clarify the requirement.

Type of general symbols for dimensioning and their designations are in Table 4-1.

Table 4-1 Specification of general symbols for dimensioning and their designations

General symbol	Meaning	Designation in Japanese
ϕ	Diameter of circular arc exceeding 180° or diameter of circle	"MARU" or "FWAI"
R	Radius	"AˆRU"
CR	Control radius	"SHIˆAˆRU"
Sϕ	Diameter of circular arc of sphere exceeding 180° or diameter of sphere	"ESUMARU" or "ESU FWAI"
SR	Sphere radius	"ESUAˆRU"
C	Chamfering of 45°	"SHII"
□	Side of square	"KAKU"
⌒	Length of circular arc	"ENKO"
t	Thickness	"TEI"
▽	Hole-depth	"ANAFUKASA"
⊔	Spot face	"ZAGURI"… surface as casting is shaved a little bit.
⊔	Counter-bore	"FUKAZAGURI"
∨	Countersink	"SARAZAGURI"

Type of the other symbols is in Table 4-2.

Table 4-2 Type of the other symbols

Symbol	Meaning	Designation in Japanese
◁	Slope	"KOUBAI"
▷	Taper	"TEˆPA"

1）直径の表し方

a) 対象とする部分の断面が円形であるとき、円の形状を表さずに示す場合には、記号「φ」を寸法数値の前に記入します（**図4-10**）。
使われる記号は、ギリシャ文字のファイです。

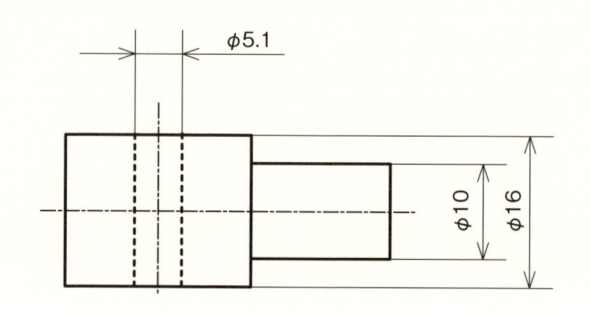

図4-10　円の形状を表さずに示す直径の指示

b) 円弧の投影図において、寸法線の端末記号が片側の場合は、半径と誤解しないように直径の寸法数値の前に「φ」を記入します。同様に、引出線を用いて記入する場合にも、直径の記号「φ」を記入します（**図4-11**）。

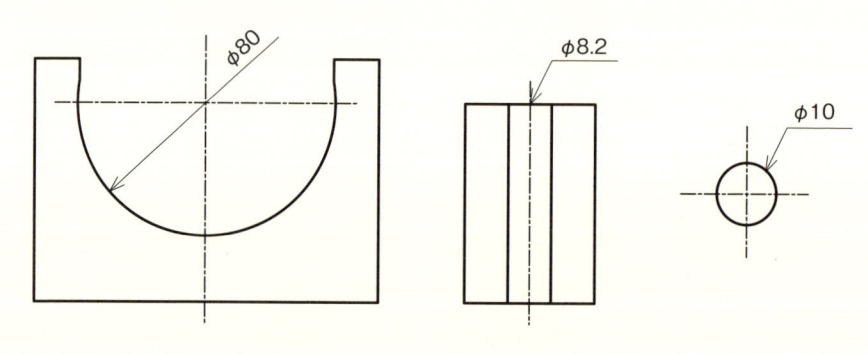

図4-11　様々な直径の表し方

-1. Expression of diameter

a) When the section of the object is circular shape, for indicating that is circular without expressing circular shape, the symbol "ø" is put before the dimension value. Refer to Figure 4-10.

The symbol used is the Greek letter phi.

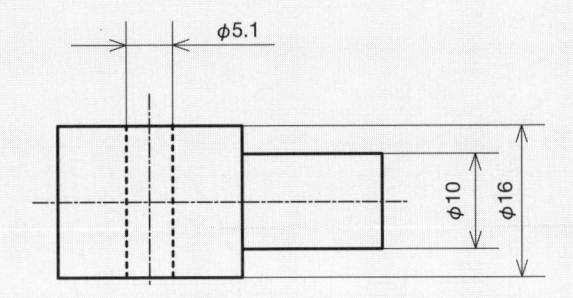

Figure 4-10 Expression of diameter dimensions without expressing circular shape

b) In a view of an circular arc, when the end symbol of dimension line is given only on one end, in order to avoid confusion with radius, symbol "ø" should be put before the dimension value on diameter. Also when the dimension is placed using the leader line, symbol "ø" should be put. Refer to Figure 4-11.

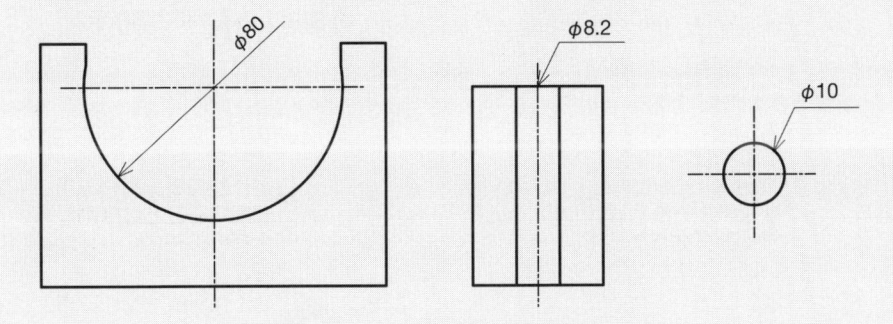

Figure 4-11 Expressions of various diameter

Chapter4 What is the rule of dimensioning!

c) 180°を超える円弧または円の中に直径の寸法を記入する場合、寸法線の両端に端末記号が付く場合は、寸法数値の前に記号「φ」は記入しません。
しかし、記号「φ」を記入する企業が多く、実務上の標準となっています（図4-12）。

さらに、記号「×」は形体が繰り返される時の数を示すために使用されます。繰り返しの数に、記号「×」とスペースを入れた直径数値を続けます。P140を参照してください。

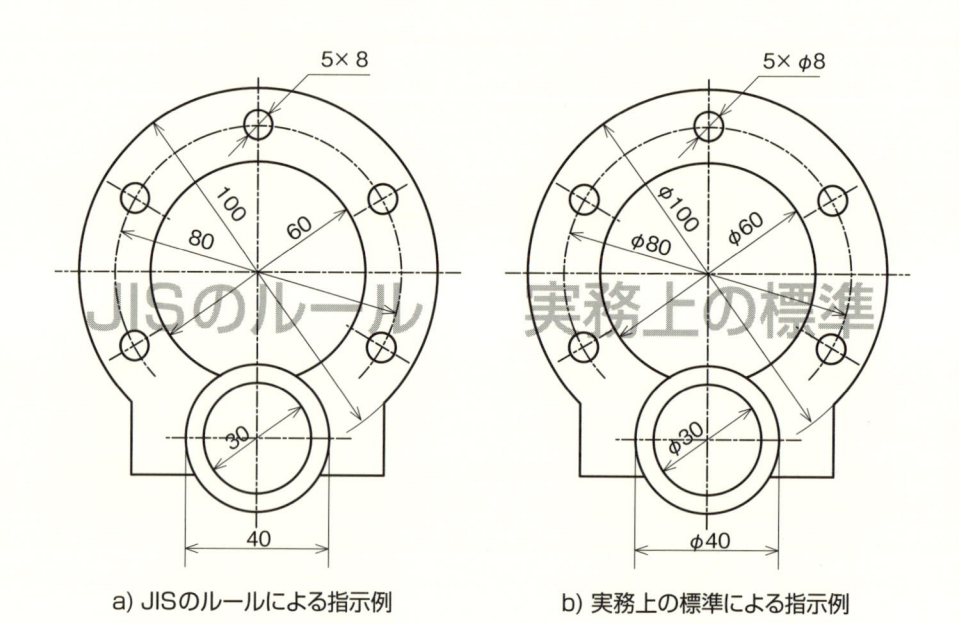

a) JISのルールによる指示例　　　b) 実務上の標準による指示例

図4-12　180°を超える円弧および全円の直径の指示

この場合、JISのルールと会社のルールでは、どっちに従ったらいいんですか！？

そやな‥どっちの表記も読み手が誤解することはないから、会社のルールに従えばええで！

第4章　寸法記入のルールって、なんやねん！

c) When the dimension of diameter is placed in a circular arc over 180° or in a circle, and end symbols are put to both ends of the dimension line, the symbol "ø" is not put before the dimension value.

However, there are many companies which express the symbol "ø" and it is positioned as the de facto standard of the design industry. Refer to Figure 4-12.

In addition, the symbol "X" is used to indicate the number when features are to be repeated. Repetitions number with symbol "X" and a space are put before the dimension value. Refer to page 141.

a) Expression by the JIS rule

b) Expression by de facto standard

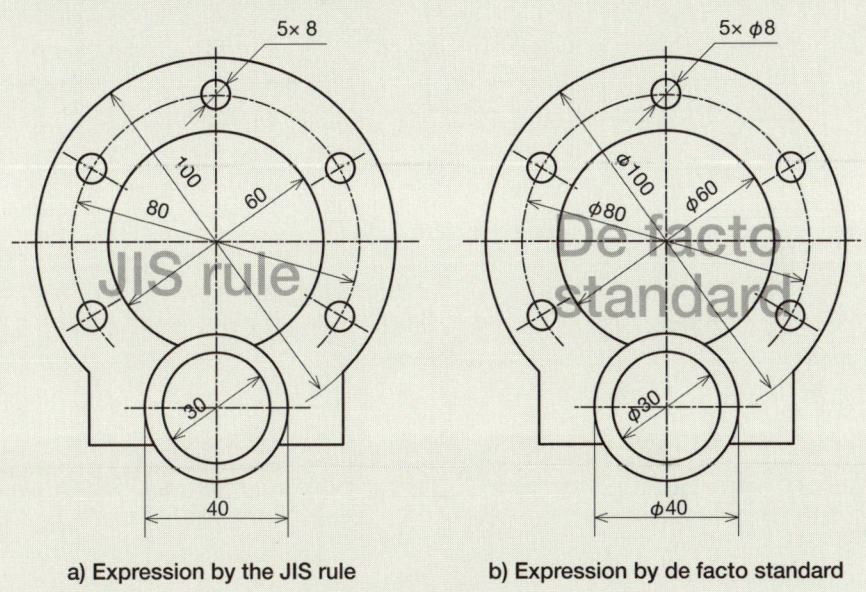

Figure 4-12 Expressions of diameter of circular arc over 180° or diameter of circle

In this case, which should I follow with JIS rule or company rule!?

Yes you do.
Readers do not misunderstand both expressions.
You should follow your company rule.

Chapter4 What is the rule of dimensioning!

2) 半径の表し方

a) 一般的に、180°以下の円弧は半径で表します。半径の寸法は、半径の記号「R」を寸法数値の前につけます。半径の寸法線は、円弧の側だけに矢をつけます（図4-13）。

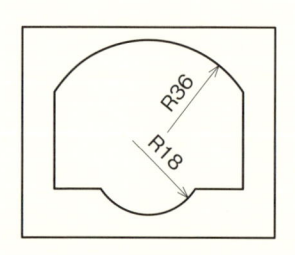

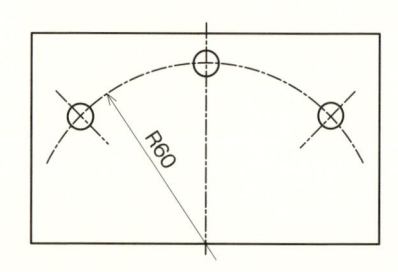

図4-13　半径の表し方

b) 矢の向きや数値は次の例のうちの一つを使用して、図面に表すことができます。（図4-14）。

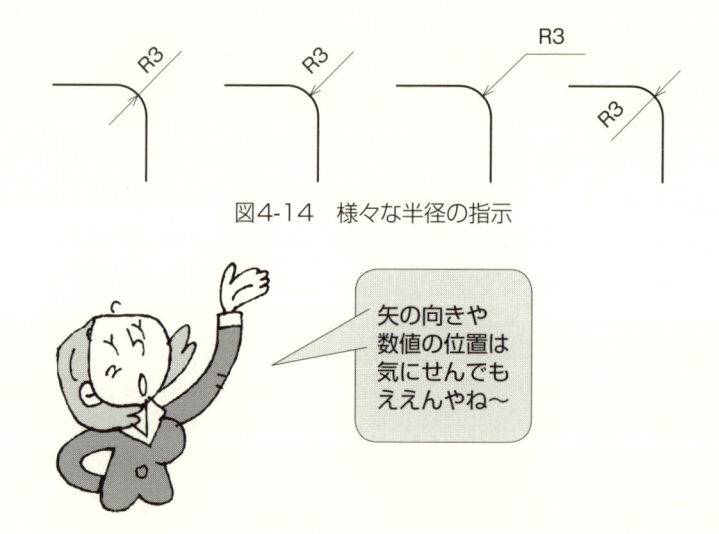

図4-14　様々な半径の指示

矢の向きや
数値の位置は
気にせんでも
ええんやね〜

-2. Expression of radius

a) Generally, circular arc of less than 180° is expressed by radius. When expressing radius dimension, place the symbol of radius "R" before the radius dimension value. Radius dimension line attaches an arrow to only circular arc side. Refer to Figure 4-13.

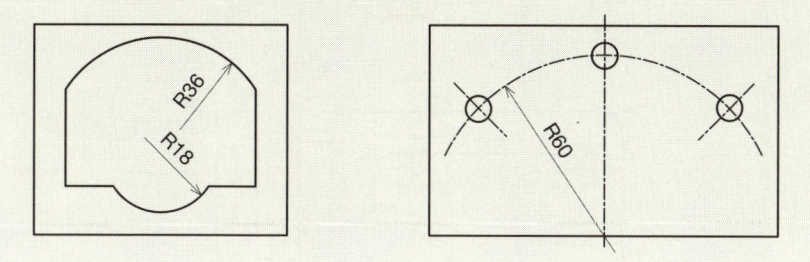

Figure 4-13 Expression of radius

b) Direction of an arrow or dimension value can be expressed on the drawing using one of following examples. Refer to Figure 4-14.

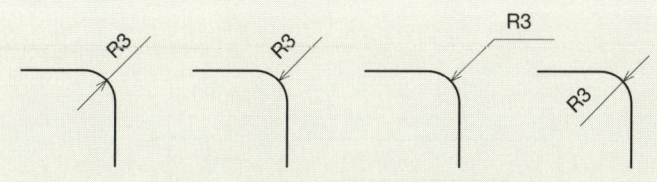

Figure 4-14 Various expressions of radius

It is necessary to care about neither direction of an arrow nor a value position, isn't it!

Chapter4 What is the rule of dimensioning!

c) 同じ大きさの2つ以上のかどや隅の丸みなどに半径を表す場合には、関連する形状の半径は数を記入しません。しかし、半径の数を記入する企業が多く、設計業界の実務上の標準となっています（図4-15）。

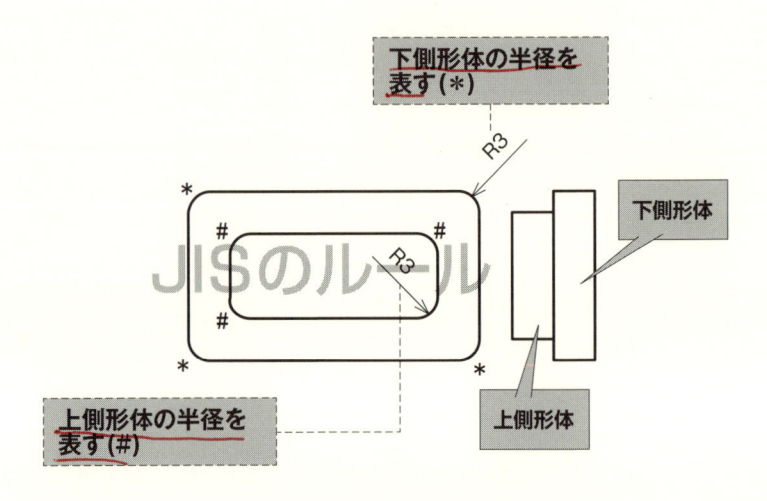

下側形体の半径を表す(*)

下側形体

上側形体の半径を表す(#)

上側形体

a) 半径の数(JISのルール)

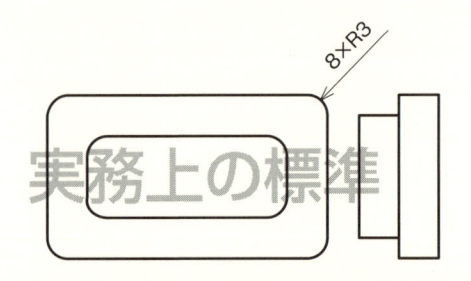

8×R3

b) 半径の数(実務上の標準)

図4-15 半径の数の表し方

c) When two or more corner roundness of the same size is expressed by radius, related number of radius is not placed. However, there are many companies which express the number of radius, and it is positioned as the de facto standard of the design industry. Refer to Figure 4-15.

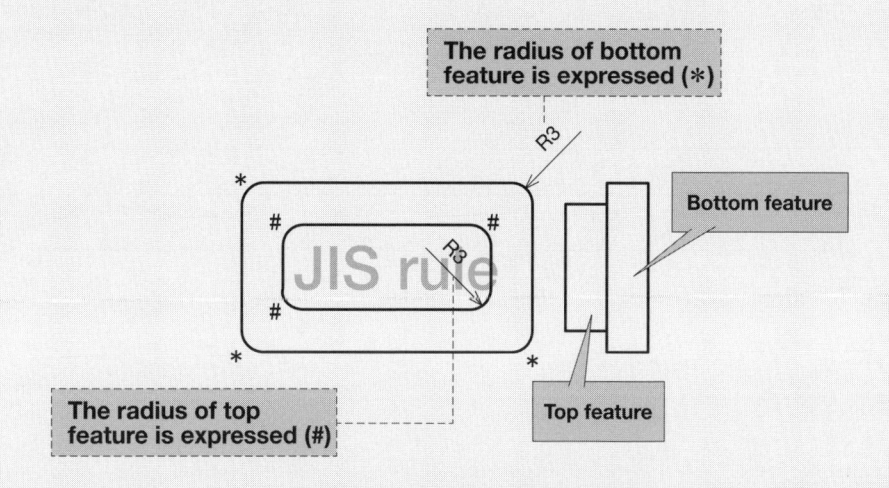

a) Number of radius (by JIS rule)

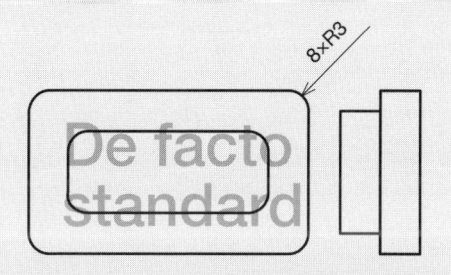

b) Number of radius (by de facto standard)

Figure 4-15 Expression of number of radius

d) 半径の寸法を指示するために円弧の中心位置を表す必要がある場合、中心位置は次の例のうちの一つを使用して図面に表します。

　一般的に、十字の中心線または黒丸で中心位置を示します（**図4-16 a**）。

　紙面などに制約があり、円弧の半径が大きい場合、その中心位置は折り曲げた寸法線で投影図の近くに置くことができます。この場合、寸法線の矢印のついた部分は、正しい中心位置に向いていなければいけません（図4-16 b）。

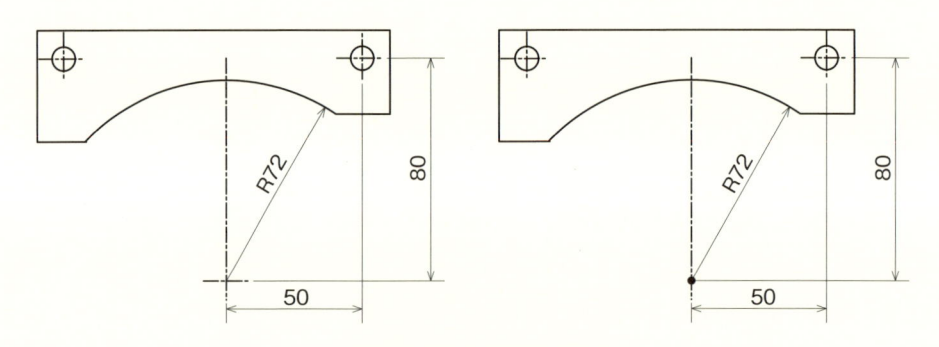

a) 円弧の中心位置を示す場合

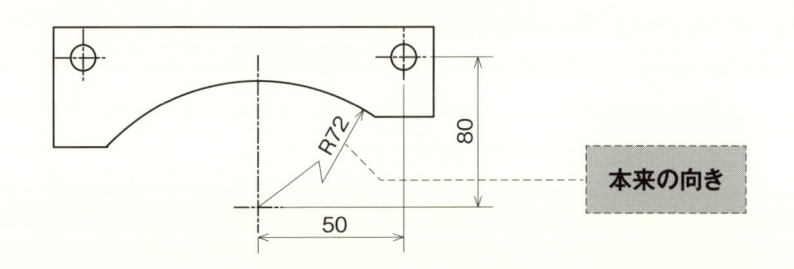

b) 紙面に制約がある例

図4-16　円弧の中心位置を表す半径寸法

d) When it is required to indicate the center position of circular arc in order to express the radius, the center position can be expressed on the drawing using one of the following examples.

Typically, center position is expressed by cross center lines or black dot. Refer to Figure 4-16 a.

When the space is limited or the radius of circular arc is large, its center position can be put to near view by the folding dimension line. In this case, the part of the dimension line with arrow should be directed to the correct center position. Refer to Figure 4-16 b.

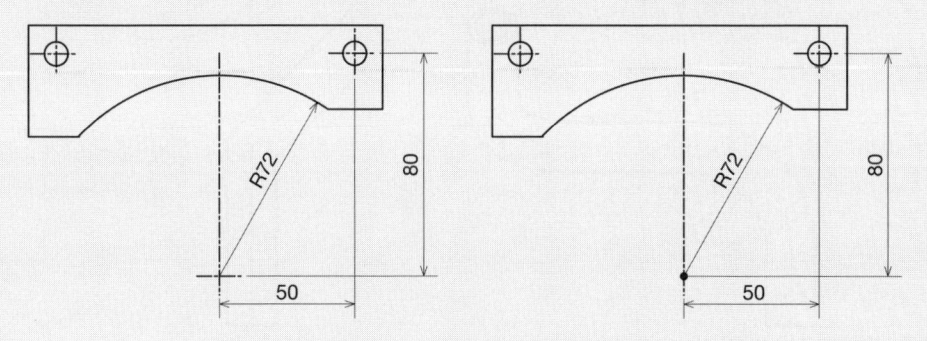

a) Radius dimension expressing an arc center position

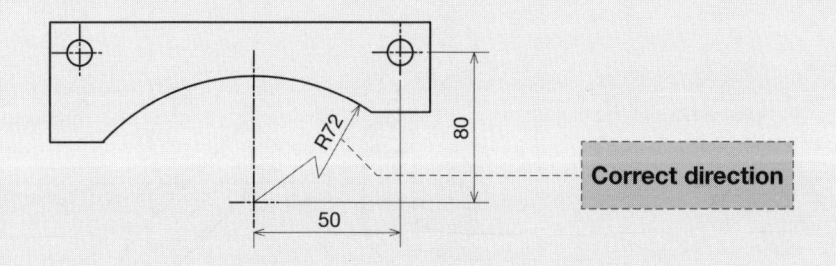

b) Example of the limited space

Figure 4-16 Radius dimension expressing center position of circular arc

e) 実形を表していない投影図に実際の半径を指示する場合には、寸法数値の前に「実R」の文字記号を記入します（**図4-17**）。

　同様に、展開した状態の半径を指示する場合には、「展開R」の文字記号を数値の前に記入します。

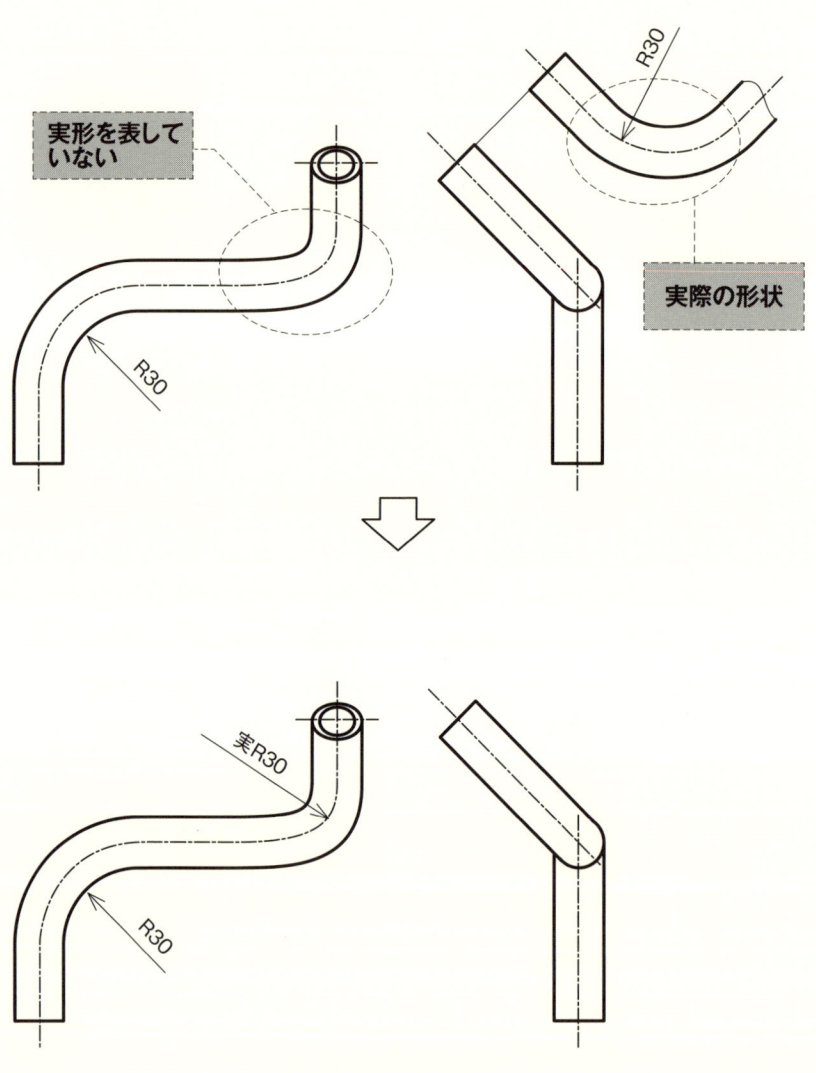

図4-17　実際の半径の表し方

e) When the actual radius is to be expressed on the view indicating no true shape, place the letter symbol "True R ("JITSU-R" in Japanese)" before the dimension value. Refer to Figure 4-17.

 Also when the developed radius is to be indicated, place "Developed R ("TENKAI-R" in Japanese)" before the dimension value.

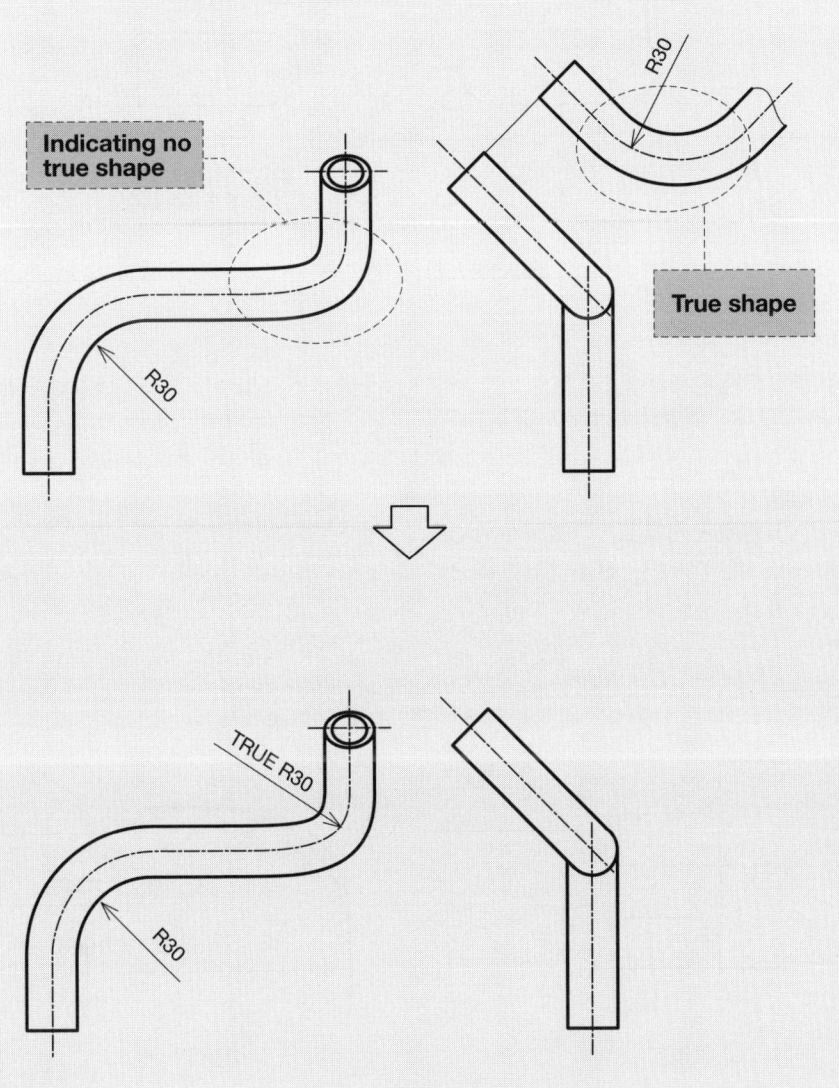

Figure 4-17 Expression of true radius

f) 半径の寸法が他から導かれる場合には、数値のない記号「(R)」によって表します（図4-18）。

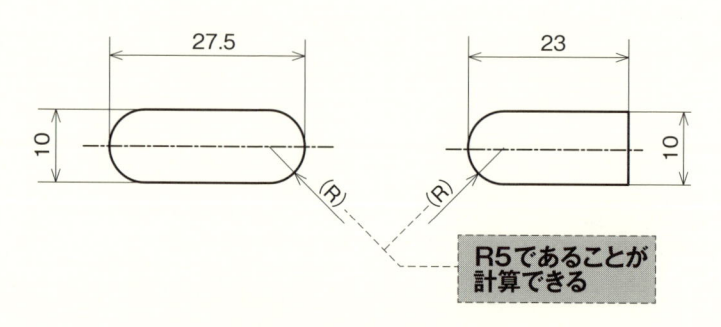

図4-18　記号(R)の表し方

g) かどや隅の丸みなどにコントロール半径を要求する場合には、半径数値の前に記号「CR」を指示します（図4-19 a）。
　　コントロール半径とは、次に示すようにコントロールされた半径です（図4-19 b）。
- 直線部と曲線部との接続が滑らかにつながる。
- 最大許容半径と最小許容半径との間（2つの曲面に接する公差域）に円弧が存在する。

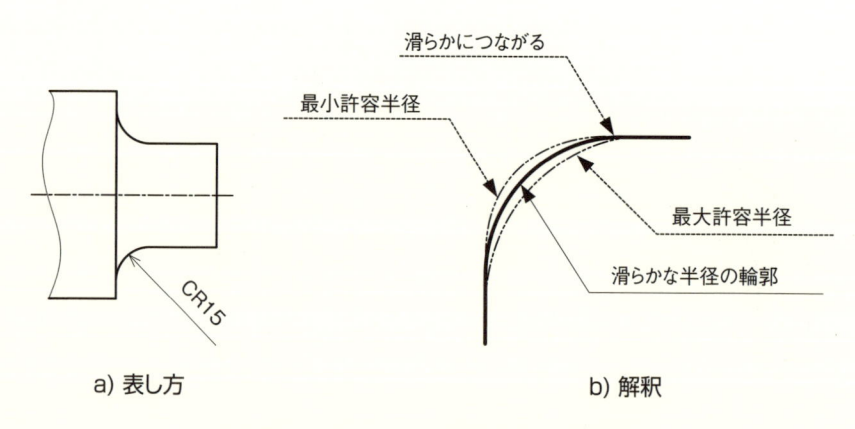

a) 表し方　　　　b) 解釈

図4-19　コントロール半径の表し方と解釈

f) When the dimension of radius can be lead to the other dimensions, the radius is expressed by symbol "(R)" without dimension value. Refer to Figure 4-18.

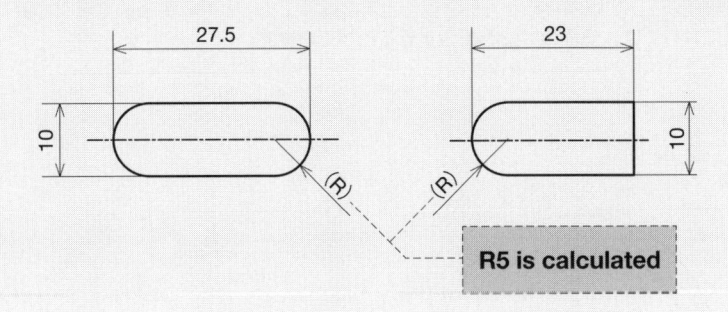

R5 is calculated

Figure 4-18 Expression of symbol (R)

g) When a control radius is required for the roundness of a corner etc., put the symbol "CR" before the radius dimension. Refer to Figure 4-19 a.
Control radius means radius controlled as follows: Refer to Figure 4-19 b.
 - The junction between a straight section and a curved section should be smoothly connected.
 - Circular arc should be exists in area between a maximum permissible radius and minimum permissible radius (tolerance zone which is tangent to two curved surfaces).

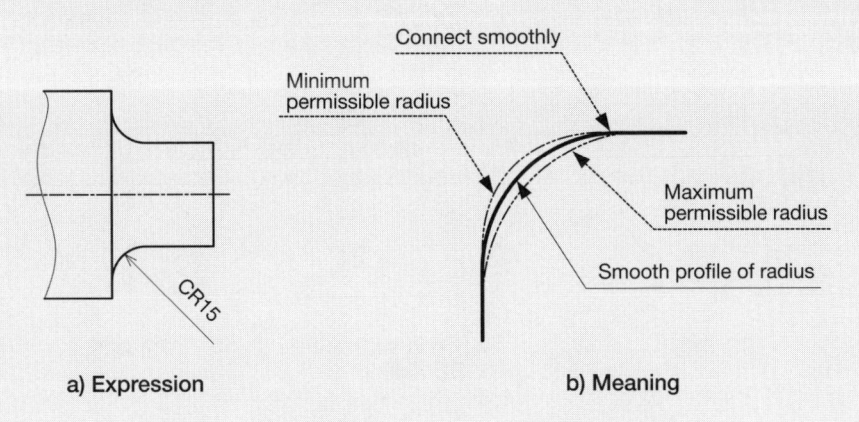

a) Expression   b) Meaning

Figure 4-19 Expression and meaning of control radius

φ(@°▽°@)　メモメモ

直径と半径の使い分け

　円弧の寸法は、円弧が180°までは半径で表し、それを超える場合には直径で表します。
　ただし、円弧が180°以内であっても、機能上または加工上、とくに直径の寸法を必要とするものに対しては、直径の寸法を記入することができます。

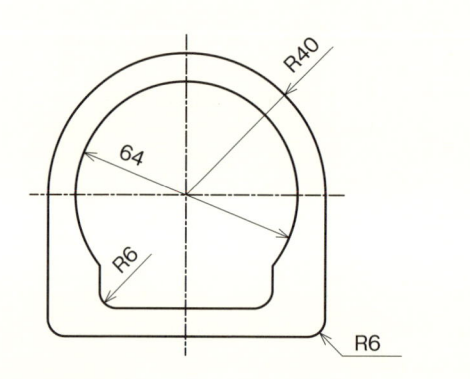

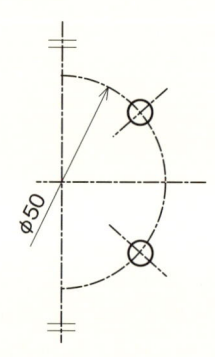

角度寸法の省略

　いくつかの形体がピッチ円上に等間隔で整列する場合、角度寸法は示しません。

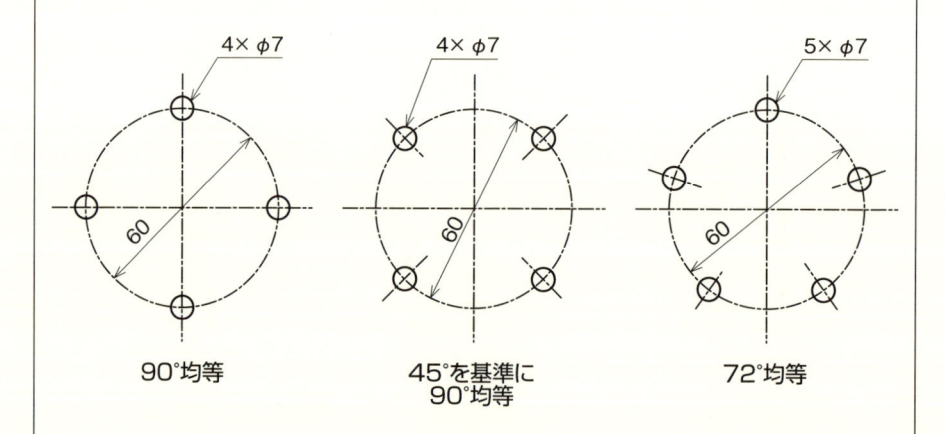

90°均等　　　45°を基準に90°均等　　　72°均等

Note ;-)

To use diameter or radius

When the angle of the circular arc is up to 180°, the dimension is expressed by the radius. When the angle of the circular arc is over 180°, dimension is expressed by the diameter.

However, even when the angle of the circular arc is up to 180°, when the dimension of diameter specially requires function or processing, the diameter dimension can be placed.

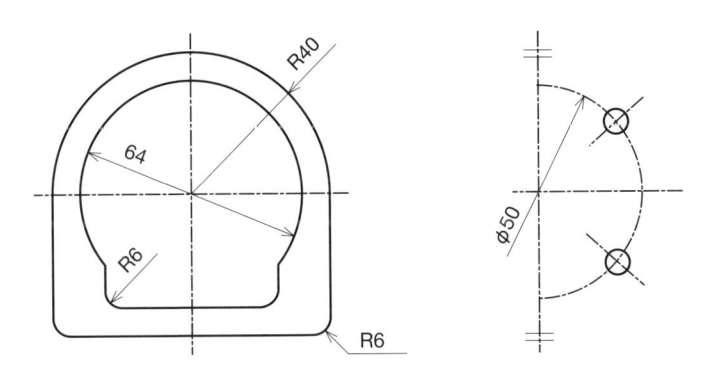

Omission of angular dimensions

When several features are arranged equal interval around a pitch circle, there is no requirement to indicate an angular dimension.

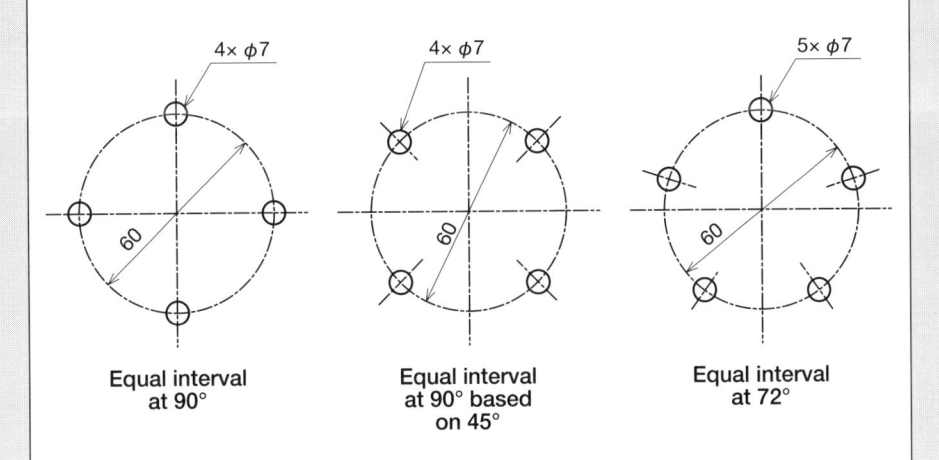

Equal interval at 90°

Equal interval at 90° based on 45°

Equal interval at 72°

Chapter4 What is the rule of dimensioning!

3) 球の直径または半径の表し方

a) 球の直径または半径の寸法は、その寸法数値の前に記号「Sφ」または「SR」を記入します（**図4-20**）。

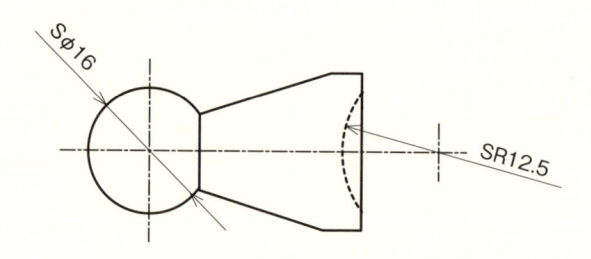

図4-20　球の直径または半径

b) 球の半径の寸法が他から導かれる場合には、数値のない記号「(SR)」によって表します（**図4-21**）。

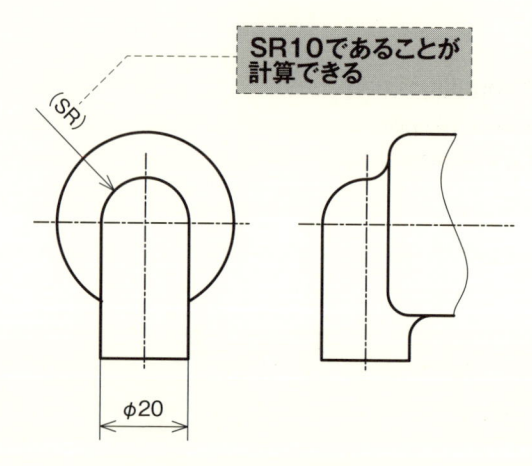

SR10であることが計算できる

図4-21　記号(SR)の表し方

-3. Expression of diameter or radius of sphere

a) The dimension of diameter or radius of sphere is put by symbol "Sϕ" or "SR" before the dimension value. Refer to Figure 4-20.

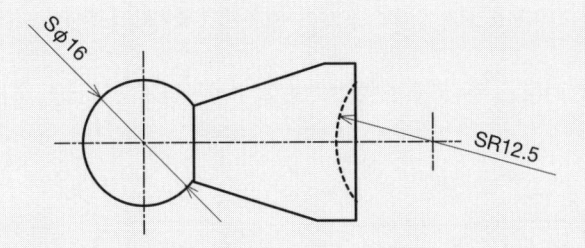

Figure 4-20 Expression of diameter or radius of sphere

b) When the dimension of radius of a sphere can be lead to the other dimensions, the radius of a sphere is expressed by symbol "(SR)" without dimension value. Refer to Figure 4-21.

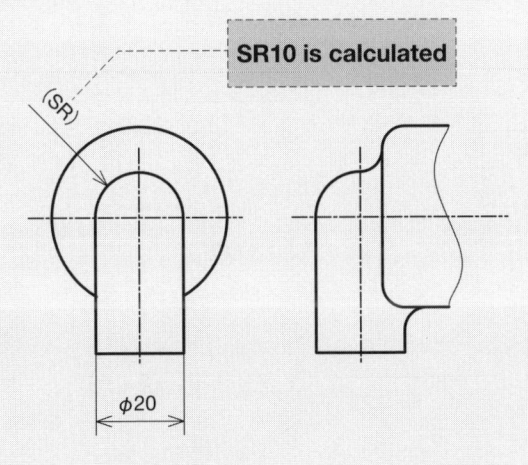

Figure 4-21 Expression of symbol (SR)

Chapter4 What is the rule of dimensioning!

4）面取りの表し方

　45°の面取りの場合には、記号「C」を寸法数値の前に記入します。

　かどの丸みと同様に、同じ大きさの2つ以上のかどにある45°の面取りを記号「C」で指示する場合には、関連する形状の面取りは個数で表しません。

　しかし、面取りの数を記入する企業が多く、設計業界の実務上の標準となっています（**図4-22 a**）。

　また、「面取りの寸法×45°」を代用できますが、日本ではほとんど使われません（図4-22 b）。

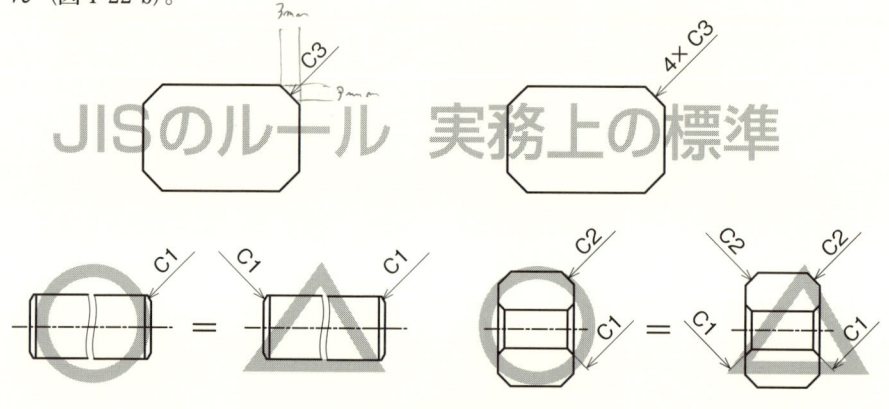

a) 個数の省略

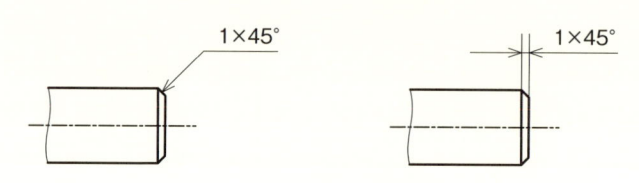

b) 他の表し方

図4-22　45°面取りの表し方

　45°以外の面取りは、角度寸法によって表します（**図4-23**）。

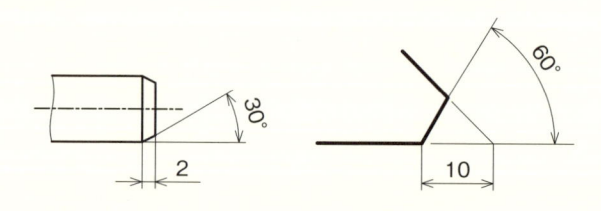

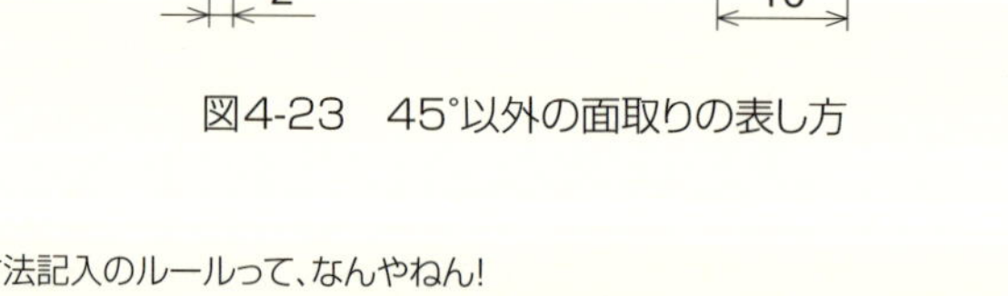

図4-23　45°以外の面取りの表し方

-4. Expression of chamfer

In case of 45° chamfer, symbol "C" is put before the dimension value.

Same as corner roundness, when two or more 45° chamfers of the same size is expressed by symbol "C", the number of chamfer that related is not placed.

However, there are many companies which express the number of chamfer, and it is positioned as the de facto standard of the design industry. Refer to Figure 4-22 a.

And "value of chamfer × 45°" can be used instead of symbol "C". But this is hardly used in Japan. Refer to Figure 4-22 b.

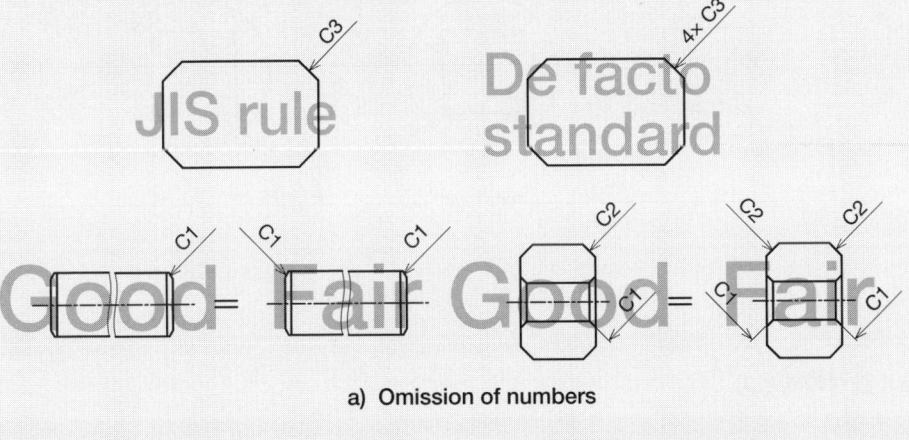

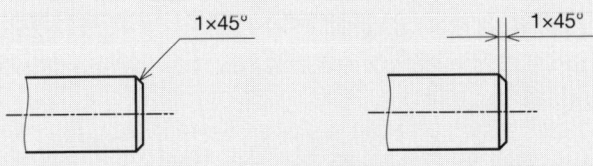

a) Omission of numbers

b) Other expression

Figure 4-22 Expression of 45° chamfer

Chamfer except 45° is expressed by angular dimension. Refer to Figure 4-23.

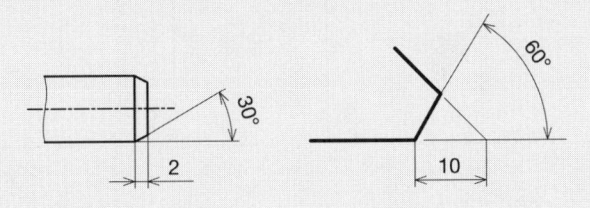

Figure 4-23 Expression of chamfer except 45°

5）正方形の辺の表し方

対象物の断面が正方形であるとき、正方形の形状を表さずに示す場合には、記号「□」を寸法数値の前に記入します（**図4-24 a**）。

正方形が投影図にあらわれる場合には、記号「□」を使うことはできず、両辺の寸法を記入しなければいけません（図4-24 b）。

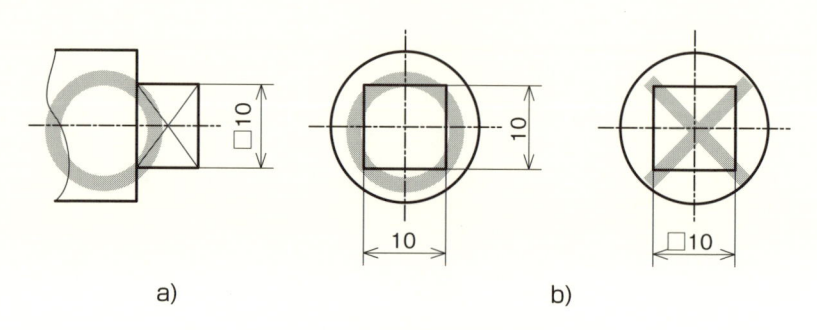

　　　　　　　　　　　　a)　　　　　　　　　　　　　　　　　b)

図4-24　角柱の辺の指示例

6）板厚の表し方

厚みの寸法を表す場合、厚さを表す寸法数値の前に、厚さを示す記号「t」を記入します。それは正面図の付近または正面図の中に記入します（**図4-25**）。

冷間圧延鋼板（SPCCなど）、購入品の樹脂製の板など、製品公差が規定されている板材の厚さ指示には特に便利です。

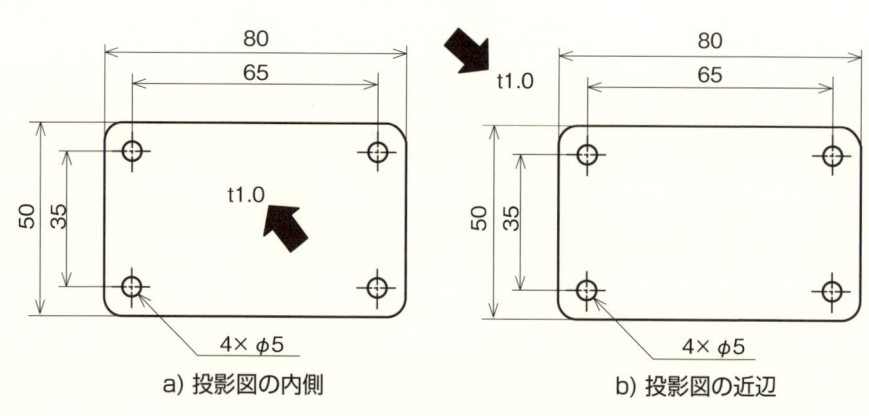

a) 投影図の内側　　　　　　　　　b) 投影図の近辺

図4-25　板厚の指示例

-5. Expression of side of square

When the section of the object is square shape, for indicating that is square without expressing square shape, the symbol " □ " is put before the dimension value. Refer to Figure 4-24 a.

When the square shape is visible in the view, the symbol " □ " cannot use. So, dimension of each side length of the square should be placed. Refer to Figure 4-24 b.

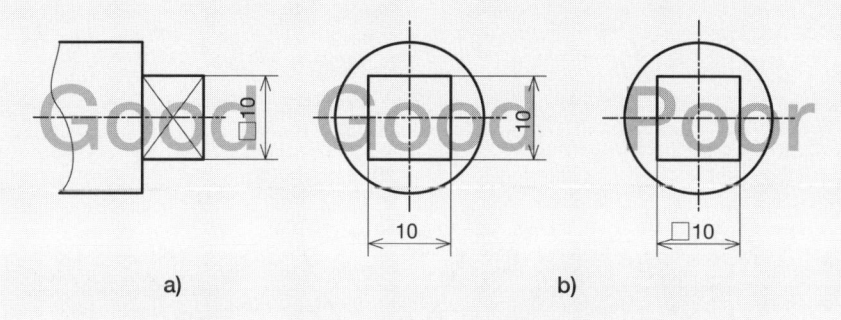

Figure 4-24 Expression of square shape

-6. Expression of plate thickness

When the dimension of plate thickness is to be expressed, place the symbol "t" before the dimension value of thickness. It should be put in the near or inside the front view. Refer to Figure 4-25.

This is especially convenient for expression of thickness of plate materials such as cold-rolled steel (SPCC etc.) or purchased resin plate of which the product tolerance is specified.

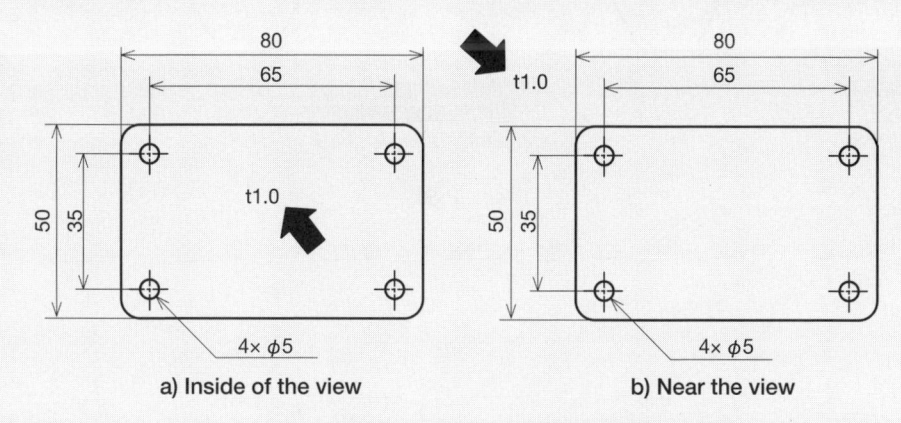

a) Inside of the view   b) Near the view

Figure 4-25 Expression of plate thickness

7) 弦および円弧の長さの表し方

a) 弦の長さは、弦に垂直な寸法補助線と、弦に平行な寸法線を用いてあらわします。（図4-26 a）。

弧の長さは、同心の円弧を寸法線として引き、寸法数値の前に円弧の長さの記号をつけます。弦の場合と同様な寸法補助線を描きます（図4-26 b）。

※弦とは、曲線状の2点間を結んだ直線をいう。

円弧とは、円の一部のような曲線または形状をいう。

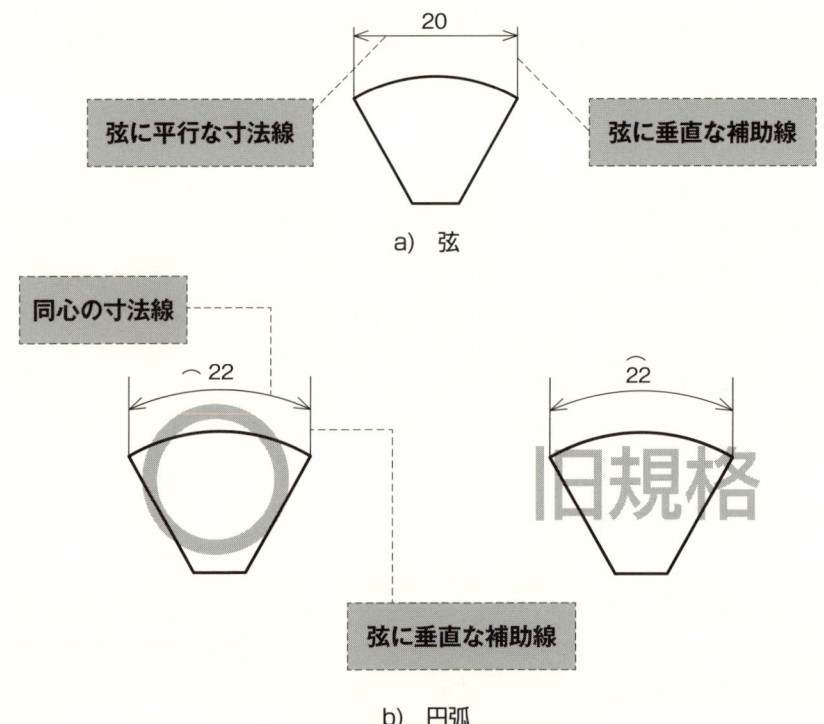

図4-26　弦、円弧の長さ寸法の例

-7. Expression of length of chord or circular arc

a) The length of chord is expressed by perpendicular projection lines to the chord and by using the dimension line parallel to the chord. Refer to Figure 4-26 a.

The length of circular arc is expressed by using the concentric dimension line and place the symbol of length of circular arc before the dimension value.

And the projection lines should be drawn same as in case of chord. Refer to Figure 4-26 b.

*A chord means a straight line joining two points on a curve.

A circular arc means a line or shape that is curved like part of a circle.

Dimension lines parallel to chord

Perpendicular projection lines to chord

a) Chord

Concentric dimension line

Perpendicular projection lines to chord

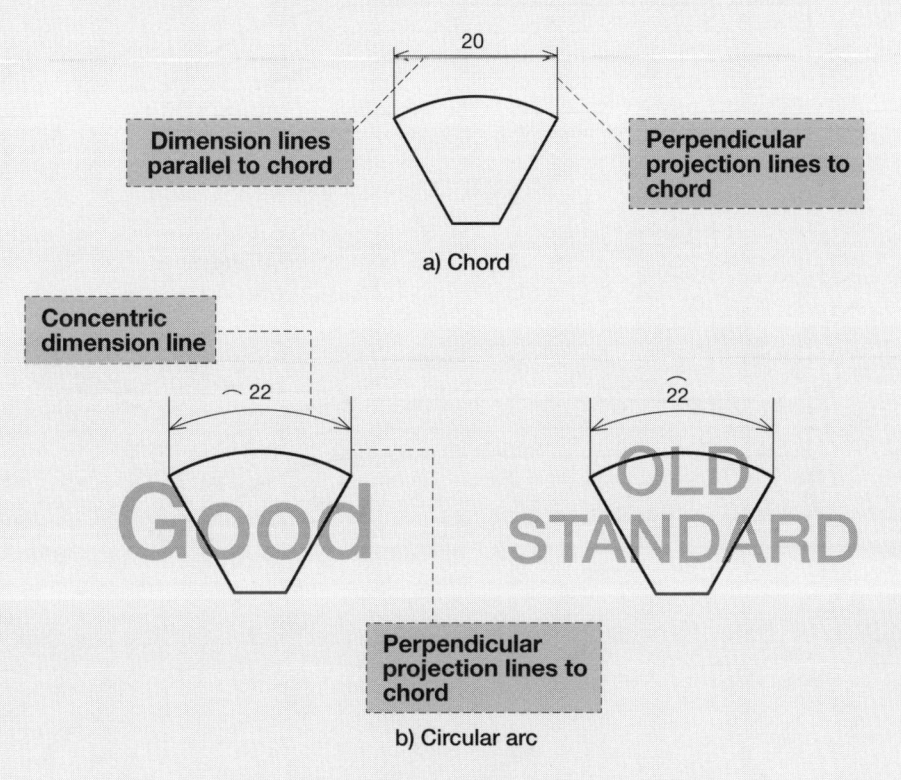

b) Circular arc

Figure 4-26 Chord and circular arc length dimensions

b) 2つ以上の同心の円弧のうち、1つの円弧の長さを明確に表す必要があるときには、次のどちらかによります。
- 円弧の寸法数値から引出線を描き矢印をつける（**図4-27** a）。
- 円弧の長さを表す寸法数値の後に、円弧の半径を括弧に入れて示す。この場合、円弧の長さに記号をつけてはならない（図4-27 b）。

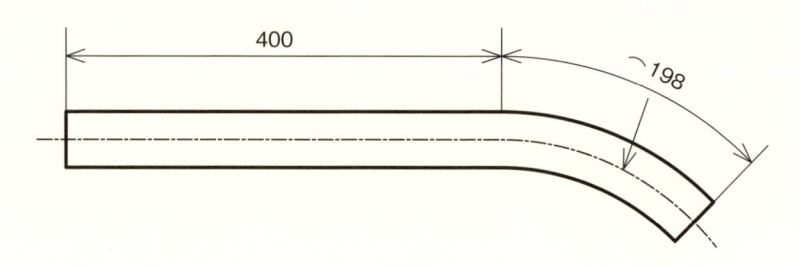

a) 寸法数値から矢

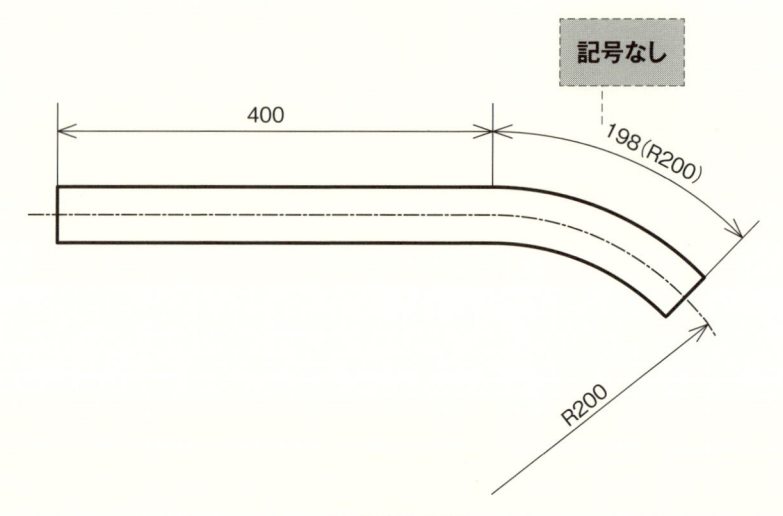

記号なし

b) 半径と併記

図4-27　円弧長さの表し方

b) When it is required to express clearly the length of one circular arc out of two or more concentric arcs, it is in accordance with either of the following:
- Draw a leader line from the dimension value of circular arc, and attach an arrow to the circular arc side. Refer to Figure 4-27 a.
- The radius of circular arc is put in parentheses, and is put after the dimension value of the circular arc length. In this case, the symbol of arc length shall not be attached. Refer to Figure 4-27 b.

a) Arrow from dimension value

b) With radius dimension

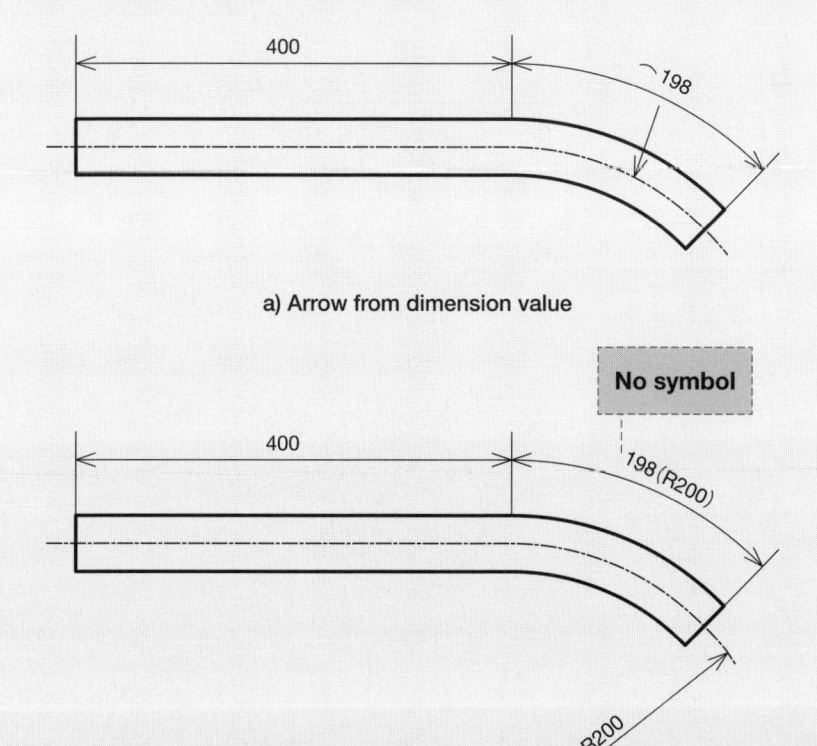

Figure 4-27 Expression of length of circular arc

8) 一群の同一寸法の表し方

　一つの面上またはピッチ円上に配置される一群の同一寸法のボルト穴、ねじ穴、ピン穴、リベット穴などの寸法は、形体の一つから引き出し線を引き出して、寸法数値の前にその総数に「×」をつけて表します（**図4-28** a）。

　この場合、一つの投影図にある形体の総数を記入します。例えば、両側にフランジをもつ管継手にいくつかの穴がある場合、片方のフランジの穴の数を記入します（図4-28 b）。

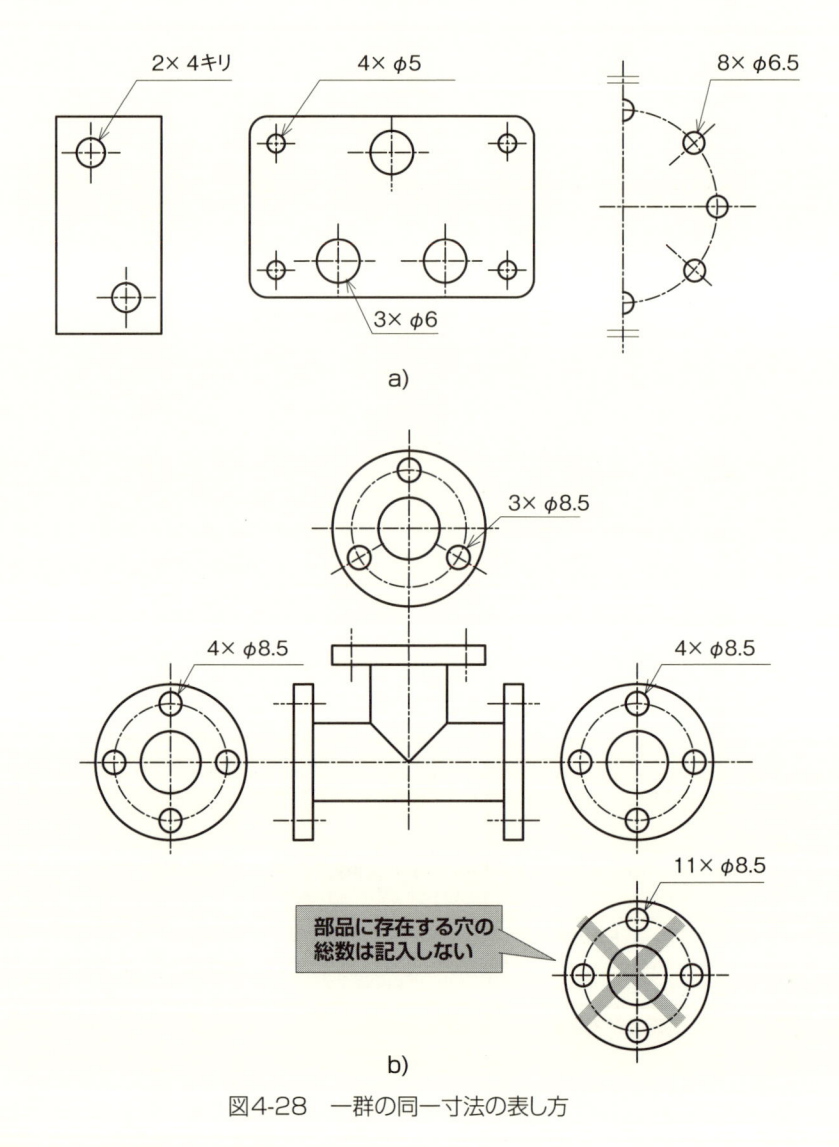

2× 4キリ　　4× φ5　　8× φ6.5

3× φ6

a)

3× φ8.5

4× φ8.5　　　　　4× φ8.5

部品に存在する穴の
総数は記入しない

11× φ8.5

b)

図4-28　一群の同一寸法の表し方

-8. Expression of a group of same dimensions

For a group of the same dimensions such as of bolt holes, machine screw holes, pin holes and rivet holes that are aligned on the one face or a pitch circle, the total number of the features are expressed together with "×" before dimension value by drawing the leader lines from one of feature. Refer to Figure 4-28 a.

In this case, the total number of features in one view is placed. For example, when there are some holes in a pipe joint having both side flanges, the number of holes of only one side flange is placed. Refer to Figure 4-28 b.

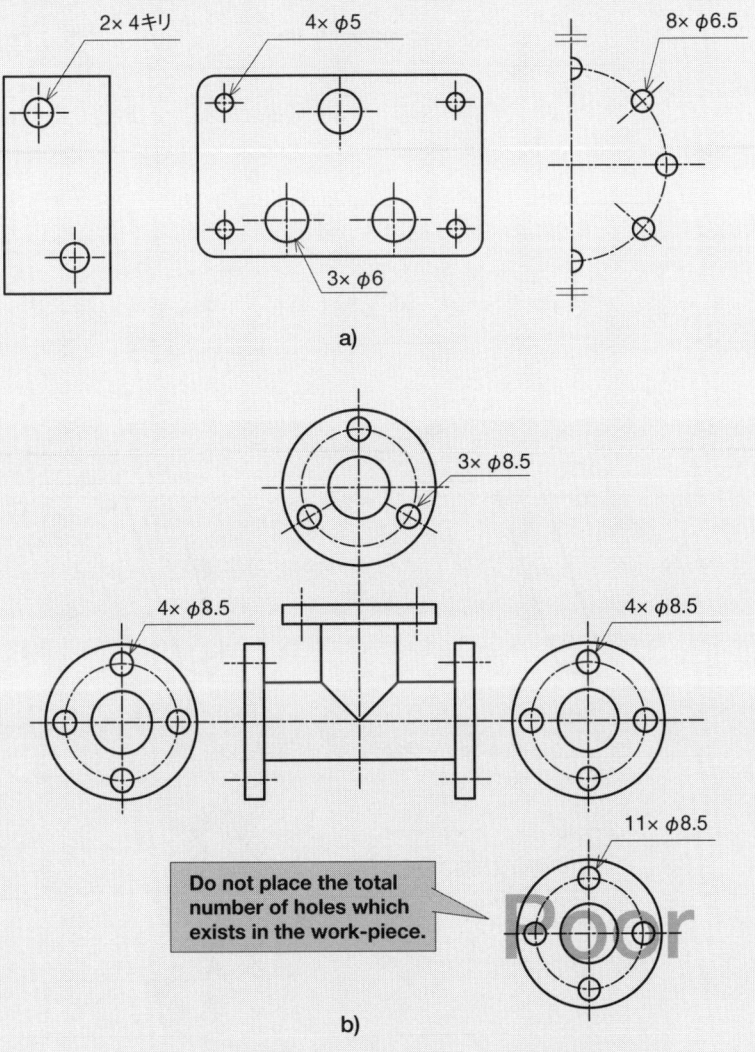

a)

b)

Figure 4-28 Expression of a group of same dimensions

Chapter4 What is the rule of dimensioning!

9) 穴深さの表し方

穴の深さを示す記号「▽」は、直径の後に記入し、穴や深ざぐりの深さ寸法の前に記入します。(図4-29)。

しかし、深さの指定がない場合は、貫通穴とみなします。

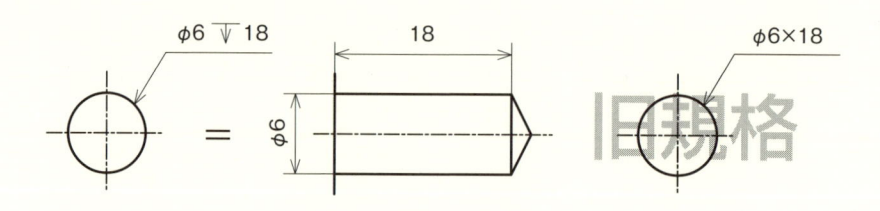

図4-29　穴深さの表し方

穴の深さとは、ドリルの先端で加工される円すい部分などを含まない円筒部の深さのことです。

また、傾斜した穴の深さは、穴の中心線上の長さで表します (図4-30)。

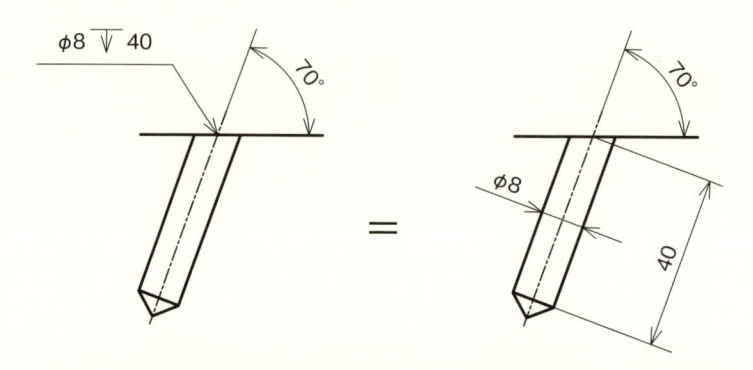

図4-30　傾斜した穴の深さの表し方

深さの指示がない場合、穴は貫通穴とみなすんやで！

-9. Expression of hole-depth

The hole-depth symbol " ▽ " is put after the diameter of the hole and is put before the hole or counter-bore depth value. Refer to **Figure 4-29**.

However, unless otherwise specified, holes are considered as through holes.

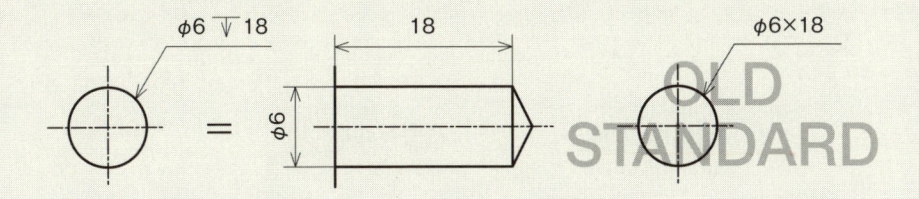

Figure 4-29 Expression of hole-depth

The hole-depth means the depth of the cylindrical part not including the conical part generated by a drill tip.

Furthermore, an inclined hole-depth means the length on the center line of the hole. Refer to **Figure 4-30**.

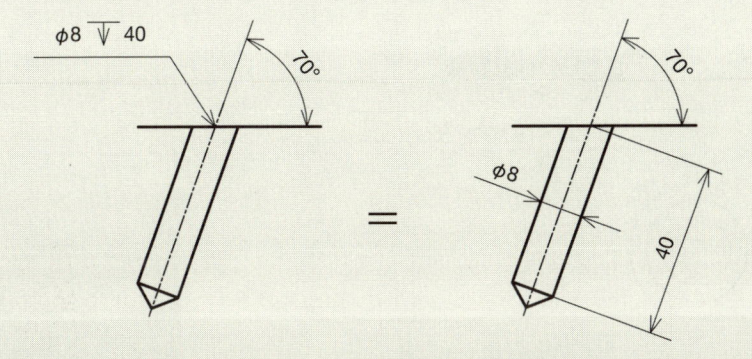

Figure 4-30 Expression of depth of inclining hole

When there are no indication of depth, holes are considered as through holes!

Chapter4 What is the rule of dimensioning!

10）ざぐりまたは深ざぐりの表し方

　ざぐりまたは深ざぐりの記号「⌴」は穴の直径の後に記入し、ざぐりまたは深ざぐりの直径と深さの前に記入します（**図4-31**）。

　なお、一般に平面を得るために鋳造品、鍛造品などの表面を削り取る程度の場合でも、その深さを指示しなければいけません。

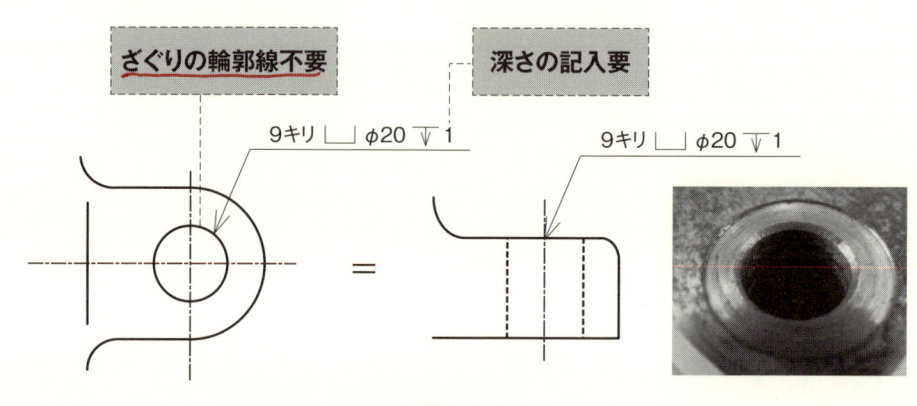

ざぐりの輪郭線不要

深さの記入要

9キリ ⌴ φ20 ▽1

a）（浅い）ざぐり

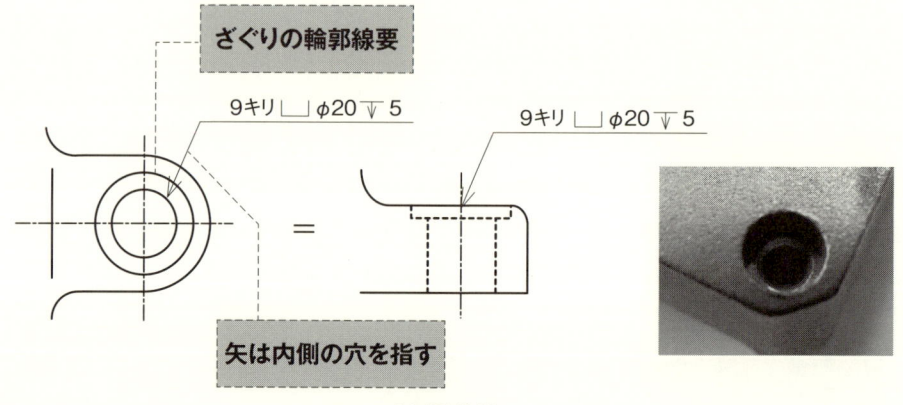

ざぐりの輪郭線要

9キリ ⌴ φ20 ▽5

矢は内側の穴を指す

b）深ざぐり

図4-31　ざぐりと深ざぐりの表し方

-10. Expression of spot face or counter-bore

The spot face or counter-bore symbol " ⊔ " is put after the diameter of the hole, and is put before the spot face or counter-bore diameter and depth value. Refer to Figure 4-31.

Furthermore, when even for chipping of casted or forged work-pieces performed to get a flat, the depth should be placed.

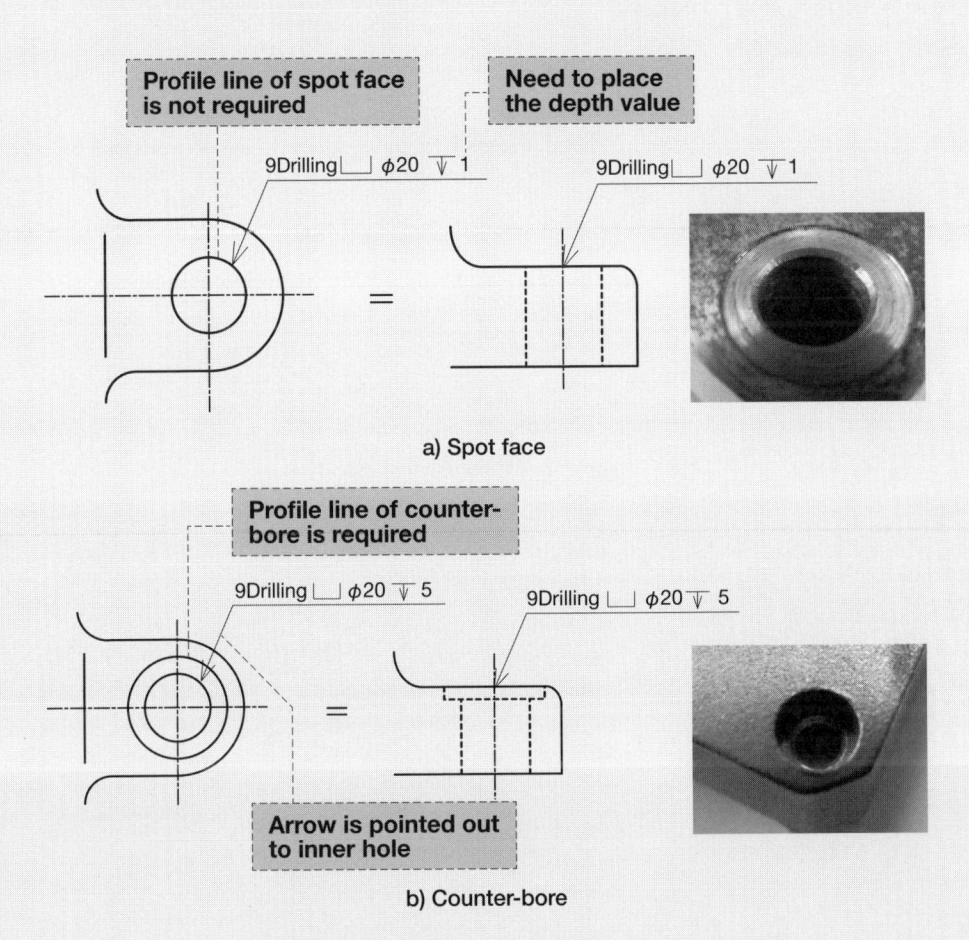

Profile line of spot face is not required

Need to place the depth value

9Drilling⊔ φ20 ▽1

9Drilling⊔ φ20 ▽1

a) Spot face

Profile line of counter-bore is required

9Drilling⊔ φ20 ▽5

9Drilling⊔ φ20 ▽5

Arrow is pointed out to inner hole

b) Counter-bore

Figure 4-31 Expression of spot face or counter-bore

11) 皿ざぐりの表し方

　皿ざぐりの記号「 ∨ 」は穴の直径の後に記入し、皿穴の入口の直径寸法の前に記入します。(**図4-32 a**)。

　皿ざぐり穴が円形形状で描かれている投影図に指示する場合、内側の穴から引き出し線を引き出します（図4-32 b）。

　皿ざぐり穴の深さを規制する要求がある場合は、皿ざぐり穴の開き角度と深さを分けて記入します（図4-32 c）。

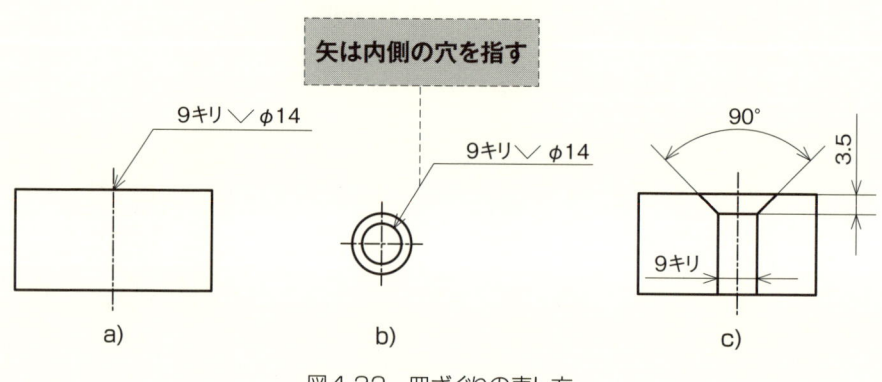

図4-32　皿ざぐりの表し方

-11. Expression of countersink

The countersink symbol "∨" is put after the diameter of the hole, and is put before the inlet diameter of countersink value. Refer to Figure 4-32 a.

When the countersink is drawn to see a circular shape in the view, draw the leader line from inner hole. Refer to Figure 4-32 b.

When the value of the countersink depth is required to control, place the opening angle and the depth of countersink individually. Refer to Figure 4-32 c.

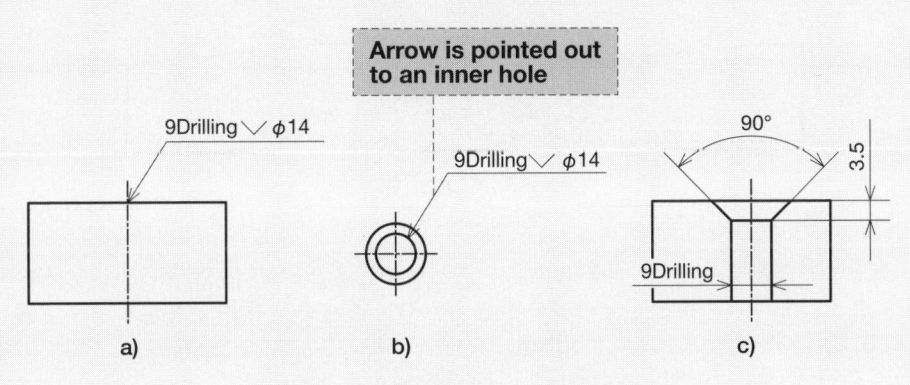

Figure 4-32 Expression of countersink

12) 傾斜の表し方

a) こう配の記号は、その傾斜と向きを合わせて数値とともに引出線の上に記入します。傾斜の値は単位長さあたりの変化の量を示します（図4-33）。

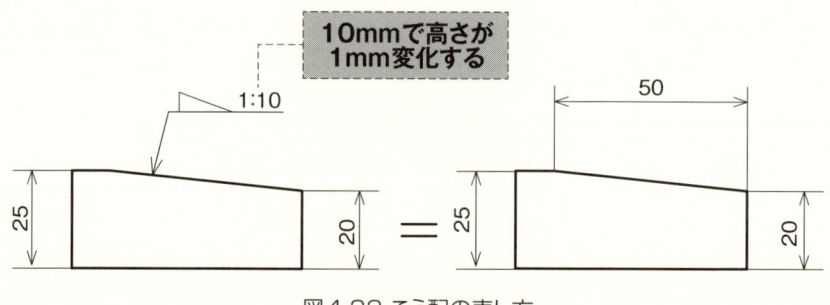

図4-33 こう配の表し方

b) テーパの記号は、その傾斜と向きを合わせて数値とともに引出線の上に記入します。テーパの値は、軸の単位長さあたりの直径の変化の量を示します（図4-34）。

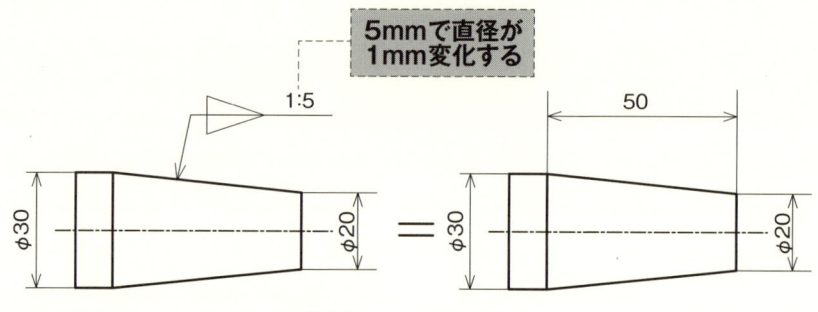

図4-34 テーパの表し方

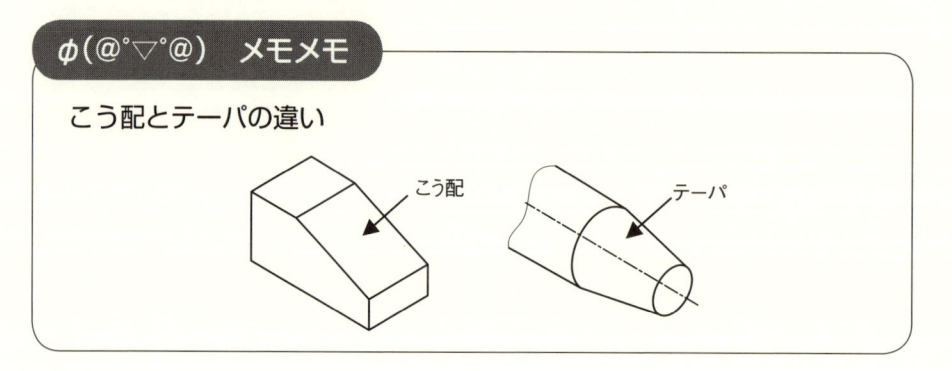

φ(@°▽°@) メモメモ

こう配とテーパの違い

-12. Expression of inclination

a) The slope symbol that synchronized a direction is put with an inclination value on leader line. The slope value is specified as the amount of change of height per unit of length. Refer to Figure 4-33.

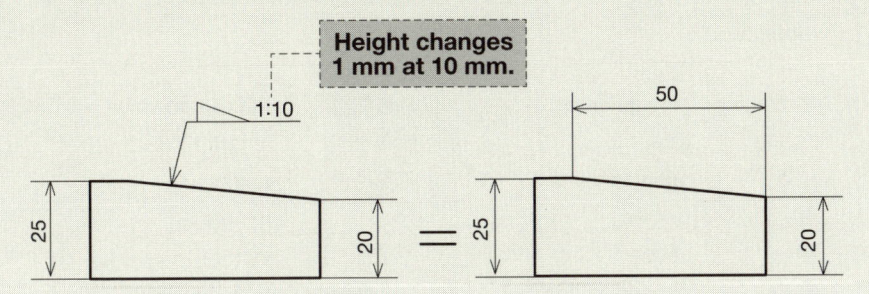

Figure 4-33 Expression of slope

b) The conical taper symbol that synchronized a direction is put with a taper value on leader line. The taper value is specified as the amount of change of diameter per unit of length. Refer to Figure 4-34.

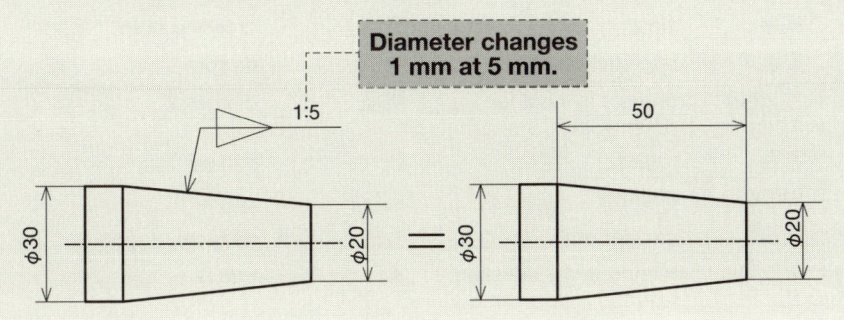

Figure 4-34 Expression of conical taper

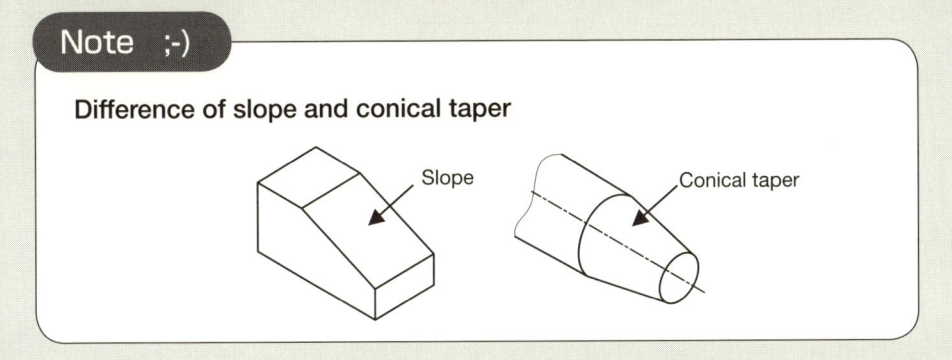

Note ;-)

Difference of slope and conical taper

Chapter4 What is the rule of dimensioning!

用語集　glossary

本章で使用した単語です。他の用語も使うことができます。
The list of words used for this chapter is shown. You can use another words, too.

寸法記入	dimensioning	寸法数値	dimension value
大きさ	size	長さ	length
位置	position	角度	angle
度	degree	分	minute
秒	second	インチ	inch
寸法線	dimension line	寸法補助線	projection line
端末記号	end of symbol	引出線	leader line
仕上がり寸法	finished dimension	鋳放し図	as-cast drawing
前加工図	pre-machining drawing	最終機械加工図	final machining drawing
隙間	gap	黒丸	black dot
斜線	slash	交点	crossing point
交差	intersection	狭い	narrow
寸法補助記号	general symbol for dimension	直径	diameter
半径	radius	球	sphere
面取り	chamfer	正方形	square
円弧	circular arc	実務上の標準	de facto standard
折り曲げた寸法線	folding dimension line	実R	true R
コントロール半径	control radius	輪郭	profile
許容半径	permissible radius	板厚	plate thickness
弦	chord	フランジ	flange
管継手	pipe joint	穴深さ	hole-depth
ドリル先端	drill tip	鋳造品	casted work-piece
鍛造品	forged work-piece	(浅い)ざぐり	spot face
深ざぐり	counter-bore	皿ざぐり	countersink
こう配	slope	テーパ	conical taper

第5章
設計意図って、どない伝えんねん！

Chapter5
In what way should I communicate my design intentions!

この章を学習すると、次の知識を得ることができます。
1）設計意図を伝える表現
2）寸法記入のコツ
3）寸法公差
4）はめあい公差の意味
5）公差解析
6）面の肌記号の指示

After studying this chapter, you will be able to get knowledge as follows:
-1. Expression to communicate design intentions
-2. Point of dimensioning
-3. Dimension tolerances
-4. Meaning of fit tolerance
-5. Tolerance analysis
-6. Expression of surface texture symbols

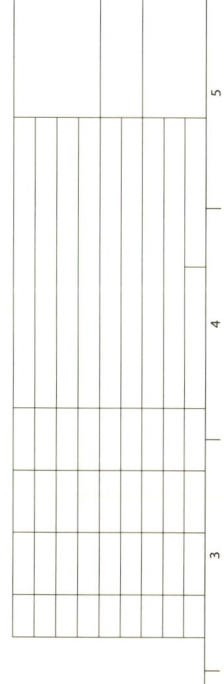

第5章 1 設計意図を伝える表現

5-1-1　特殊な加工の穴の指定

a) ドリル加工、打抜き、鋳抜きなど、穴の加工方法を指示する必要がある場合には、加工方法の前に工具の呼び寸法を使用します（**図**5-1）。
　加工方法の簡略名称は日本工業規格（JIS）によります（**表**5-1）。

※この場合、指示した加工方法に対する寸法の普通公差を適用します。ただし、リーマ穴に限ってははめあい公差を適用します。

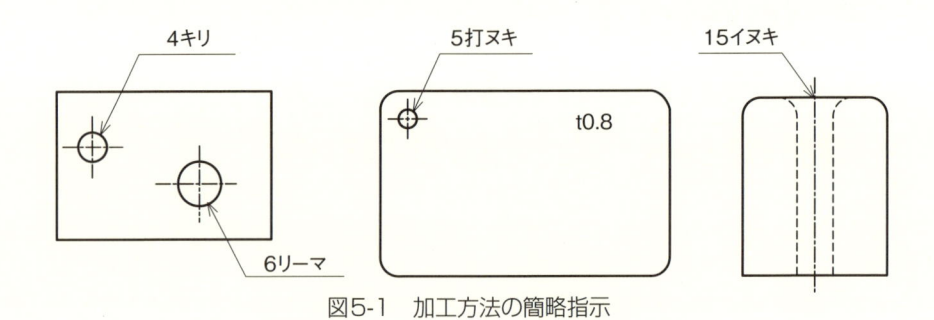

図5-1　加工方法の簡略指示

表5-1　JISによる穴の加工方法の簡略名称

加工方法	簡略名称
鋳放し	イヌキ
プレス打抜き	打ヌキ
ドリル加工	キリ
リーマ加工	リーマ

全ての穴に簡略名称を指示せなあきませんか？

まー、たまに使うのは「キリ」くらいやな。それから、薄板の部品はプレス加工やから、「キリ」を指示したらアカンで！

| Chapter5 | 1 | **Expression to communicate design intentions** |

5-1-1 Indication of the specific processing hole

a) When the processing method of holes such as drilling or punching or casting etc. is needed to indicate, use the nominal dimension value of the tool before processing method. Refer to Figure 5-1.

  The simplified designation of the processing method is given in the Japanese Industrial Standard (JIS). Refer to Table 5-1.

* In this case, the general dimension tolerance on relating to the indicated processing method is applied. However, fit tolerance is applied only reaming hole.

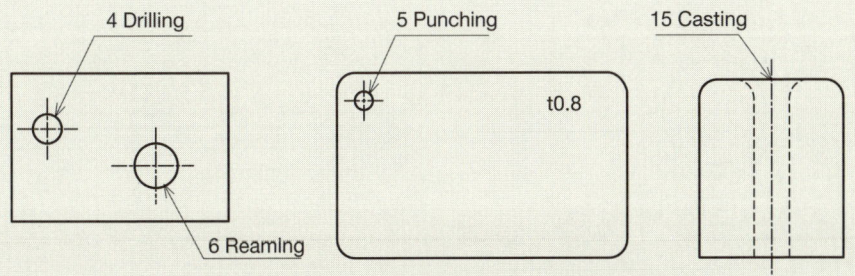

4 Drilling 5 Punching 15 Casting

t0.8

6 Reaming

Figure 5-1 Simplified designation of processing method

Table 5-1 Simplified designation of processing method of hole by JIS

Processing method	Simplified designation
Casting	イヌキ (INUKI)
Punching	打ヌキ (UCHINUKI)
Drilling	キリ (KIRI)
Reaming	リーマ (RI^MA)

Is it necessary to indicate simplified designation to all holes?

Hmmm, "KIRI" is often used. However, do not indicate "KIRI" to the thin plate because punching.

Chapter5 In what way should I communicate my design intension!

5-1-2 加工・処理範囲の限定

　対象物の面の一部分に特殊な加工を施す場合には、その範囲を、外形線に平行にわずかに離して引いた太い一点鎖線で示すことができます。また、投影図中の特定の範囲を指示する場合には、その範囲を太い一点鎖線で囲みます。
　これらの場合、特殊な加工に関する要求事項を寸法と合わせて指示しなければけません（図5-2）。

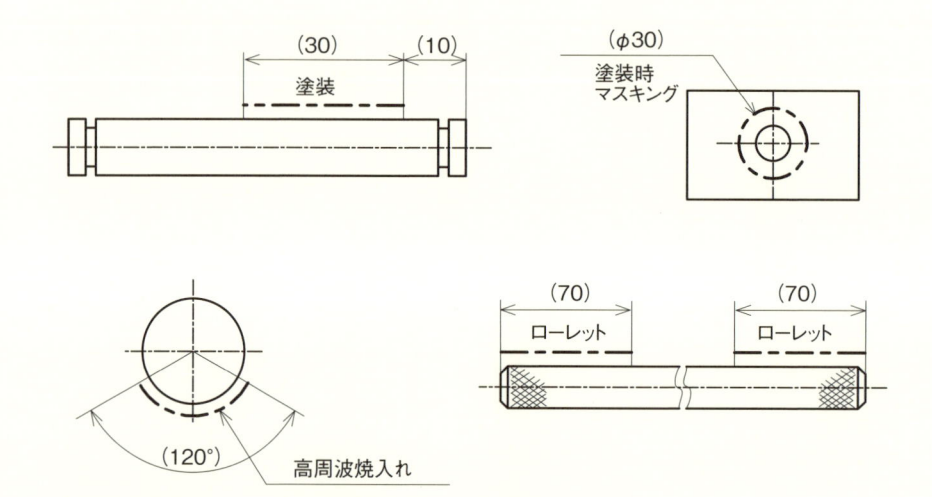

図5-2　加工・処理範囲の限定

φ(@°▽°@)　メモメモ

高周波焼入れ

　高周波焼入れは、歯車や軸の一部分の特別な領域を強くするために処理するものです。結果として生じる硬化した領域は、強度特性とともに摩耗と疲労抵抗力を向上させます。銅製の誘導コイルを使用する電磁気によって、特定の周波数と電力で電流を流します。

適用できる材質：
・炭素鋼…S45C、S55Cなど
・合金鋼…SCM435、SCM440など
・ステンレス鋼（マルテンサイト）…SUS420、SUS440など
・鋳鉄…FCD450、FCD600など
・粉末冶金

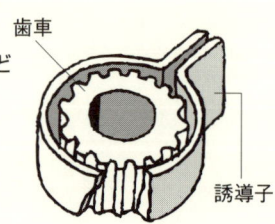

5-1-2 Limitation of processing and treatment area

When special processing is applied to a part of surface of the object, the area can be indicated by thick long-dashed dotted line drawn in parallel to visible outline and slightly apart from it. Further, when specific area in the view is indicated, enclose the area with thick long-dashed dotted line.

In these cases, the requirements related to special processing have to be indicated with dimensions. Refer to Figure 5-2.

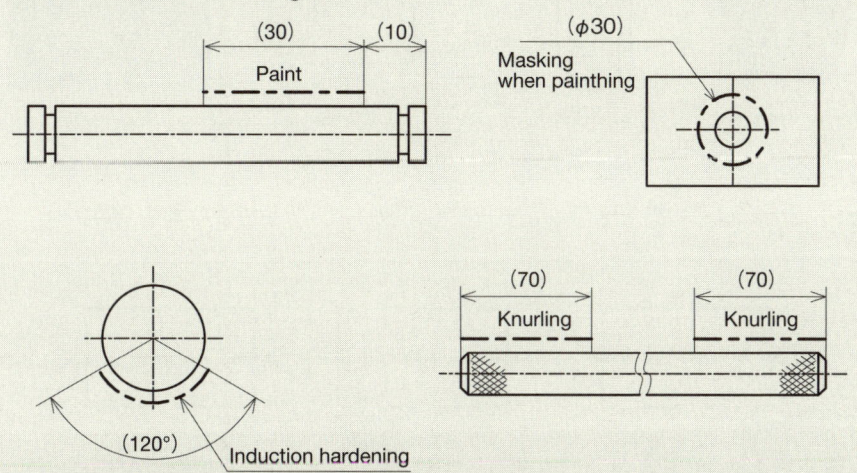

Figure 5-2 Limitation of processing and treatment area

Note ;-)

Induction hardening

Induction hardening is applied to strengthen a specific area of a part of gears or shafts.

The resultant hardened area improves the wear and fatigue resistances along with strength characteristics.

It is an electromagnetic process using a copper inductor coil, which is fed a current at a specific frequency and power level.

Applicable materials:
- Carbon steels ... S45C, S55C, etc.
- Alloy steels ... SCM435, SCM440, etc.
- Stainless steels (martensitic) ... SUS420, SUS440, etc.
- Cast iron ... FCD450, FCD600, etc.
- Powder metallurgy

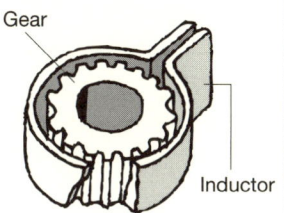

5-1-3 特殊な加工の指示

・ローレット加工する部分を表面の一部分に描いて表すことができます（**図5-3**）。

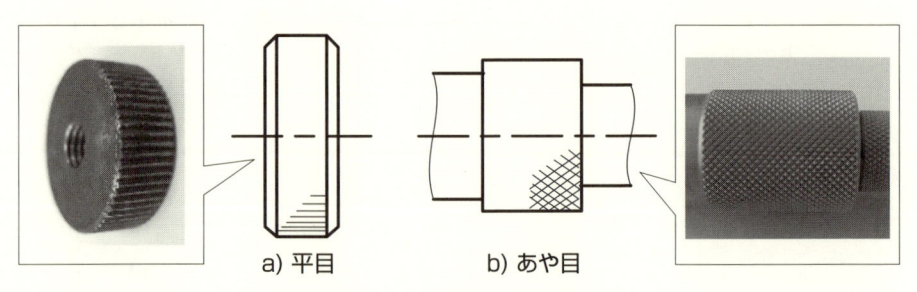

a) 平目　　b) あや目

図5-3　ローレット加工の指示

・センター穴はその形を図に表さず、簡略記号によって指示されます（**図5-4**）。

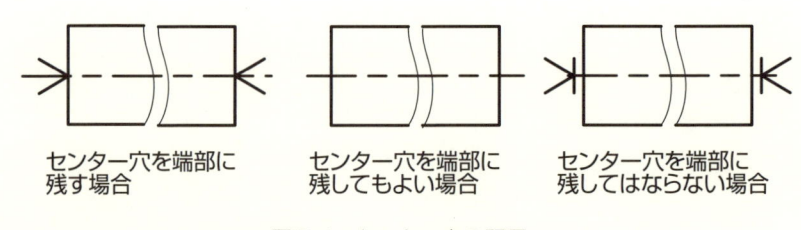

センター穴を端部に残す場合　　センター穴を端部に残してもよい場合　　センター穴を端部に残してはならない場合

図5-4　センター穴の記号

φ(@˚▽˚@)　メモメモ

センター穴

センター穴とは、旋盤や円筒研削盤などで加工するための基準穴をいいます。真直度や同軸度、振れ公差を指示する部品に適用します。

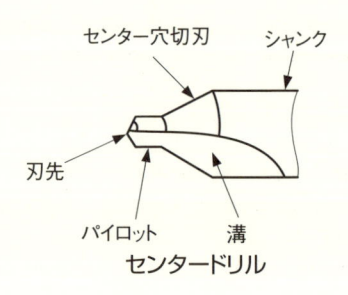

センター穴切刃　シャンク
刃先
パイロット　溝
センタードリル

5-1-3 Indications of special processing

- Knurled part can be indicated as graphic pattern on a part of the surface. Refer to Figure 5-3.

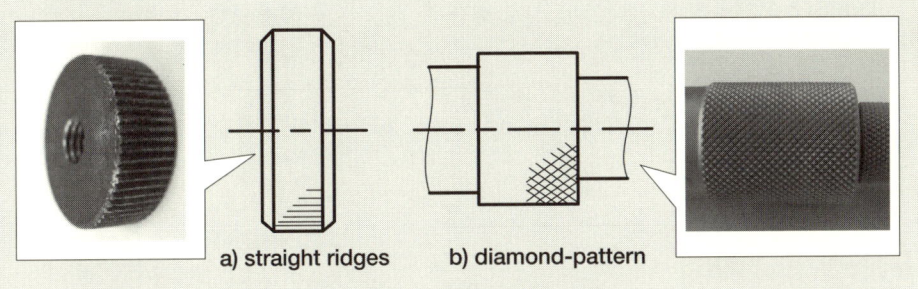

a) straight ridges   b) diamond-pattern

Figure 5-3 Indication of knurling

- Center drill holes are not indicated as shape to the views, but are indicated by a simple sign. Refer to Figure 5-4.

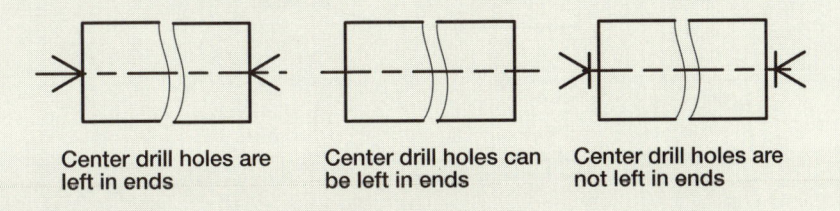

Center drill holes are left in ends

Center drill holes can be left in ends

Center drill holes are not left in ends

Figure 5-4 Simple sign of the center drill hole

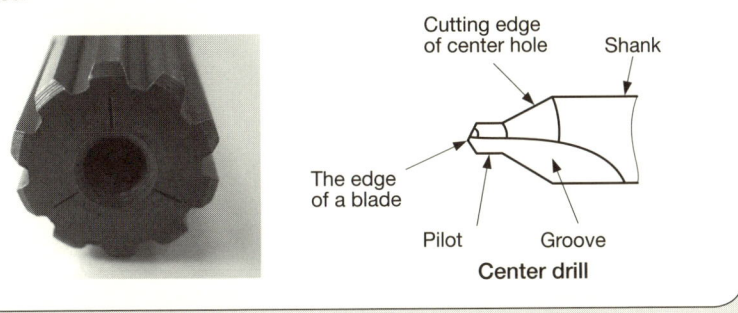

> **Note ;-)**
>
> ### Center drill hole
>
> The center drill hole means a processing datum hole in a lathe or a cylindrical grinder. It is applied to the work-pieces which indicate straightness, concentricity, and run-out tolerance.
>
> Cutting edge of center hole Shank
>
> The edge of a blade
>
> Pilot   Groove
>
> **Center drill**

Chapter5 In what way should I communicate my design intension! 157

5-1-4　その他の特殊な情報の指示

　その他の特殊な図示法は、次によります（図5-5）。
・加工前の形、または粗材寸法を示す場合には、細い二点鎖線で図示します。
・隣接する部品を参考として図示する場合は、細い二点鎖線で図示します。しかし投影図は隣接する部品に隠されていてもかくれ線としてはいけません。隣接部品が断面図になる場合、ハッチングを施しません。
・部品図に用いられることはごくまれですが、参考情報として動作範囲などを表す場合もあります。

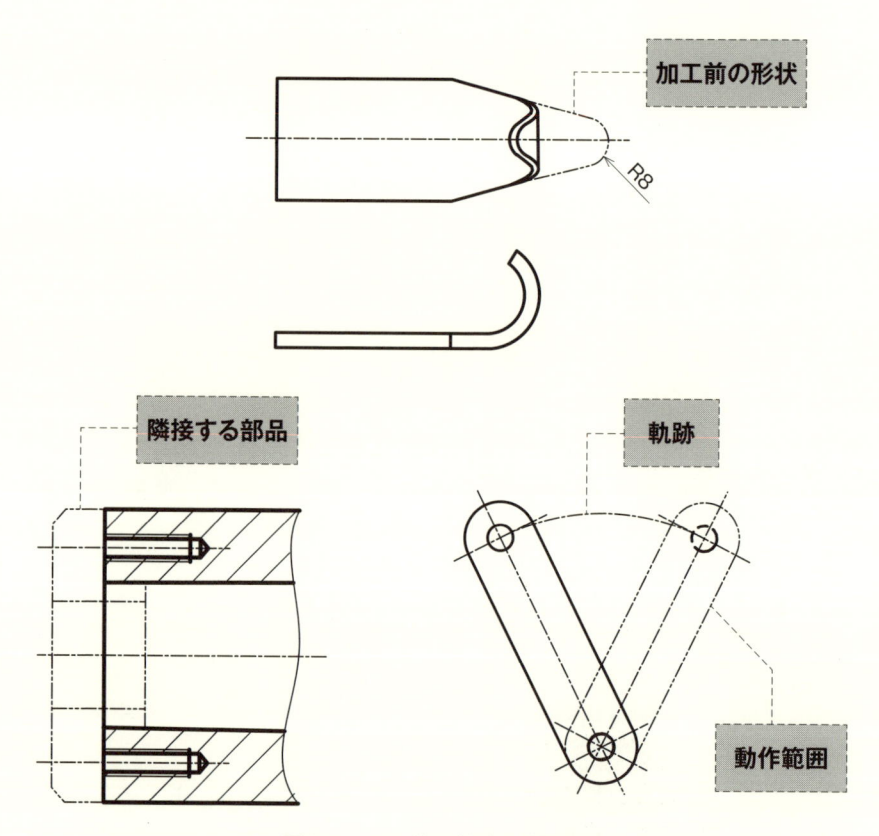

図5-5　その他の特殊な図示法

5-1-4 Indication of the other special information

Indication of the other special methods is as follows: Refer to Figure 5-5.
- When the shapes before processing or rough material dimensions are indicated, the thin long-dashed double-dotted lines should be used.
- When adjacent work-piece to the object as a reference is indicated, the thin long-dashed double-dotted line should be used. However, even if the view is hidden by the adjacent work-piece, do not use as a hidden line. Hatching is not given when the adjacent work-piece is indicated by sectional view.
- Although being used for a work-piece drawing is very rare, a working area can be indicated as reference information.

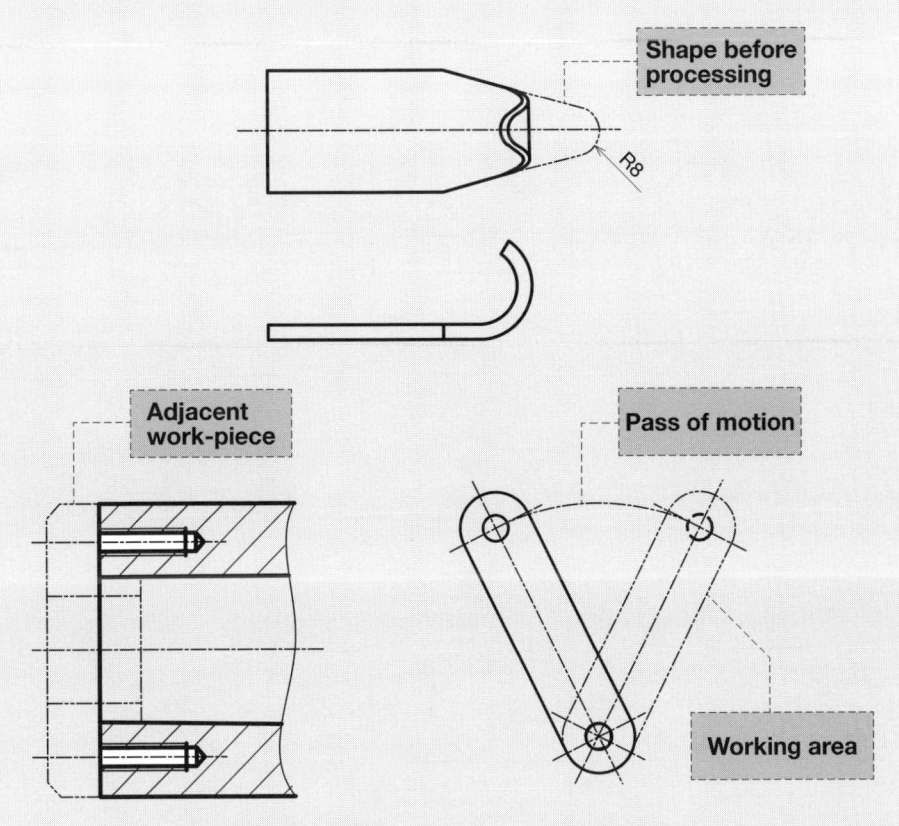

Figure 5-5 Indication of the other special methods

5-1-5 同一形状であることの指示

　T形管継手のフランジのように、1つの部品に全く同一寸法の部分が2つ以上ある場合は、寸法はそのうちの一つだけに記入します。この場合、寸法を記入しない部分に注意書きをします（図5-6）。

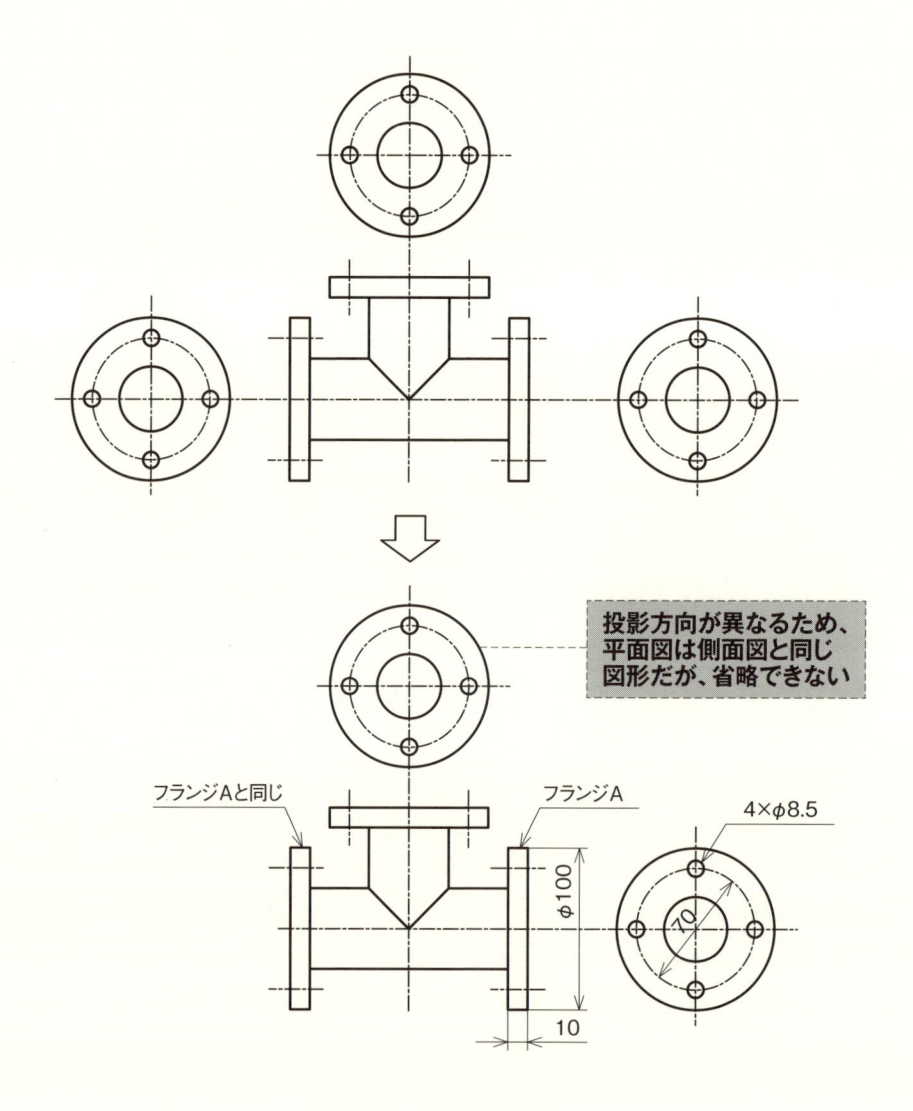

投影方向が異なるため、平面図は側面図と同じ図形だが、省略できない

フランジAと同じ　　フランジA　　4×φ8.5

φ100

10

図5-6　同一形状の指示

5-1-5 Indications of same shapes

Like the flange of T-pipe joint etc., when one work-piece completely has two or more parts of the same dimension, the dimensions are placed only in one of them.

In this case, notes are indicated the parts that dimensions are not placed. Refer to Figure 5-6.

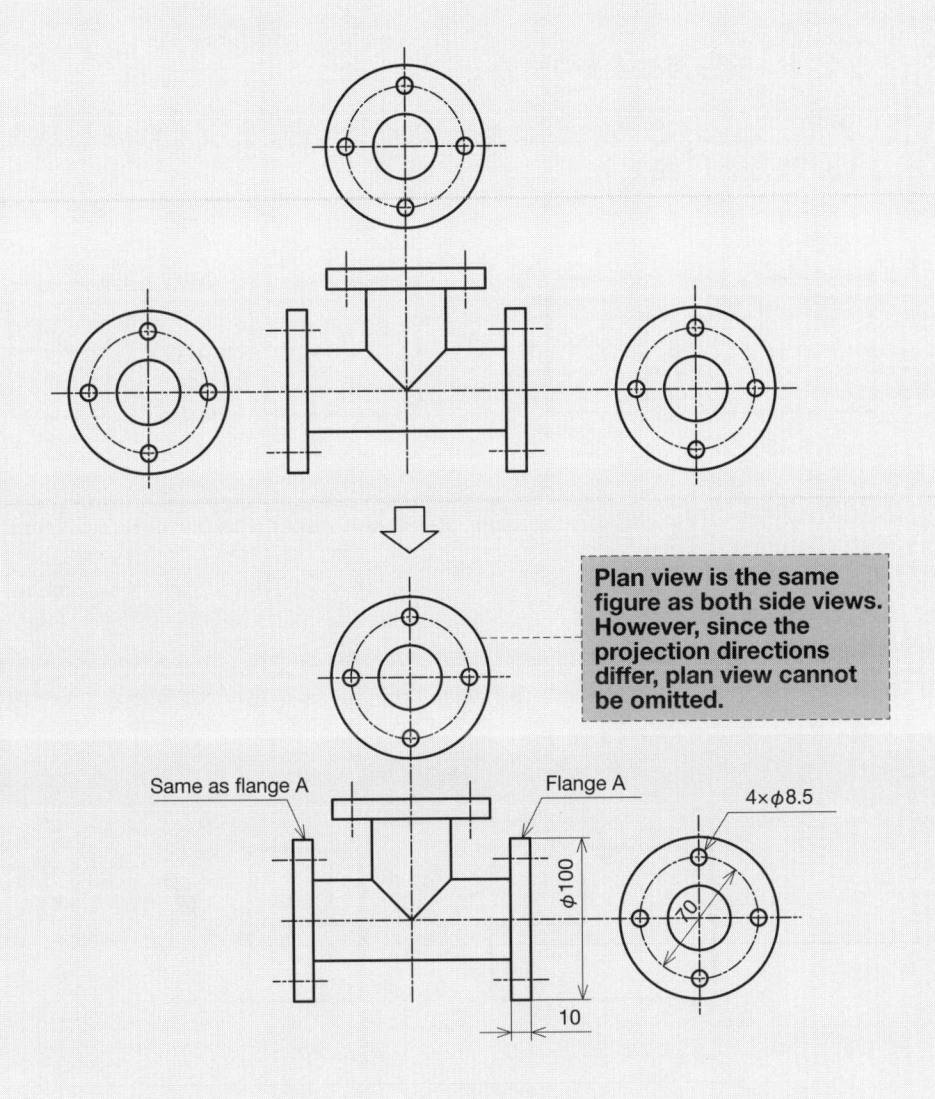

Plan view is the same figure as both side views. However, since the projection directions differ, plan view cannot be omitted.

Figure 5-6 Indication of same shapes

Chapter5 In what way should I communicate my design intension!

| 第5章 | 2 | 寸法記入のコツ |

> 対象物の大きさ、形状および位置を最も明瞭に表すのに必要で十分な寸法を記入します。

5-2-1 投影図と寸法の関係

　寸法は、なるべく正面図に集中して指示します。なぜなら、正面図は最も特徴を表す図だからです（図5-7）。

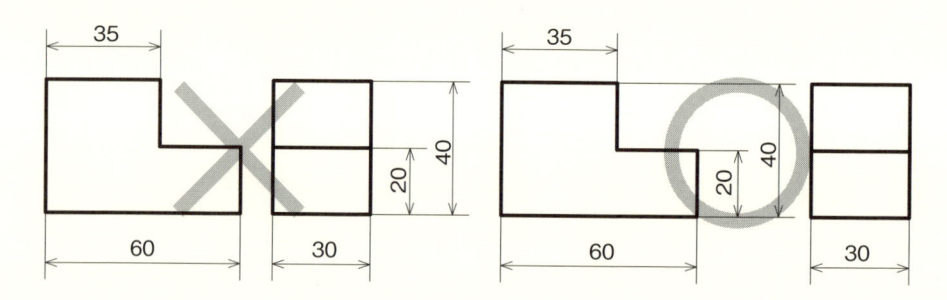

図5-7　正面図に集中させた寸法記入

5-2-2 重複寸法

a) 寸法は、重複記入を避けなければいけません（図5-8）。
　ただし、複数枚に分割される一連の図面では、寸法を重複して記入することで理解しやすくなるため許されます。

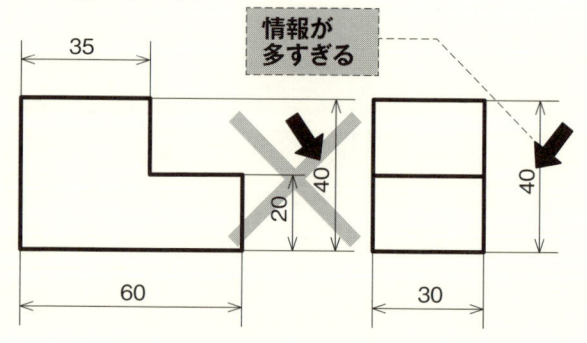

図5-8　重複する寸法記入

Chapter 5 — 2 — Point of dimensioning

> Dimensions that are necessary and sufficient to express the size, shape and position of the object most clearly are placed.

5-2-1 Relation between views and dimensions

Dimension should be placed collectively on the front view, as far as possible. You know the front view expresses a characteristic most. Refer to **Figure 5-7**.

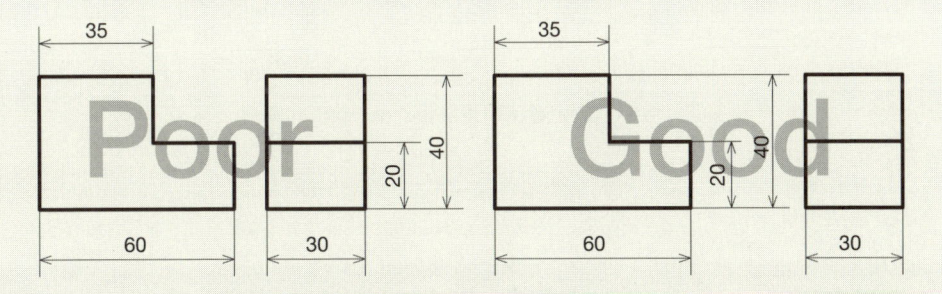

Figure 5-7 Collectively dimensioning on the front view

5-2-2 Duplicated dimensions

a) Duplicated dimensions shall be avoided. Refer to **Figure 5-8**.

However, duplicated dimensions are permitted to easy understand in the multi-sheet drawing.

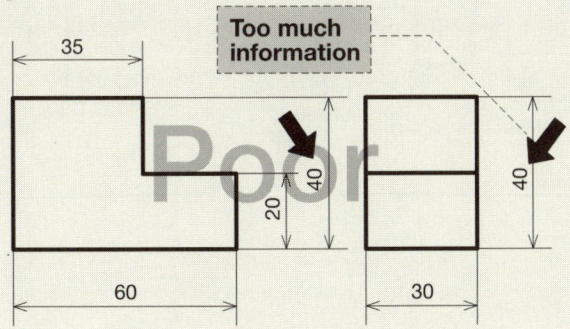

Figure 5-8 Duplicated dimensioning

b) 機能上あるいは加工上、特に注意を促したい場合は重複寸法を使うことができます。この場合、重複するいくつかの寸法数値の前に黒丸をつけ、重複寸法を示したことを図面に注記します（**図5-9**）。

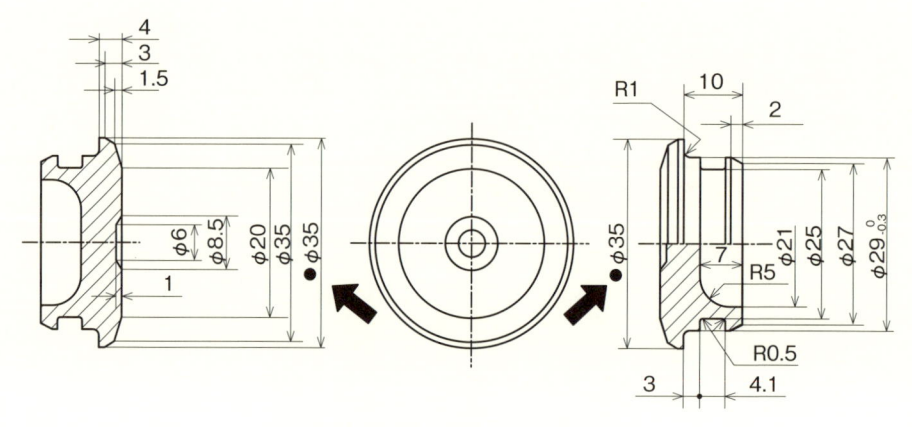

注記　黒丸（●）印は重複寸法をあらわす

図5-9　黒丸を使った重複する寸法記入

c) 加工に配慮して、寸法を計算して求める必要がないように記入する場合、重複する寸法を参考寸法として示すことができます（**図5-10**）。
　参考寸法は数値を括弧で囲み、参考情報が提供されるだけで、計測や組立てで使用されません。

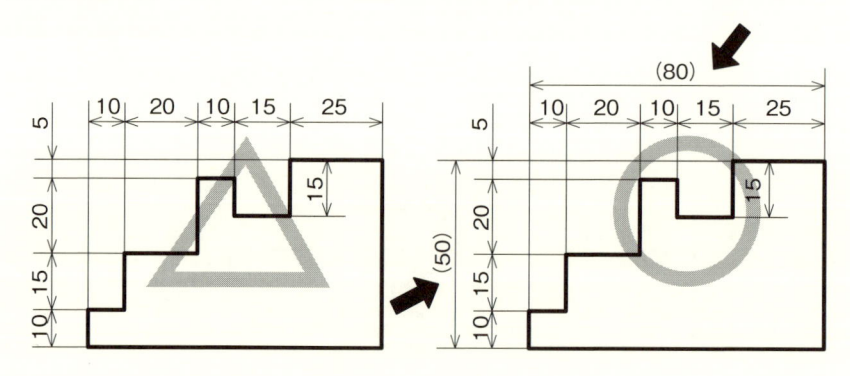

図5-10　参考寸法

b) When there is special attention on function or processing, duplicated dimensions can be used. In this case black dots are put before duplicated dimensions, and a note to explain that it is duplicated dimensions are given. Refer to Figure 5-9.

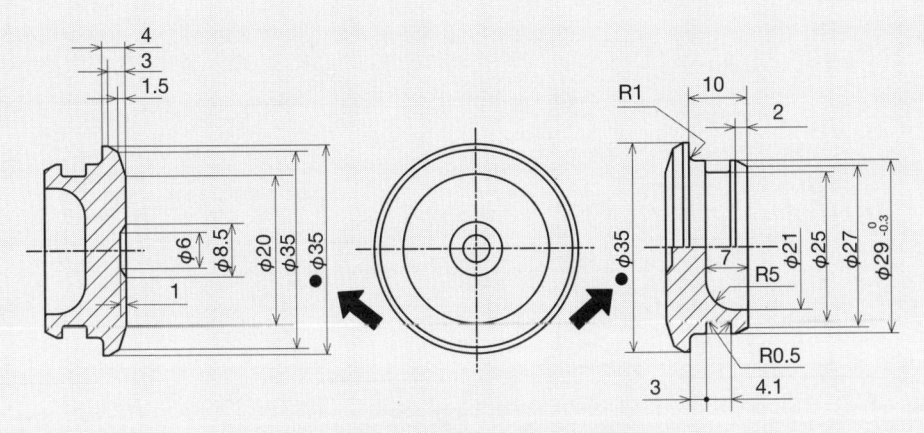

NOTE   Black dots indicate duplicated dimensions.

Figure 5-9 Duplicated dimensioning with black dots

c) For convenience of the processing, the reference dimension can be indicated so that no calculation is necessary. Refer to Figure 5-10.

Reference dimension is the numerical value enclosed in parentheses provided for information only and it is not used in the measurement or assembly.

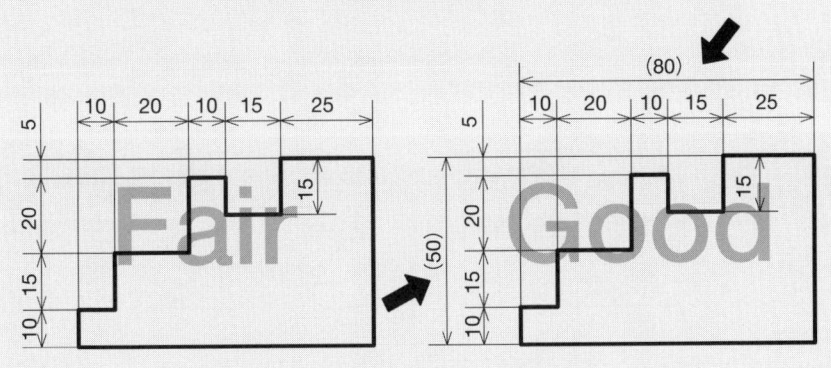

Figure 5-10 Reference dimension

Chapter5   In what way should I communicate my design intension!

5-2-3　基準面や基準軸をもとにした一連の寸法

a) 基準とする形体（面または線）がある場合には、その形体をもとにして寸法を記入します（図5-11）。

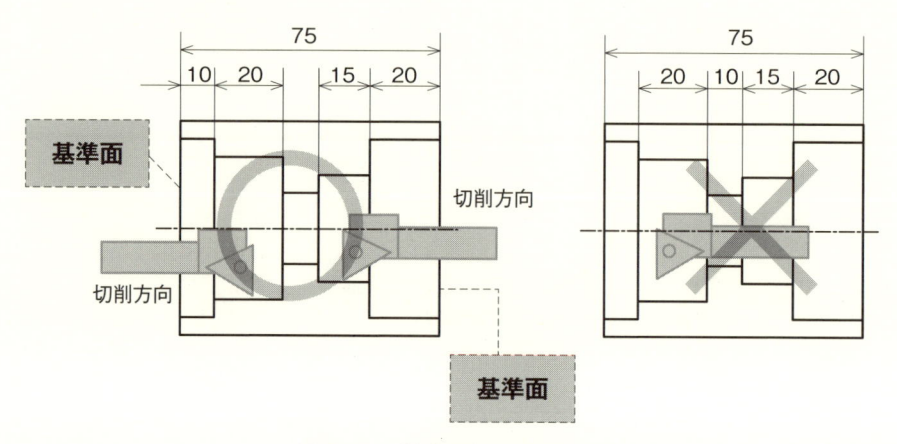

図5-11　基準面をもとにした寸法

b) 原則として、対称形状の対象物の寸法は、中心線あるいは中心平面を基準に振り分けます（図5-12）。

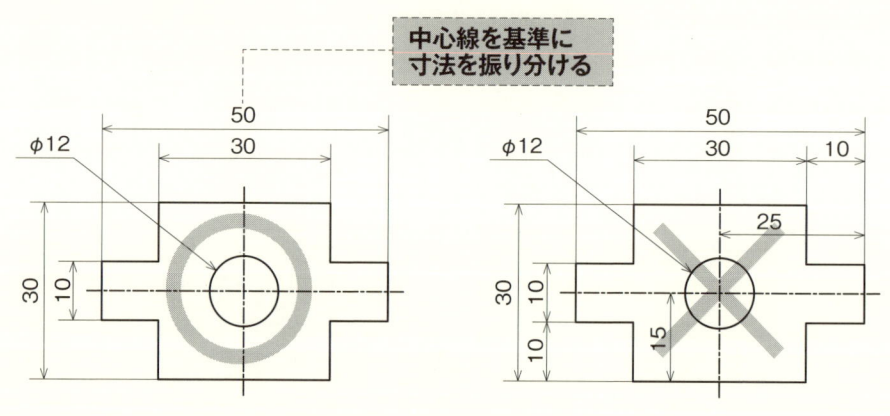

図5-12　中心振分け寸法

5-2-3 Related dimensions based on datum surface or datum axis

a) When there is a datum feature (surface or axis), the dimensions based on the datum feature should be indicated. Refer to Figure 5-11.

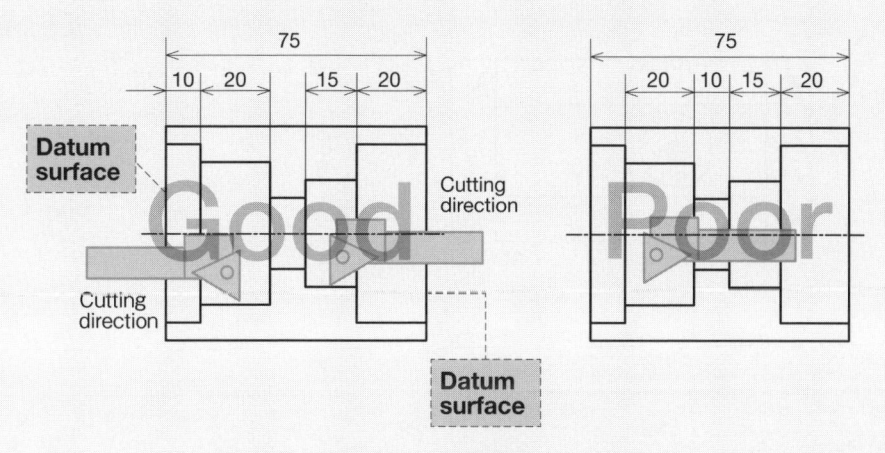

Figure 5-11 Dimensioning based on datum plane

b) As a rule, dimensions of symmetrical object are divided as the datum center line or center plane. Refer to Figure 5-12.

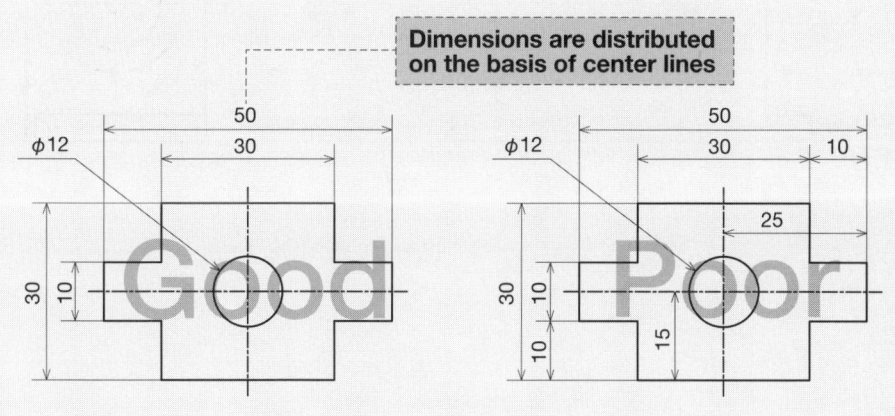

Figure 5-12 Center divided dimensions

c) 寸法はなるべく工程ごとに配列を分けて記入します。また、関連する寸法は、なるべく近くにまとめて記入します。(**図5-13**)。

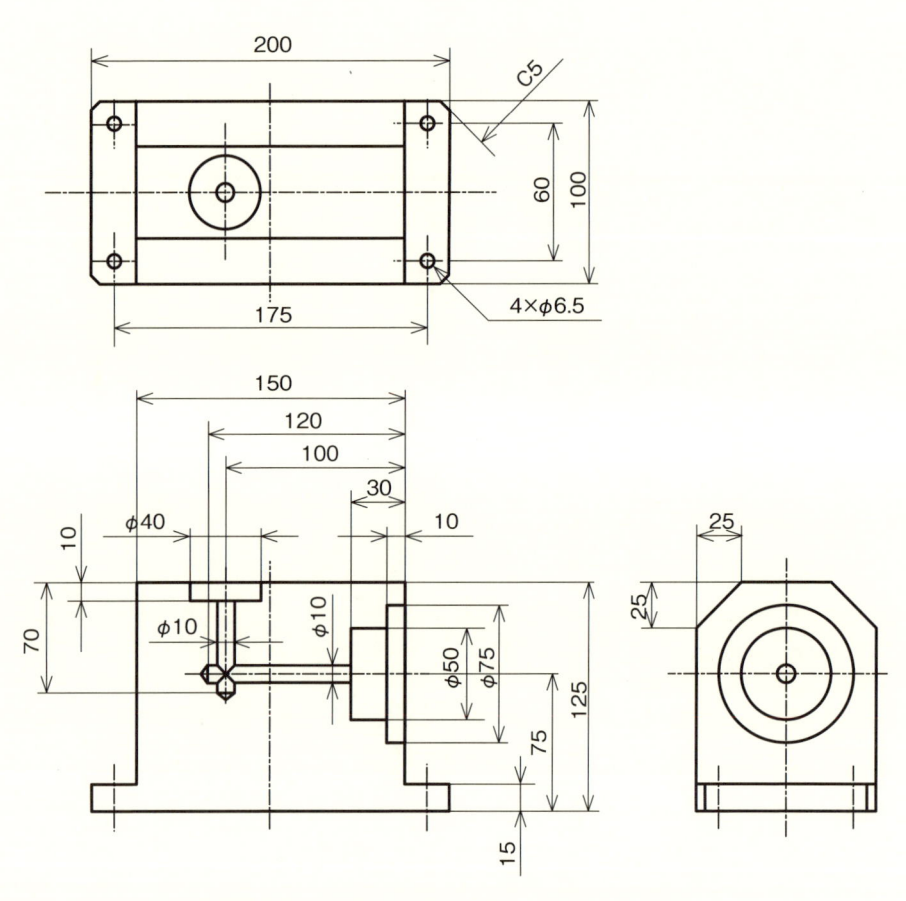

図5-13　工程ごとに配列を分けて記入した寸法例

c) Dimensions should be divided arrangement according to each process as far as possible. In addition, related dimensions should be placed collectively to near place. Refer to Figure 5-13.

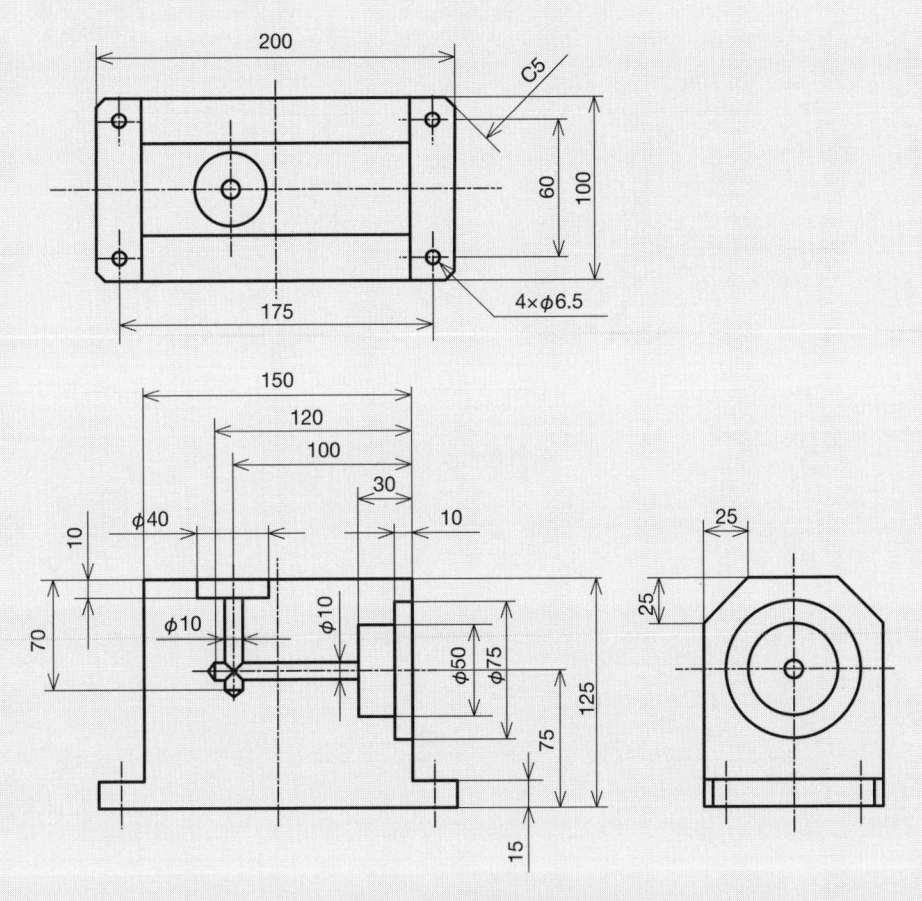

Figure 5-13 Dimensioning that divided arrangement for each process

寸法記入の解説を図5-14に示します。

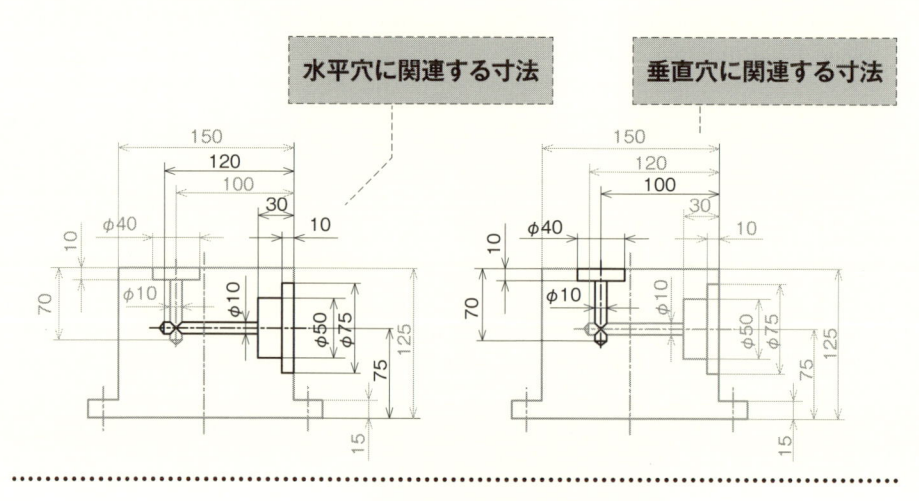

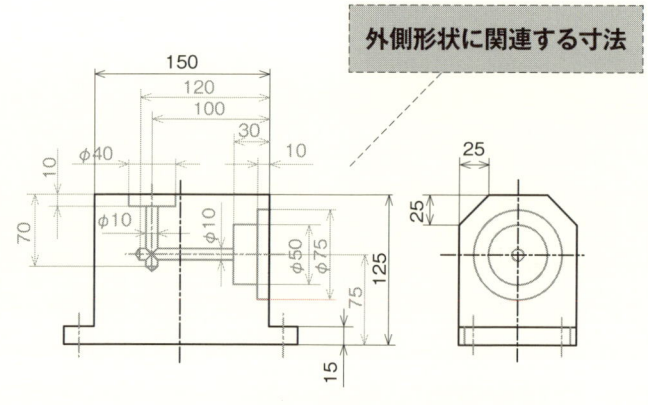

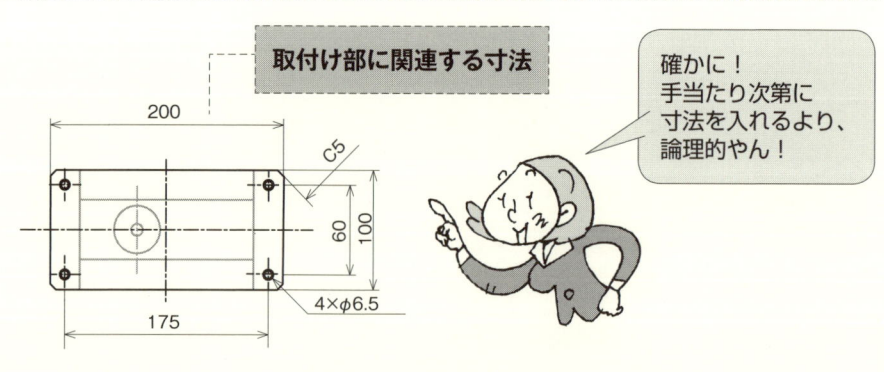

水平穴に関連する寸法

垂直穴に関連する寸法

外側形状に関連する寸法

取付け部に関連する寸法

確かに！
手当たり次第に
寸法を入れるより、
論理的やん！

図5-14　寸法記入の分解

Break down of dimensioning is shown in Figure 5-14.

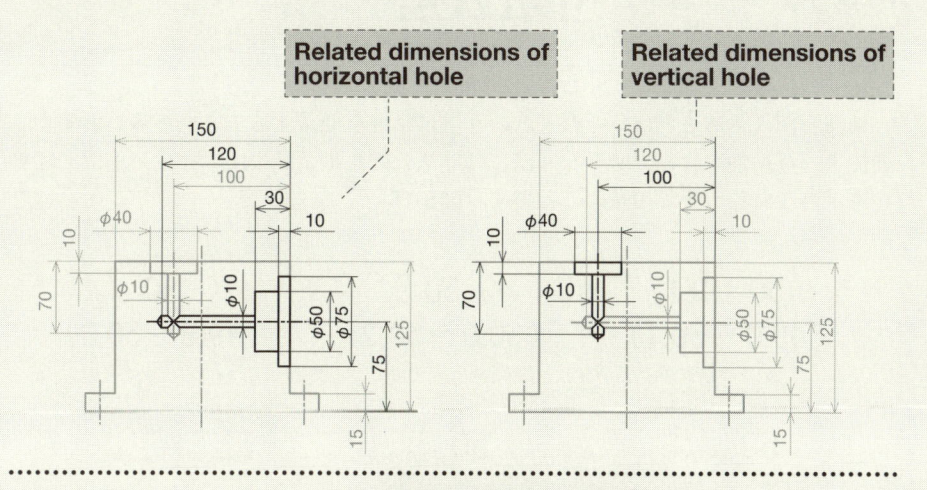

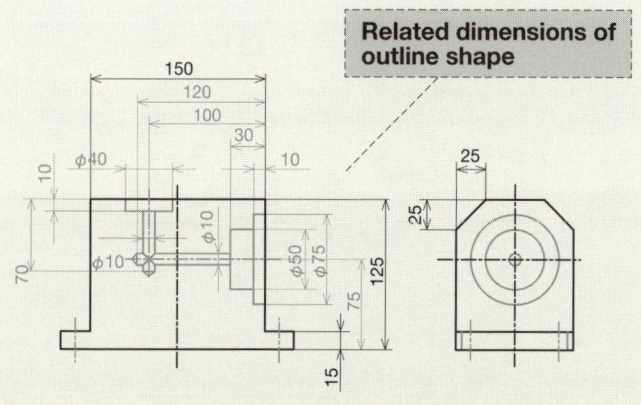

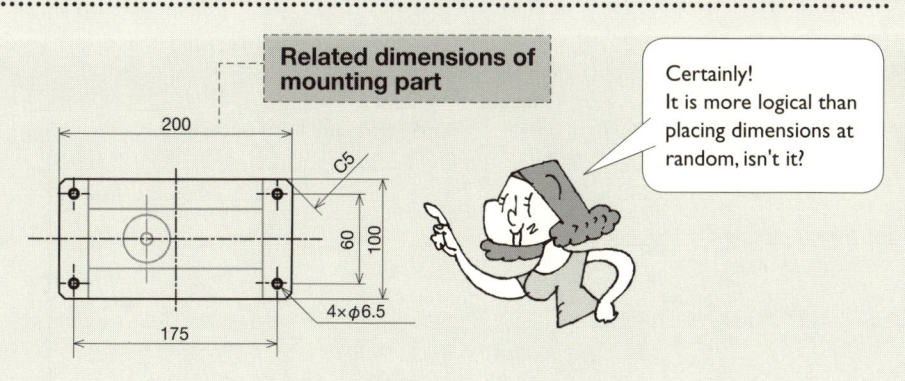

Figure 5-14 Breakdown of dimensioning

Certainly! It is more logical than placing dimensions at random, isn't it?

Chapter5 In what way should I communicate my design intension!

第5章 3 寸法公差

寸法を記入することは、それぞれの寸法に精度を要求することを意味します。機能上（互換性を含む）必要な寸法には寸法公差を指示します。寸法公差は、ばらつきが許される大きさの量のことです（**図5-15**）。

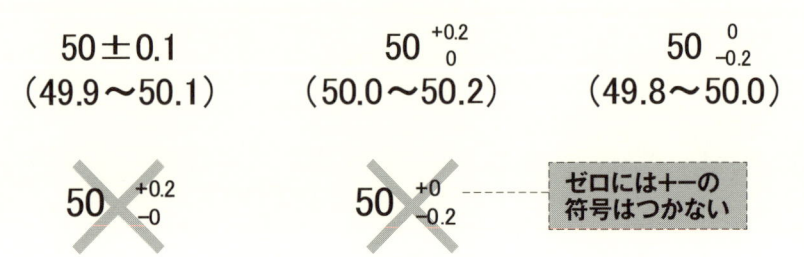

$$50 \pm 0.1 \quad\quad 50 {}^{+0.2}_{0} \quad\quad 50 {}^{0}_{-0.2}$$
$$(49.9 \sim 50.1) \quad (50.0 \sim 50.2) \quad (49.8 \sim 50.0)$$

$$50 \;\cancel{{}^{+0.2}_{-0}} \quad\quad 50 \;\cancel{{}^{+0}_{-0.2}}$$

ゼロには+－の符号はつかない

図5-15　寸法公差の表し方

しかし、理論的に正しい寸法および参考寸法を除きます。理論的に正しい寸法は、寸法数値を長方形の枠で囲み、幾何公差と合わせて使用します（**図5-16**）。

幾何公差と一緒に使う　　$\boxed{50}$　＝　50.0000

図5-16　理論寸法の表し方と意味

それでは、寸法公差の指示がない場合はどう考えればよいのでしょう？

この場合は、個々に規定する普通公差を適用します。適用するJISの規格番号および等級記号または普通公差の表を表題欄の中またはその付近に指示するとよいでしょう（**図5-17**）。

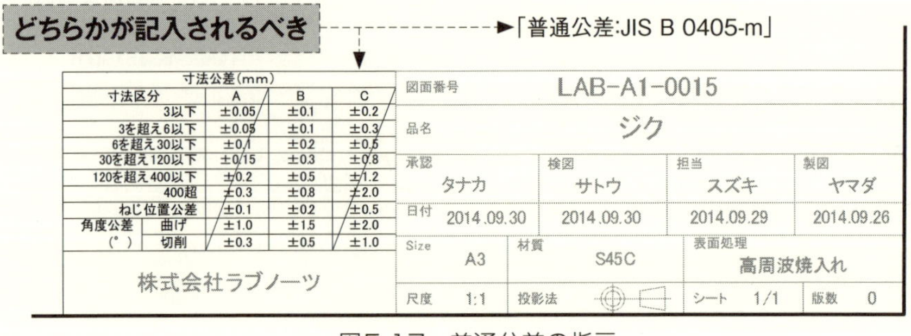

どちらかが記入されるべき　→　「普通公差:JIS B 0405-m」

図5-17　普通公差の指示

| Chapter5 | 3 | **Dimension tolerances** |

Dimensioning means the accuracy required for each dimension.

For dimensions required due to the function (including interchangeability), the dimension tolerance must be indicated. Dimension tolerance is the quantity of a size allowed variation. Refer to **Figure 5-15**.

50 ± 0.1 $50^{+0.2}_{\ 0}$ $50^{\ 0}_{-0.2}$
(49.9〜50.1)   (50.0〜50.2)   (49.8〜50.0)

~~$50^{+0.2}_{\ 0}$~~ Poor   ~~$50^{+0}_{-0.2}$~~ Poor   ───── No sign (+/−) to zero

Figure 5-15 Expression of tolerance

However, "theoretically exact dimensions" and "reference dimensions" are excepted. Theoretically exact dimension is put within rectangular frames, and it is used together with geometrical tolerance. Refer to **Figure 5-16**.

Use with geometrical tolerance ───── $\boxed{50}$ = 50.0000

Figure 5-16 Expression and meaning of theoretically exact dimensions

Well then, when the dimension tolerance is not indicated, how do we think?

In this case, the general dimension tolerances individually specified are applied. The number of the JIS standard and class symbol or table of general dimension tolerance value may be indicated within or near the title block. Refer to **Figure 5-17**.

Either should be placed. ─────▶ "general dimension tolerance:JIS B 0405-m"

Tolerance(mm)				Drawing Number	LAB-A1−0015		
Dimension	A	B	C				
Up to 3	±0.05	±0.1	±0.2	Parts name	SHAFT		
Over 3 up to 6	±0.05	±0.1	±0.3				
Over 6 up to 30	±0.1	±0.2	±0.5				
Over 30 up to 120	±0.15	±0.3	±0.8	APPROVALS	CHECK	IN CHARGE	DRAWN
Over 120 up to 400	±0.2	±0.5	±1.2	Tanaka	Satou	Suzuki	Yamada
Over 400	±0.3	±0.8	±2.0				
Screw location	±0.1	±0.2	±0.5	Date 30/09/2014	30/09/2014	29/09/2014	26/09/2014
Angle Tolerance bending	±1.0	±1.5	±2.0				
(°) machining	±0.3	±0.5	±1.0	Size A3	Material S45C	Finish induction hardening	
Labnotes Co., Ltd.				Scale 1:1	Projection	Sheet 1/1	Revision 0

Figure 5-17 Indication of general dimension tolerance

JIS B 0405によって規定される、切削加工の普通公差を**表5-2**に示します。

表5-2　面取り部分を除く長さ寸法の普通公差

公差等級	基準寸法の区分							
	0.5(*)以上 3以下	3を超え 6以下	6を超え 30以下	30を超え 120以下	120を超え 400以下	400を超え 1000以下	1000を超え 2000以下	2000を超え 4000以下
	許容差							
精級(f)	±0.05	±0.05	±0.1	±0.15	±0.2	±0.3	±0.5	-
中級(m)	±0.1	±0.1	±0.2	±0.3	±0.5	±0.8	±1.2	±2
粗級(c)	±0.2	±0.3	±0.5	±0.8	±1.2	±2	±3	±4
極粗級(v)	-	±0.5	±1	±1.5	±2.5	±4	±6	±8

*) 0.5mm未満の基準寸法に対しては、その基準寸法に続けて公差を個々に指示する

したがって、公差等級が中級の場合、次のような解釈になります（**図5-18**）。表5-2の灰色の部分を確認してみてください。

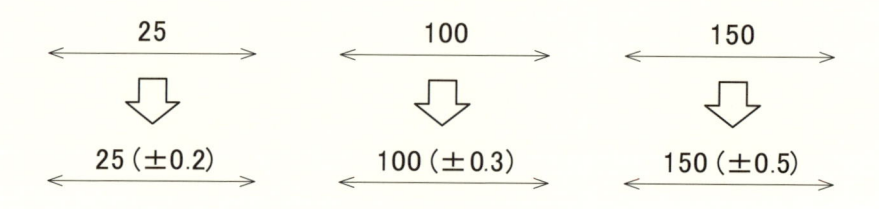

```
 25              100             150
  ↓               ↓               ↓
25(±0.2)      100(±0.3)      150(±0.5)
```

図5-18　寸法と普通公差（中級の場合）

General dimension tolerance for cutting specified by JIS B 0405 is shown in Table 5-2.

Table 5-2 General dimension tolerance of the length without chamfer

Tolerance grade	Basic size range							
	From 0.5(*) To 3	Over 3 up to 6	Over 6 up to 30	Over 30 up to 120	Over 120 up to 400	Over 400 up to 1000	Over 1000 up to 2000	Over 2000 up to 4000
	Tolerance							
Fine(f)	±0.05	±0.05	±0.1	±0.15	±0.2	±0.3	±0.5	-
Medium (m)	±0.1	±0.1	±0.2	±0.3	±0.5	±0.8	±1.2	±2
Coarse (C)	±0.2	±0.3	±0.5	±0.8	±1.2	±2	±3	±4
Very coarse(v)	-	±0.5	±1	±1.5	±2.5	±4	±6	±8

*) For nominal size below 0.5 mm, tolerance is indicated individually to relevant nominal size.

So, when the general dimension tolerance is a medium class, dimensions mean as follows: Refer to Figure 5-18.

Check gray zone in Table 5-2.

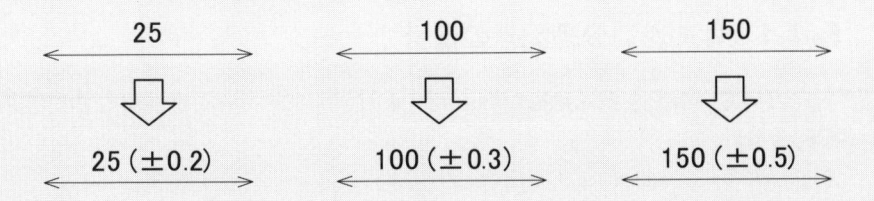

25 ⇩ 25 (±0.2)

100 ⇩ 100 (±0.3)

150 ⇩ 150 (±0.5)

Figure 5-18 Dimensions and general dimension tolerance (In case of medium grade)

第5章 4 はめあい公差の指示とその意味

　はめあい公差は、切削工具、在庫部材、ゲージなどに使用される穴と軸の寸法公差の決めごとです。
　これらの寸法公差が標準化されることで、切削工具、材料、ゲージは世界的に共通化が図れるのです。

　はめあいには、3種類があります。
- すきまばめは、穴と軸を組み立てたとき、2つのはめ合い部が常に隙間がある場合をいいます。
- しまりばめは、穴と軸を組み立てたとき、2つのはめ合い部が常に干渉いている場合をいいます。
- 中間ばめは、穴と軸を組み立てたとき、2つのはめ合い部に隙間があるときや干渉しているときがある場合をいいます。

5-4-1　はめあい公差記号の指示

　穴にはめあい公差を指示する場合、大文字のアルファベットを使用します（図5-19 a）。
　軸にはめあい公差を指示する場合、小文字のアルファベットを使用します（図5-19 b）。

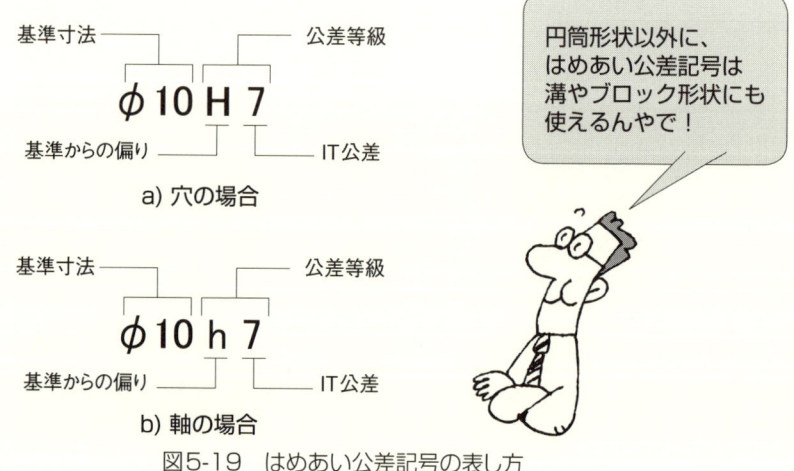

```
    基準寸法        公差等級
       φ 10 H 7
    基準からの偏り    IT公差
         a) 穴の場合

    基準寸法        公差等級
       φ 10 h 7
    基準からの偏り    IT公差
         b) 軸の場合
```

円筒形状以外に、はめあい公差記号は溝やブロック形状にも使えるんやで！

図5-19　はめあい公差記号の表し方

Chapter 5 — 4. Indication and meaning of fit tolerance

Fits tolerance is a tolerance system of hole or shaft that is used for cutting tools, material stock, gages, etc.

If these tolerances standardize, cutting tools, material, and gages are generally available throughout the world.

A fit has three types.
- Clearance fit ("SUKIMA-BAME" in Japanese) is when two mating parts will always have space when the hole and shaft assembled.
- Interference fit ("SHIMARI-BAME" in Japanese) is when two mating parts will always interfere when the hole and shaft assembled.
- Transition fit ("TYUKAN-BAME" in Japanese) is when two mating parts will sometimes have a clearance or sometimes interfere when the hole and shaft assembled.

5-4-1 Indication of fit tolerance symbol

When indicate fit tolerance in a hole, the Latin alphabet capital letter is used. Refer to Figure 5-19 a.

When indicate fit tolerance on a shaft, the Latin alphabet small letter is used. Refer to Figure 5-19 b.

ϕ 10 H 7

- Nominal size
- Fundamaental deviation
- Tolerance grade
- IT grade

a) In case of hole

ϕ 10 h 7

- Nominal size
- Fundamaental deviation
- Tolerance grade
- IT grade

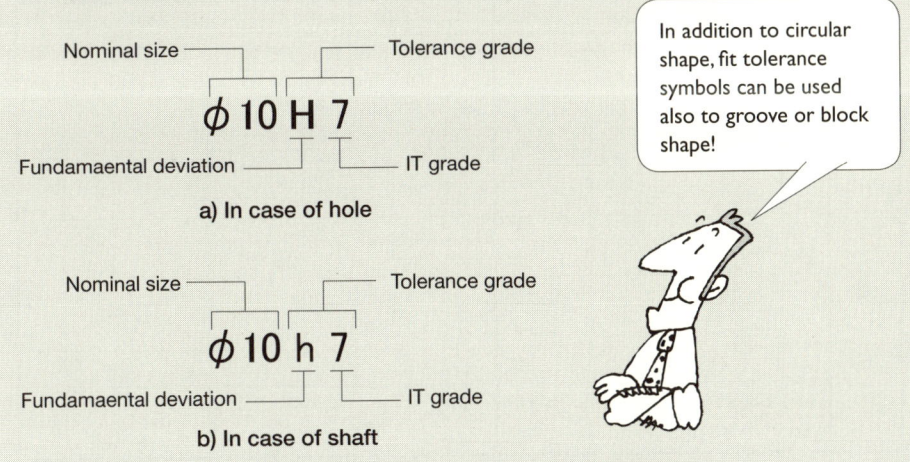

b) In case of shaft

Figure5-19 Expression of fit tolerance symbol

In addition to circular shape, fit tolerance symbols can be used also to groove or block shape!

5-4-2　はめあい基準方式

はめあいの基準方式には、穴基準方式と軸基準方式の2つの種類があります。

1) 穴基準方式

表5-3　穴基準方式

基準穴	軸の公差域クラス
	すきまばめ / 中間ばめ / しまりばめ
H6	g5, h5 / js5, k5, m5
H6	f6, g6, h6 / js6, k6, m6 / n6(*), p6(*)
H7	f6, g6, h6 / js6, k6, m6 / n6, p6(*), r6(*), s6, t6, u6, x6
H7	e7, f7, h7 / js7
H8	f7, h7
H8	e8, f8, h8
	d9, e9
H9	d8, e8, h8
H9	c9, d9, e9, h9
H10	b9, c9, d9

* これらのはめあいは、寸法の区分によっては例外を生じる

2) 軸基準方式

表5-4　軸基準方式

基準軸	穴の公差域クラス
	すきまばめ / 中間ばめ / しまりばめ
h5	H6 / Js6, K6, M6 / N6(*), P6
h6	F6, G6, H6 / Js6, K6, M6 / N6, P6(*)
	F7, G7, H7 / Js7, K7, M7 / N7, P7(*), R7, S7, T7, U7, X7
h7	E7, F7, H7
	F8, H8
h8	D8, E8, F8, H8
	D9, E9, H9
h9	D8, E8, H8
	C9, D9, E9, H9
	B10, C10, D10

* これらのはめあいは、寸法の区分によっては例外を生じる

5-4-2 General fit system combinations

General fit system combinations have two types of "Hole basis system of fits" and "Shaft basis system of fits".

-1. Hole basis system of fits

Table 5-3 Hole basis system of fits

Base hole	Tolerance class of shaft																
	Clearance fit								Transition fit			Interference fit					
	b	c	d	e	f	g	h	js	k	m	n	p	r	s	t	u	x
H6						g5	h5	js5	k5	m5							
H6					f6	g6	h6	js6	k6	m6	n6 (*)	p6 (*)					
H7					f6	g6	h6	js6	k6	m6	n6	p6 (*)	r6 (*)	s6	t6	u6	x6
H7				e7	f7		h7	js7									
H8					f7		h7										
H8				e8	f8		h8										
H8			d9	e9													
H9			d8	e8			h8										
H9		c9	d9	e9			h9										
H10	b9	c9	d9														

* Exceptions occur in some classifications of dimensions.

-2. Shaft basis system of fits

Table 5-4 Shaft basis system of fits

Base shaft	Tolerance class of hole																
	Clearance fit								Transition fit			Interference fit					
	B	C	D	E	F	G	H	Js	K	M	N	P	R	S	T	U	X
h5							H6	Js6	K6	M6	N6 (*)	P6					
h6					F6	G6	H6	Js6	K6	M6	N6	P6 (*)					
h6					F7	G7	H7	Js7	K7	M7	N7	P7 (*)	R7	S7	T7	U7	X7
h7				E7	F7		H7										
h7					F8		H8										
h8			D8	E8	F8		H8										
h8			D9	E9			H9										
h9			D8	E8			H8										
h9		C9	D9	E9			H9										
h9	B10	C10	D10														

* Exceptions occur in some classifications of dimensions.

表5-5 穴の公差域クラス
Table 5-5 Tolerance band of hole

基準寸法の区分(mm) Nominal size range(mm)		穴の公差域クラス (μm) Tolerance band of hole(μm)																									基準寸法の区分(mm) Nominal size range(mm)	
超 Over	以下 Up to	F6	F7	F8	G6	G7	H5	H6	H7	H8	H9	H10	Js5	Js6	Js7	K5	K6	K7	M5	M6	M7	N6	N7	P6	P7	R7	S7	超 Over	以下 Up to
—	3	+12 / +6	+16 / +6	+20 / +6	+8 / +2	+12 / +2	+4 / 0	+6 / 0	+10 / 0	+14 / 0	+25 / 0	+40 / 0	±2	±3	±5	0 / -4	0 / -6	0 / -10	-2 / -6	-2 / -8	-2 / -12	-4 / -10	-4 / -14	-6 / -12	-6 / -16	-10 / -20	-14 / -24	—	3
3	6	+18 / +10	+22 / +10	+28 / +10	+12 / +4	+16 / +4	+5 / 0	+8 / 0	+12 / 0	+18 / 0	+30 / 0	+48 / 0	±2.5	±4	±6	0 / -5	+2 / -6	+3 / -9	-3 / -8	-1 / -9	0 / -12	-5 / -13	-4 / -16	-9 / -17	-8 / -20	-11 / -23	-15 / -27	3	6
6	10	+22 / +13	+28 / +13	+35 / +13	+14 / +5	+20 / +5	+6 / 0	+9 / 0	+15 / 0	+22 / 0	+36 / 0	+58 / 0	±3	±4.5	±7.5	+1 / -5	+2 / -7	+5 / -10	-4 / -10	-3 / -12	0 / -15	-7 / -16	-4 / -19	-12 / -21	-9 / -24	-13 / -28	-17 / -32	6	10
10	18	+27 / +16	+34 / +16	+43 / +16	+17 / +6	+24 / +6	+8 / 0	+11 / 0	+18 / 0	+27 / 0	+43 / 0	+70 / 0	±4	±5.5	±9	+2 / -6	+2 / -9	+6 / -12	-4 / -12	-4 / -15	0 / -18	-9 / -20	-5 / -23	-15 / -26	-11 / -29	-16 / -34	-21 / -39	10	18
18	30	+33 / +20	+41 / +20	+53 / +20	+20 / +7	+28 / +7	+9 / 0	+13 / 0	+21 / 0	+33 / 0	+52 / 0	+84 / 0	±4.5	±6.5	±10.5	+1 / -8	+2 / -11	+6 / -15	-5 / -14	-4 / -17	0 / -21	-11 / -24	-7 / -28	-18 / -31	-14 / -35	-20 / -41	-27 / -48	18	30
30	50	+41 / +25	+50 / +25	+64 / +25	+25 / +9	+34 / +9	+11 / 0	+16 / 0	+25 / 0	+39 / 0	+62 / 0	+100 / 0	±5.5	±8	±12.5	+2 / -9	+3 / -13	+7 / -18	-5 / -16	-4 / -20	0 / -25	-12 / -28	-8 / -33	-21 / -37	-17 / -42	-25 / -50	-34 / -59	30	50
50	80	+49 / +30	+60 / +30	+76 / +30	+29 / +10	+40 / +10	+13 / 0	+19 / 0	+30 / 0	+46 / 0	+74 / 0	+120 / 0	±6.5	±9.5	±15	+3 / -10	+4 / -15	+9 / -21	-6 / -19	-5 / -24	0 / -30	-14 / -33	-9 / -39	-26 / -45	-21 / -51	-30 / -60 -32 / -62	-42 / -72 -48 / -78	50	80
80	120	+58 / +36	+71 / +36	+90 / +36	+34 / +12	+47 / +12	+15 / 0	+22 / 0	+35 / 0	+54 / 0	+87 / 0	+140 / 0		±11	±17.5	+2 / -13	+4 / -18	+10 / -25	-8 / -23	-6 / -28	0 / -35	-16 / -38	-10 / -45	-30 / -52	-24 / -59	-38 / -73 -41 / -76	-58 / -93 -66 / -101	80	120
120	180	+68 / +43	+83 / +43	+106 / +43	+39 / +14	+54 / +14	+18 / 0	+25 / 0	+40 / 0	+63 / 0	+100 / 0	+160 / 0		±12.5	±20	+3 / -15	+4 / -21	+12 / -28	-9 / -27	-8 / -33	0 / -40	-20 / -45	-12 / -52	-36 / -61	-28 / -68	-48 / -83 -50 / -90 -53 / -93	-77 / -117 -85 / -125 -93 / -133	120	180
180	250	+79 / +50	+96 / +50	+122 / +50	+44 / +15	+61 / +15	+20 / 0	+29 / 0	+46 / 0	+72 / 0	+115 / 0	+185 / 0		±14.5	±23	+2 / -18	+5 / -24	+13 / -33	-11 / -31	-8 / -37	0 / -46	-22 / -51	-14 / -60	-41 / -70	-33 / -79	-60 / -106 -63 / -109 -67 / -113	-105 / -151 -113 / -159 -123 / -169	180	250
250	315	+88 / +56	+108 / +56	+137 / +56	+49 / +17	+69 / +17	+23 / 0	+32 / 0	+52 / 0	+81 / 0	+130 / 0	+210 / 0		±16	±26	+3 / -20	+5 / -27	+16 / -36	-13 / -36	-9 / -41	0 / -52	-25 / -57	-14 / -66	-47 / -79	-36 / -88	-74 / -126 -78 / -130		250	280
315	400	+98 / +62	+119 / +62	+151 / +62	+54 / +18	+75 / +18	+25 / 0	+36 / 0	+57 / 0	+89 / 0	+140 / 0	+230 / 0		±18	±28.5	+3 / -22	+7 / -29	+17 / -40	-14 / -39	-10 / -46	0 / -57	-26 / -62	-16 / -73	-51 / -87	-41 / -98	-87 / -144		315	355
400	500	+108 / +68	+131 / +68	+165 / +68	+60 / +20	+83 / +20	+27 / 0	+40 / 0	+63 / 0	+97 / 0	+155 / 0	+250 / 0		±20	±31.5	+2 / -25	+8 / -32	+18 / -45	-16 / -43	-10 / -50	0 / -63	-27 / -67	-17 / -80	-55 / -95	-45 / -108	-93 / -150 -103 / -166 -109 / -172		355	400 450 500

同上 Same as above

表5-6 軸の公差域クラス
Table 5-6 Tolerance band of shaft

基準寸法の区分 (mm) / Nominal size range(mm)

軸の公差域クラス (μm) / Tolerance band of shaft(μm)

Over	Up to	f6	f7	f8	g4	g5	g6	h4	h5	h6	h7	h8	h9	js4	js5	js6	js7	k4	k5	k6	m4	m5	m6	n6	p6	r6	s6
—	3	−6/−12	−6/−16	−6/−20	−2/−5	−2/−6	−2/−8	0/−3	0/−4	0/−6	0/−10	0/−14	0/−25	±1.5	±2	±3	±5	+3/0	+4/0	+6/0	+5/+2	+6/+2	+8/+2	+10/+4	+12/+6	+16/+10	+20/+14
3	6	−10/−18	−10/−22	−10/−28	−4/−8	−4/−9	−4/−12	0/−4	0/−5	0/−8	0/−12	0/−18	0/−30	±2	±2.5	±4	±6	+5/+1	+6/+1	+9/+1	+8/+4	+9/+4	+12/+4	+16/+8	+20/+12	+23/+15	+27/+19
6	10	−13/−22	−13/−28	−13/−35	−5/−9	−5/−11	−5/−14	0/−4	0/−6	0/−9	0/−15	0/−22	0/−36	±2	±3	±4.5	±7.5	+5/+1	+7/+1	+10/+1	+10/+6	+12/+6	+15/+6	+19/+10	+24/+15	+28/+19	+32/+23
10	18	−16/−27	−16/−34	−16/−43	−6/−11	−6/−14	−6/−17	0/−5	0/−8	0/−11	0/−18	0/−27	0/−43	±2.5	±4	±5.5	±9	+6/+1	+9/+1	+12/+1	+12/+7	+15/+7	+18/+7	+23/+12	+29/+18	+34/+23	+39/+28
18	30	−20/−33	−20/−41	−20/−53	−7/−13	−7/−16	−7/−20	0/−6	0/−9	0/−13	0/−21	0/−33	0/−52	±3	±4.5	±6.5	±10.5	+8/+2	+11/+2	+15/+2	+14/+8	+17/+8	+21/+8	+28/+15	+35/+22	+41/+28	+48/+35
30	50	−25/−41	−25/−50	−25/−64	−9/−16	−9/−20	−9/−25	0/−7	0/−11	0/−16	0/−25	0/−39	0/−62	±3.5	±5.5	±8	±12.5	+9/+2	+13/+2	+18/+2	+16/+9	+20/+9	+25/+9	+33/+17	+42/+26	+50/+34	+59/+43
50	65	−30/−49	−30/−60	−30/−76	−10/−18	−10/−23	−10/−29	0/−8	0/−13	0/−19	0/−30	0/−46	0/−74	±4	±6.5	±9.5	±15	+10/+2	+15/+2	+21/+2	+19/+11	+24/+11	+30/+11	+39/+20	+51/+32	+60/+41	+72/+53
65	80	−30/−49	−30/−60	−30/−76	−10/−18	−10/−23	−10/−29	0/−8	0/−13	0/−19	0/−30	0/−46	0/−74	±4	±6.5	±9.5	±15	+10/+2	+15/+2	+21/+2	+19/+11	+24/+11	+30/+11	+39/+20	+51/+32	+62/+43	+78/+59
80	100	−36/−58	−36/−71	−36/−90	−12/−22	−12/−27	−12/−34	0/−10	0/−15	0/−22	0/−35	0/−54	0/−87	±5	±7.5	±11	±17.5	+13/+3	+18/+3	+25/+3	+23/+13	+28/+13	+35/+13	+45/+23	+59/+37	+73/+51	+93/+71
100	120	−36/−58	−36/−71	−36/−90	−12/−22	−12/−27	−12/−34	0/−10	0/−15	0/−22	0/−35	0/−54	0/−87	±5	±7.5	±11	±17.5	+13/+3	+18/+3	+25/+3	+23/+13	+28/+13	+35/+13	+45/+23	+59/+37	+76/+54	+101/+79
120	140	−43/−68	−43/−83	−43/−106	−14/−26	−14/−32	−14/−39	0/−12	0/−18	0/−25	0/−40	0/−63	0/−100	±6	±9	±12.5	±20	+15/+3	+21/+3	+28/+3	+27/+15	+33/+15	+40/+15	+52/+27	+68/+43	+88/+63	+117/+92
140	160	−43/−68	−43/−83	−43/−106	−14/−26	−14/−32	−14/−39	0/−12	0/−18	0/−25	0/−40	0/−63	0/−100	±6	±9	±12.5	±20	+15/+3	+21/+3	+28/+3	+27/+15	+33/+15	+40/+15	+52/+27	+68/+43	+90/+65	+125/+100
160	180	−43/−68	−43/−83	−43/−106	−14/−26	−14/−32	−14/−39	0/−12	0/−18	0/−25	0/−40	0/−63	0/−100	±6	±9	±12.5	±20	+15/+3	+21/+3	+28/+3	+27/+15	+33/+15	+40/+15	+52/+27	+68/+43	+93/+68	+133/+108
180	200	−50/−79	−50/−96	−50/−122	−15/−29	−15/−35	−15/−44	0/−14	0/−20	0/−29	0/−46	0/−72	0/−115	±7	±10	±14.5	±23	+18/+4	+24/+4	+33/+4	+31/+17	+37/+17	+46/+17	+60/+31	+79/+50	+106/+80	+151/+122
200	225	−50/−79	−50/−96	−50/−122	−15/−29	−15/−35	−15/−44	0/−14	0/−20	0/−29	0/−46	0/−72	0/−115	±7	±10	±14.5	±23	+18/+4	+24/+4	+33/+4	+31/+17	+37/+17	+46/+17	+60/+31	+79/+50	+109/+84	+159/+130
225	250	−50/−79	−50/−96	−50/−122	−15/−29	−15/−35	−15/−44	0/−14	0/−20	0/−29	0/−46	0/−72	0/−115	±7	±10	±14.5	±23	+18/+4	+24/+4	+33/+4	+31/+17	+37/+17	+46/+17	+60/+31	+79/+50	+113/+94	+169/+140
250	280	−56/−88	−56/−108	−56/−137	−17/−33	−17/−40	−17/−49	0/−16	0/−23	0/−32	0/−52	0/−81	0/−130	±8	±11.5	±16	±26	+20/+4	+27/+4	+36/+4	+36/+20	+43/+20	+52/+20	+66/+34	+88/+56	+126/+98	+190/+158
280	315	−56/−88	−56/−108	−56/−137	−17/−33	−17/−40	−17/−49	0/−16	0/−23	0/−32	0/−52	0/−81	0/−130	±8	±11.5	±16	±26	+20/+4	+27/+4	+36/+4	+36/+20	+43/+20	+52/+20	+66/+34	+88/+56	+130/+108	+202/+170
315	355	−62/−98	−62/−119	−62/−151	−18/−36	−18/−43	−18/−54	0/−18	0/−25	0/−36	0/−57	0/−89	0/−140	±9	±12.5	±18	±28.5	+22/+4	+29/+4	+40/+4	+39/+21	+46/+21	+57/+21	+73/+37	+98/+62	+144/+114	+226/+190
355	400	−62/−98	−62/−119	−62/−151	−18/−36	−18/−43	−18/−54	0/−18	0/−25	0/−36	0/−57	0/−89	0/−140	±9	±12.5	±18	±28.5	+22/+4	+29/+4	+40/+4	+39/+21	+46/+21	+57/+21	+73/+37	+98/+62	+150/+126	+244/+208
400	450	−68/−108	−68/−131	−68/−165	−20/−40	−20/−47	−20/−60	0/−20	0/−27	0/−40	0/−63	0/−97	0/−155	±10	±13.5	±20	±31.5	+25/+5	+32/+5	+45/+5	+43/+23	+50/+23	+63/+23	+80/+40	+108/+68	+166/+132	+272/+232
450	500	−68/−108	−68/−131	−68/−165	−20/−40	−20/−47	−20/−60	0/−20	0/−27	0/−40	0/−63	0/−97	0/−155	±10	±13.5	±20	±31.5	+25/+5	+32/+5	+45/+5	+43/+23	+50/+23	+63/+23	+80/+40	+108/+68	+172/+132	+292/+252

Chapter5　In what way should I communicate my design intension!

第5章 5 公差解析

最悪状態を検討するために、構成部品が最悪状態であると仮定して公差解析を行います。(図5-20)。

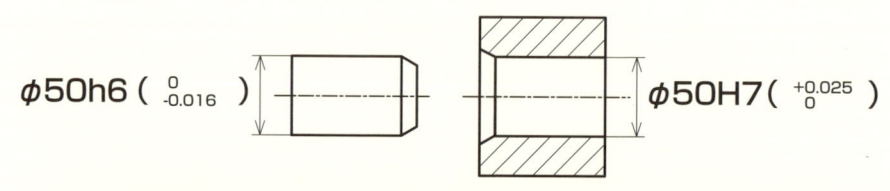

$\phi 50h6 \begin{pmatrix} 0 \\ -0.016 \end{pmatrix}$ $\phi 50H7 \begin{pmatrix} +0.025 \\ 0 \end{pmatrix}$

項目	軸(mm)	穴(mm)
基準(ノミナル)寸法	50	
最大許容寸法	50.000	50.025
最小許容寸法	49.984	50.000
寸法公差	0.016	0.025
はめあいの種類	すきまばめ	
最小すきま	0.000	
最大すきま	0.041	

図5-20　公差の管理

直列寸法は、幾何学的な閉回路をもつ、互いにつながった1セットの独立した平行寸法です。この累積する公差は、足し算によって求められます (図5-21)。

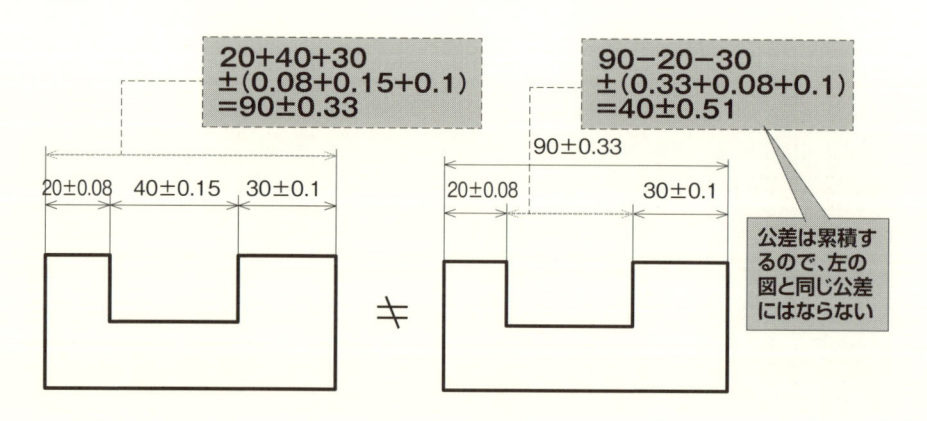

左図:
20+40+30
±(0.08+0.15+0.1)
=90±0.33

20±0.08　40±0.15　30±0.1

右図:
90−20−30
±(0.33+0.08+0.1)
=40±0.51

90±0.33
20±0.08　　30±0.1

公差は累積するので、左の図と同じ公差にはならない

図5-21　累積公差

Chapter 5 — 5 Tolerance analysis

In the worst-case approach, we analyze tolerances assuming components will be at their worst-case conditions. Refer to Figure 5-20.

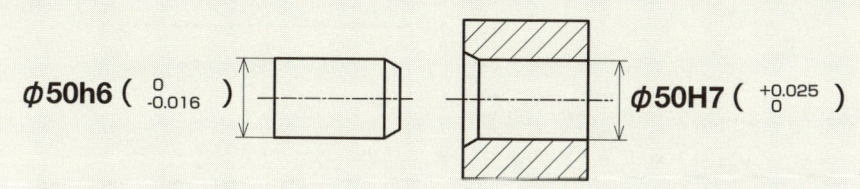

$\phi 50h6 \left(\begin{smallmatrix} 0 \\ -0.016 \end{smallmatrix} \right)$　　　$\phi 50H7 \left(\begin{smallmatrix} +0.025 \\ 0 \end{smallmatrix} \right)$

Item	Shaft (mm)	Hole (mm)
Nominal dimension	50	
Upper limits	50.000	50.025
Lower limits	49.984	50.000
Tolerance	0.016	0.025
Type of fit	Clearance fit	
Minimum clearance	0.000	
Maximum clearance	0.041	

Figure 5-20 Management of tolerance

A linear dimensional chain is a set of independent parallel dimensions which continue each other to create a geometrically closed circuit. This stacked tolerance can be gotten by addition. Refer to Figure 5-21.

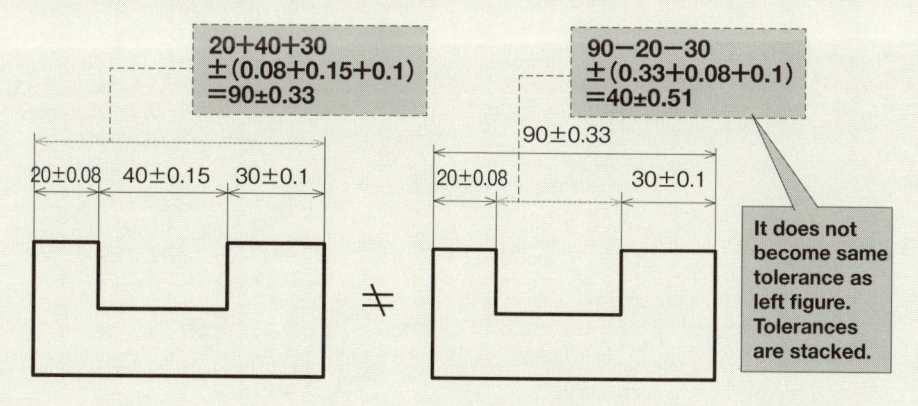

$20+40+30$
$\pm(0.08+0.15+0.1)$
$=90\pm0.33$

$90-20-30$
$\pm(0.33+0.08+0.1)$
$=40\pm0.51$

20 ± 0.08　40 ± 0.15　30 ± 0.1

90 ± 0.33

20 ± 0.08　30 ± 0.1

It does not become same tolerance as left figure. Tolerances are stacked.

Figure 5-21 Tolerance stack

4つの部品をハウジングの中に挿入する必要があるとき、それら全ての寸法公差の累積が重要となります。もし、これらの部品の隙間が機能に影響するとしたら、問題が発生することになります（図5-22）。

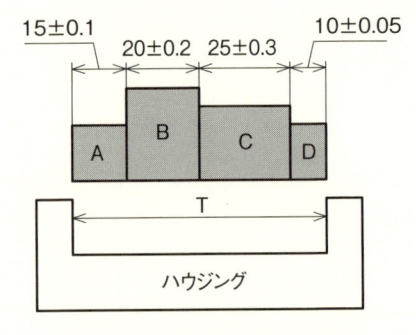

図5-22　個々の部品の組立

　そこで、次に示すような解決策があります。
・公差の合計がより小さくなるよう、各部品の公差を制限すること。
・各部品のばらつきが統計的に分散すると考えること。

　一般的に、公差解析する場合は、次の2つの方法があります。
① 最悪状態での公差解析： $T = \Sigma |T_i|$
　$T = 15+20+25+10 \pm (0.1+0.2+0.3+0.05) = 70 \pm 0.65$　（mm）
　　ハウジングの寸法Tは、70.65mm以上にしなければ、それらをはめあうことはできません。
② 統計的な公差解析（二乗和平方根）： $T = \sqrt{\Sigma |T_i|}$
　$T = 15+20+25+10 \pm \sqrt{(0.1^2+0.2^2+0.3^2+0.05^2)} = 70 \pm 0.38$　（mm）
　　ハウジングの寸法Tは、70.38mm以上でよいことになります。

　統計的な公差解析を使う場合、寸法のバラツキが正規分布に従うことを前提としなければいけません（図5-23）。
　上記から明らかなように、最悪状態による製品は、二乗和平方根よりも大きなばらつきをもちます。直列する寸法の数が増えるほど、その違いがさらに大きくなります。

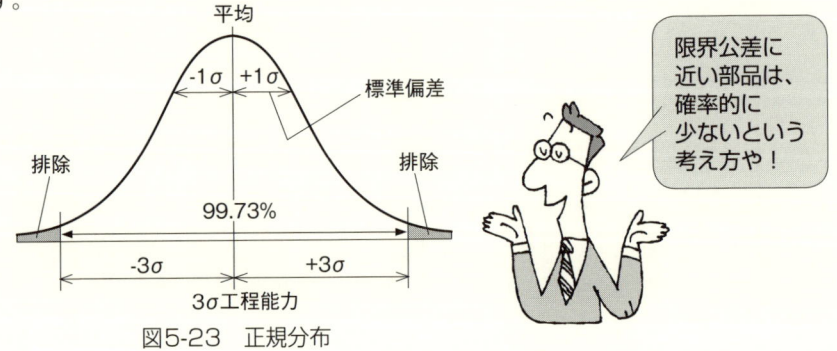

限界公差に近い部品は、確率的に少ないという考え方や！

図5-23　正規分布

When 4 work-pieces are fit into housing, stack of all those tolerances shall be significant. If controlling the spacing of these work-pieces is influence to function, a problem occurs. Refer to Figure 5-22.

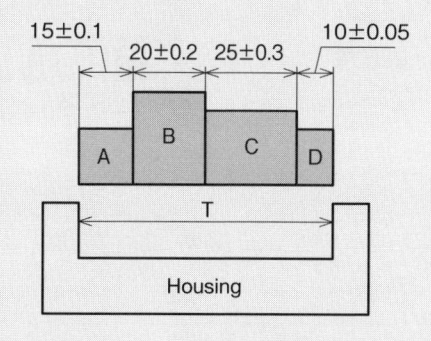

Figure 5-22 Assembly of individual work-pieces

Therefore there are solutions as follows,
- Tighten up the tolerances on each work-piece so the sum of the tolerances is lower.
- Consider that the variation in each work-piece is likely to be statistically distributed.

Generally, when carrying out tolerance analysis, there are the following two methods.
-1. Tolerance analysis of worst-case (WC): $T = \Sigma \mid T_i \mid$
   $T = 15+20+25+10 \pm (0.1+0.2+0.3+0.05) = 70 \pm 0.65$ (mm)
   If the dimension T of housing is not made over 70.65mm, they cannot fit.
-2. Tolerance analysis of statistical (root-sum-square; RSS): $T = \sqrt{\Sigma \mid T_i \mid}$
   $T = 15+20+25+10 \pm \sqrt{(0.1^2+0.2^2+0.3^2+0.05^2)} = 70 \pm 0.38$ (mm)
   Dimension T of housing will be satisfied over 70.38mm.

When statistical tolerance analysis is used, dimension variation must be premised on following a normal distribution. Refer to Figure 5-23.
Clearly from the above, WC products are much variation than RSS. The difference is even greater as the number of work-piece dimensions in the chain increase.

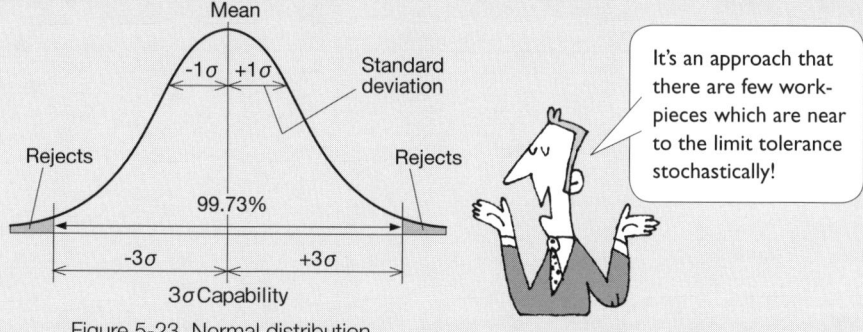

It's an approach that there are few work-pieces which are near to the limit tolerance stochastically!

Figure 5-23 Normal distribution

第5章 6 面の肌記号の指示

> 機械加工によって生じる凹凸やうねり、筋目は、表面粗さとして数値化されます。
> 通常、測定は工具痕の一般的な方向に対して垂直に走る直線上で行われます。

5-6-1 面の肌記号の新旧

　面の肌記号は、改定によって古いものから最新記号までの３種類があります（**表5-7**）。

表5-7　面の肌記号の新旧

	1992年以前の記号	2002年以前の記号 （代表数値例）	2002年以降の記号 （代表数値例）
加工の有無や方法を問わない	〜 （なみと呼ぶ）	数値 （上限の数値を記入）	数値 （上限の数値を記入）
機械加工を要求 （粗い仕上げ）	▽ （いっぱつ）	25	Ra 25
機械加工を要求 （経済的で良好な仕上げ）	▽▽ （にはつ）	6.3	Ra 6.3
機械加工を要求 （綺麗な仕上げ）	▽▽▽ （さんぱつ）	1.6	Ra 1.6
機械加工を要求 （大変綺麗な仕上げ）	▽▽▽▽ （よんぱつ）	0.4	Ra 0.4

| Chapter5 | 6 | # Indication of surface texture symbols |

Unevenness, waviness, tool marks arise by machining is converted into numerical value as surface roughness.

The measurements are usually made along a line, running at perpendicular to the general direction of tool marks on the surface.

5-6-1 Old and new of the surface texture symbol

There are three types surface texture symbol from an old symbol to the newest symbol by revision of the rules. Refer to Table 5-7.

Table 5-7 Old and new symbols of the surface texture

	Symbol up to 1992	Symbol up to 2002 (typical numerical example)	Symbol after 2002 (typical numerical example)
Surface may be produced by any method	∼ (Called NAMI)	Value (Fill in upper limit value)	Value (Fill in upper limit value)
Material removal requied (coarse finishing)	▽ (Called IPPATSU)	25	Ra 25
Material removal requied (economical fine finishing)	▽▽ (Called NIHATSU)	6.3	Ra 6.3
Material removal requied (fine finishing)	▽▽▽ (Called SANPATSU)	1.6	Ra 1.6
Material removal requied (very fine finishing)	▽▽▽▽ (Called YONPATSU)	0.4	Ra 0.4

5-6-2　面の肌を示すパラメータの記号と位置

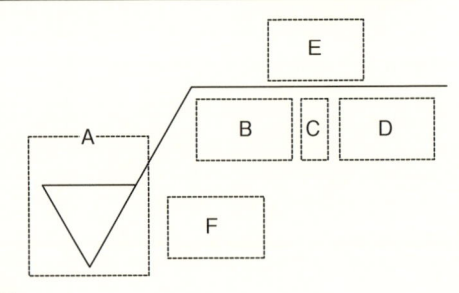

(図：A, B, C, D, E, F の配置)

A：基準となる記号

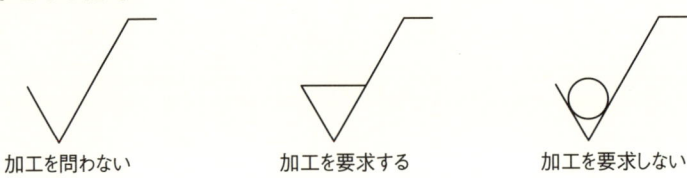

- 加工を問わない
- 加工を要求する
- 加工を要求しない

B：粗さパラメータ

　　Ra…算術平均粗さ

　　Rz…最大高さ

C：空白（半角ダブルスペース）

D：マイクロメートルによる粗さの値（一般的に標準数を使用する）

代表的なパラメータ値	0.05　0.1　0.2　0.4　0.8　1.6　3.2　6.3　12.5　25　50　　100　200　400

E：加工方法の記号

　加工方法の記号は、要求される面の肌が1つの特別な加工によって作られる場合に指示されます。

代表的な加工方法	旋削	フライス削り	研削	バフ
略号	L	M	G	FB

F：筋目方向の記号

　加工によって生じる筋目方向を指示することができます。

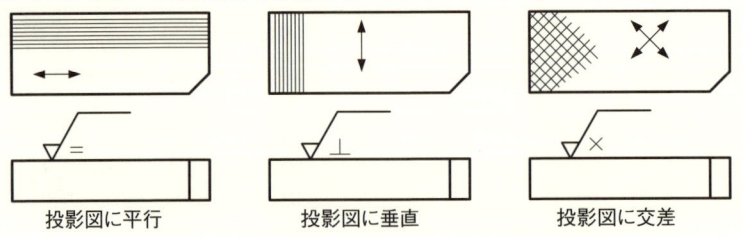

- 投影図に平行（＝）
- 投影図に垂直（⊥）
- 投影図に交差（×）

図5-24　面の肌を示すパラメータの記号と位置

5-6-2 Symbols and position of parameters indicating surface texture

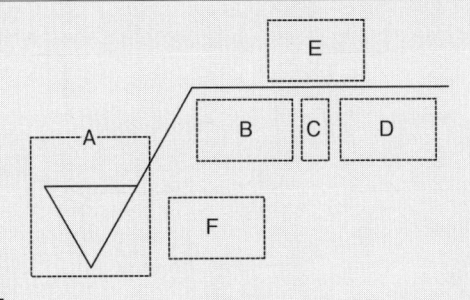

A : Basic symbol

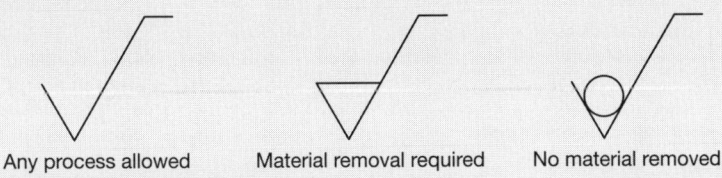

Any process allowed   Material removal required   No material removed

B : Roughness parameters
   Ra: Arithmetical mean deviation of profile
   Rz: Maximum height of profile

C : Blank(double single-byte characters)

D : Roughness value in micrometers(Use preferred numbers generally)

Typical roughness values	0.05 0.1 0.2 0.4 0.8 1.6 3.2 6.3 12.5 25 50 100 200 400

E : Processing symbols

When the required surface texture is produced by one particular processing, processing symbol is indicated.

Typical processing	Lathe turning	Milling	Grinding	Buffing
Code	L	M	G	FB

F : Surface pattern symbols

Surface pattern which produces by processing can be indicated.

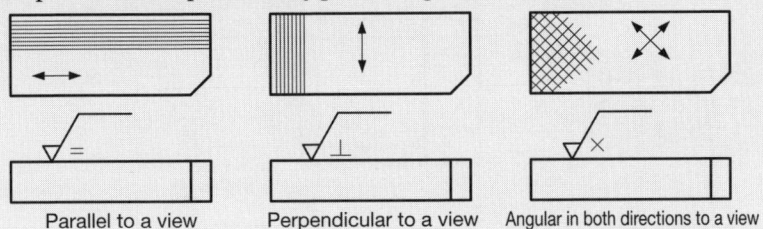

Parallel to a view   Perpendicular to a view   Angular in both directions to a view

Figure 5-24 Symbols and position of parameters indicating surface texture

Chapter5 In what way should I communicate my design intension!

5-6-3 面の肌記号の指示

面の肌記号は、外形線の上あるいは寸法補助線の上に指示することができます。

記号のエッジは、面が存在する方向に向けますが、最新の記号は、真っすぐか90°回転した方向でしか使うことができません（**図5-25**）。

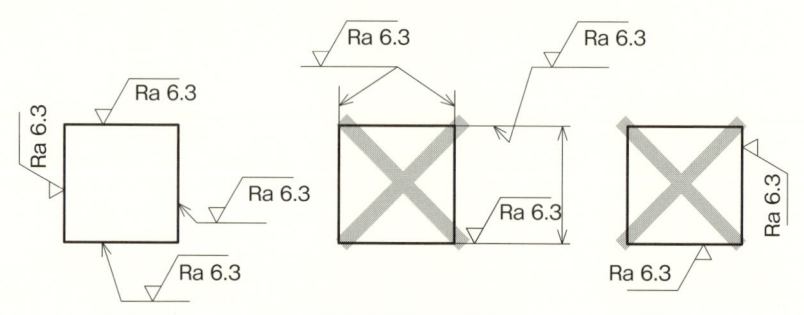

図5-25 面の肌記号の向き

基準記号に円を付けると、その投影図のすべての表面に適用されます。しかし、投影図の表裏は適用されません（**図5-26**）。

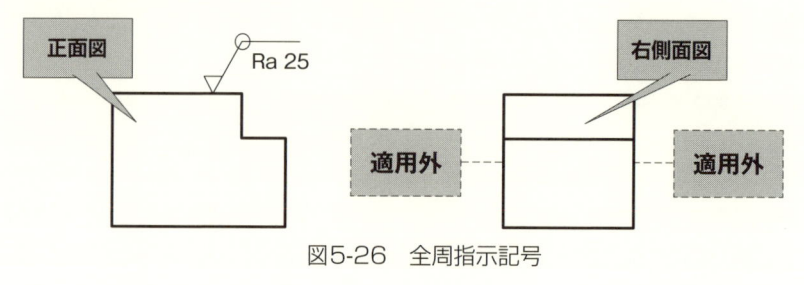

図5-26 全周指示記号

主たる面の肌記号を投影図の付近に指示することができます（**図5-27**）。

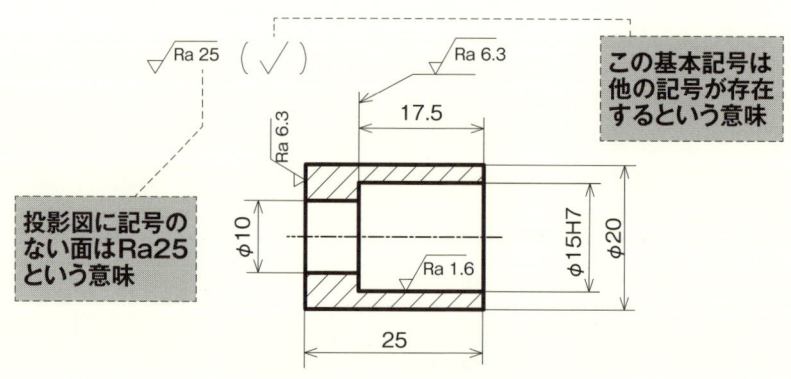

図5-27 主たる面の肌記号の指示

5-6-3 Indication of surface texture symbol

Surface texture symbol can be put on visible outlines or projection lines. The edge of a symbol is turned to the direction in which a surface exists. However, the newest symbols have to use only in upright or 90° rotation. Refer to Figure 5-25.

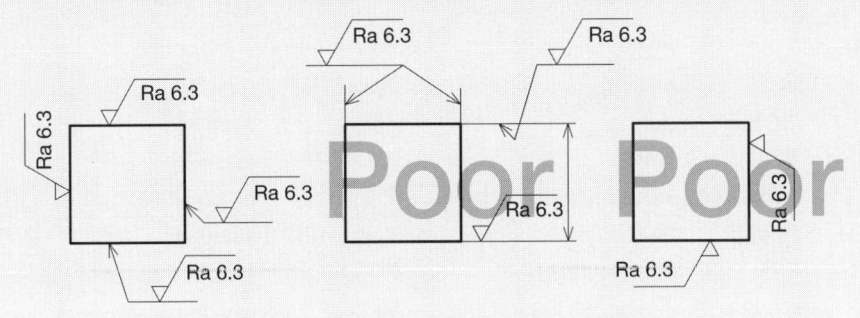

Figure 5-25 Direction of surface texture symbol

The basic symbol with circle applies to all surfaces in the view. However, the backside and front-side of the view does not apply. Refer to Figure 5-26.

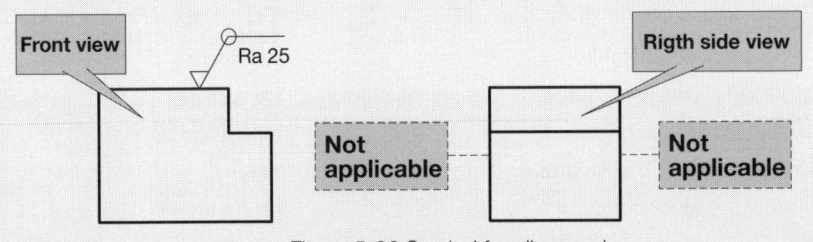

Figure 5-26 Symbol for all around

Main surface texture symbol can indicate near the views. Refer to Figure 5-27.

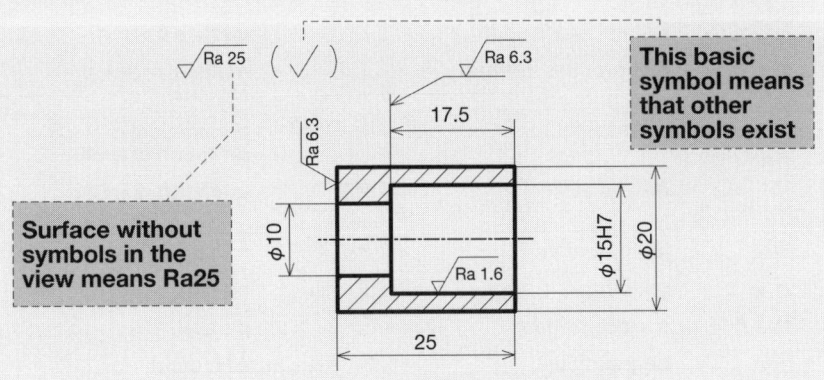

Figure 5-27 Indication of main surface texture symbol

用語集　glossary

本章で使用した単語です。他の用語も使うことができます。

The list of words used for this chapter is shown. You can use another words, too.

鋳放し	casting	プレス打抜き	punching
きりもみ	drilling	リーマ仕上げ	reaming
塗装	paint	マスキング	masking
高周波焼入れ	induction hardening	炭素鋼	carbon steel
合金鋼	alloy steel	ローレット	knurling
平目	straight ridge	あや目	diamond-pattern
センター穴	center drill hole	動作範囲	working area
隣接する部品	adjacent work-piece	重複する寸法	duplicated dimension
参考寸法	reference dimension	基準面(軸)	datum surface (axis)
中心振分け寸法	center divided dimension	寸法公差	dimension tolerance
理論寸法	theoretically exact dimensions	普通公差	general dimension tolerance
公差等級	tolerance grade	精級	fine grade
中級	medium grade	粗級	coarse grade
極粗級	very coarse grade	区分	range
はめあい公差	fit tolerance	すきまばめ	clearance fit
しまりばめ	interference fit	中間ばめ	transition fit
公差域クラス	tolerance band	公差解析	tolerance analysis
正規分布	normal distribution	面の肌	surface texture
加工を問わない	any process allowed	加工を要求する	material removal required
加工を要求しない	no material removed	算術平均粗さ	arithmetical mean deviation of profile
マイクロメートル	micrometer	標準数	preferred number
旋削	lathe turning	フライス削り	milling
研削	grinding	バフ	buffing
筋目方向	surface pattern	平行	parallel
垂直	perpendicular	全周	all around

第6章
特殊な記号って、どない使うねん！

Chapter6
In what way should I use special symbols!

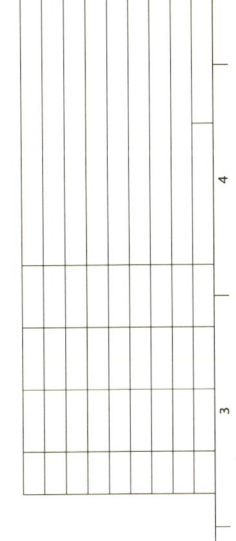

この章を学習すると、次の知識を得ることができます。
1) ねじの指示
2) 溶接記号の理解
3) 幾何公差記号の理解

After studying this chapter, you will be able to get knowledge as follows:
-1. Indication of screw thread
-2. Understanding weld symbols
-3. Understanding geometrical tolerance symbols

| 第6章 | 1 | # ねじの指示 |

ねじは回転運動を直線運動または力に変換するために使用される螺旋の形をしたものです。

6-1-1　ねじの種類

ねじの種類を表6-1に示します。

表6-1　ねじの種類

ねじの記号	ねじの名称	用途
M	メートルねじ	固定
Tr	台形ねじ	直線移動が必要なとき 大きな荷重を受けるとき
G	管用（くだよう）平行ねじ	管の固定 気密接合が必要でないとき
Rc	管用（くだよう）テーパめねじ	管の固定 気密接合が必要なとき
R	管用（くだよう）テーパおねじ	

6-1-2　並目ねじと細目ねじ、左ねじ

　メートルねじには並目ねじと細目ねじがあり、それぞれピッチが異なります（図6-1 a）。
　細目ねじは、ねじの寸法に続けて、ピッチの数値を指示します（図6-1 b）。
　また、通常のねじは右ねじです。これは右回りに回した時に締めるねじのことです。
　左ねじを指示する場合は、ねじの寸法に続けて「LH」を記入します（図6-1 c）。LHは、Left handの省略です。

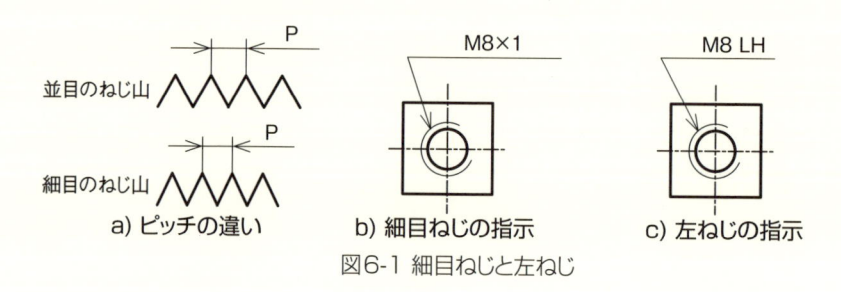

a) ピッチの違い　　b) 細目ねじの指示　　c) 左ねじの指示
図6-1　細目ねじと左ねじ

Chapter6	1	# Indication of screw thread

Screw is a helical shape used to convert rotational movement to linear movement or force.

6-1-1 Type of thread

Type of thread is shown in Table 6-1.

Table 6-1 Type of thread

Thread symbol	Name of thread	Usual use
M	Metric thread	Fixing
Tr	Acme thread	When linear movement is required. When receiving big load.
G	Parallel pipe thread	Pipe fitting When pressure-tight joints are not need on the thread.
Rc	Internal tapered pipe thread	Pipe fitting When pressure-tight joints are need on the thread.
R	External tapered pipe thread	

6-1-2 Coarse pitch thread and fine pitch thread, left-hand thread

Metric thread has coarse pitch thread and fine pitch thread, and each pitch differ. Refer to Figure 6-1 a.

Fine pitch thread is indicated continuing pitch value to the thread dimension. Refer to Figure6-1 b.

And usual thread is a right-hand thread. Right-hand thread will tighten when turned clockwise.

When indicating a left-hand thread, "LH" is put in after thread dimension. Refer to Figure 6-1 c. LH stands for Left hand.

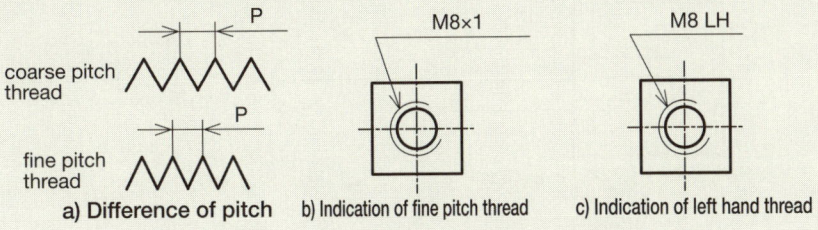

a) Difference of pitch   b) Indication of fine pitch thread   c) Indication of left hand thread

Figure 6-1 Fine pitch thread and left hand thread

ねじのピッチを**表6-2**に示します。

表6-2　ねじのピッチ

ピッチ	M3	M4	M5	M6	M8	M10	M12	M16	M20	M24
並目	0.5	0.7	0.8	1	1.25	1.5	1.75	2	2.5	3
細目	0.35	0.5	0.5	0.75	1 0.75	1.25 1 0.75	1.5 1.25 1	1.5 1	2 1.5 1	2 1.5 1

6-1-3　ねじの投影図

ねじは簡略図で表します（**図6-2**）。

丸く見える方向からおねじを見るとき、おねじの外側（外径）の線は太い実線で円を、内側の線（谷径）は細い実線で弧を描きます。

丸く見える方向からめねじを見るとき、めねじの外側の線（谷径）は細い実線で弧を、内側の線（内径）は太い実線で円を描きます。

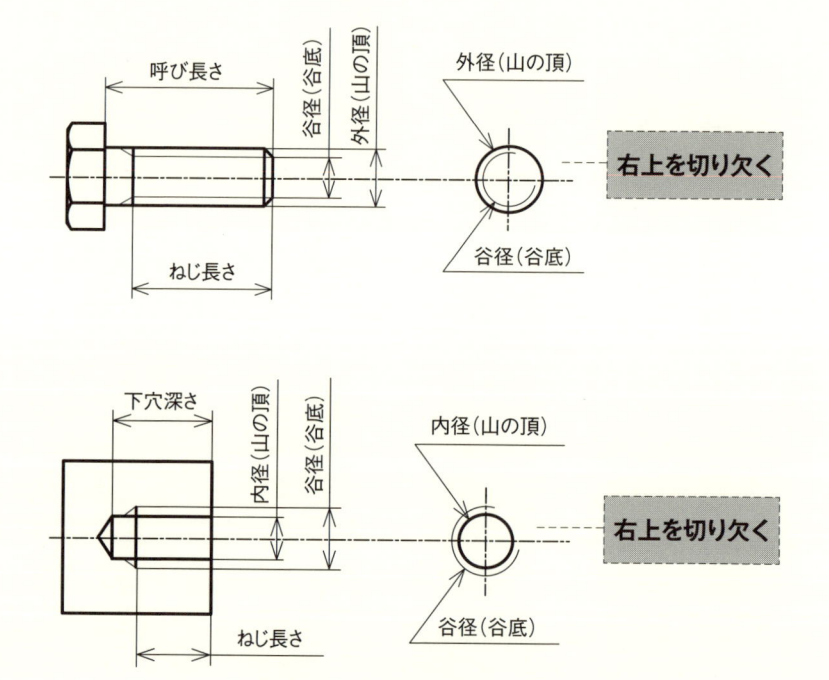

図6-2　ねじの表し方

Thread pitch is shown in Table 6-2.

Table 6-2 Thread pitch

Pitch	M3	M4	M5	M6	M8	M10	M12	M16	M20	M24
Coarse	0.5	0.7	0.8	1	1.25	1.5	1.75	2	2.5	3
Fine	0.35	0.5	0.5	0.75	1 0.75	1.25 1 0.75	1.5 1.25 1	1.5 1	2 1.5 1	2 1.5 1

6-1-3 View of thread

Thread is expressed simplified drawing. Refer to Figure 6-2.

When see a external thread from the round direction, outside line of external thread (crest) is drawn a circle as thick continuous line, and inside line (root) is drawn an arc as thin continuous line.

When see a internal thread from the round direction, outside line of internal thread (root) is drawn an arc as thin continuous line, and inside line (crest) is drawn a circle as thick continuous line.

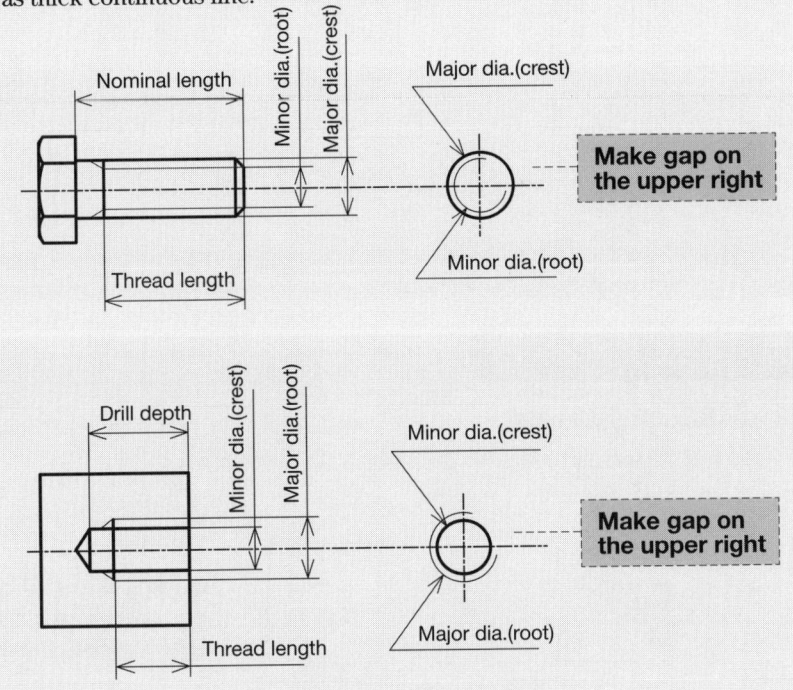

Figure 6-2 Expression of thread

Chapter6 In what way should I use special symbols!

6-1-4　ねじの寸法記入

ねじの寸法は**図6-3**のように指示することができます。

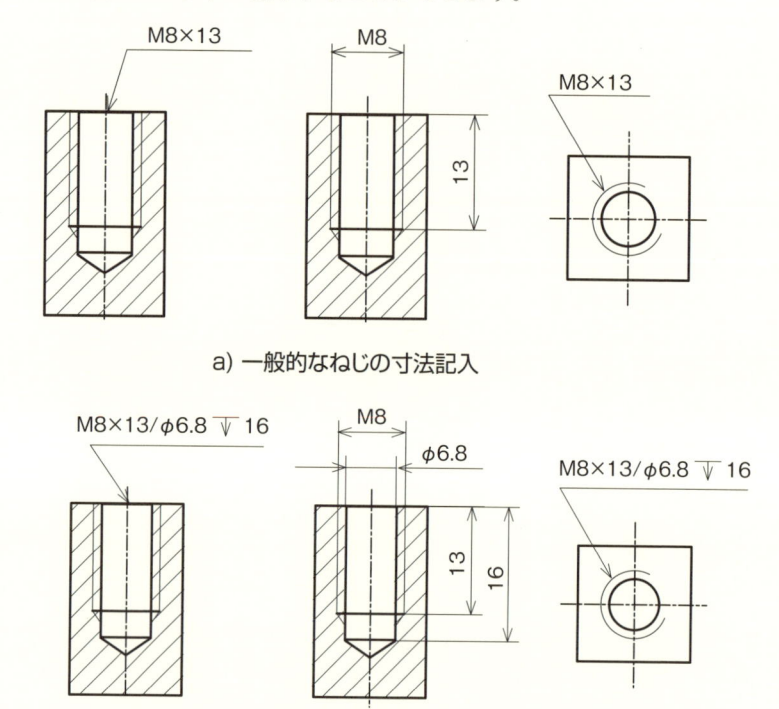

a) 一般的なねじの寸法記入

b) 下穴径と下穴深さまで示したねじの寸法記入

図6-3　ねじの寸法記入

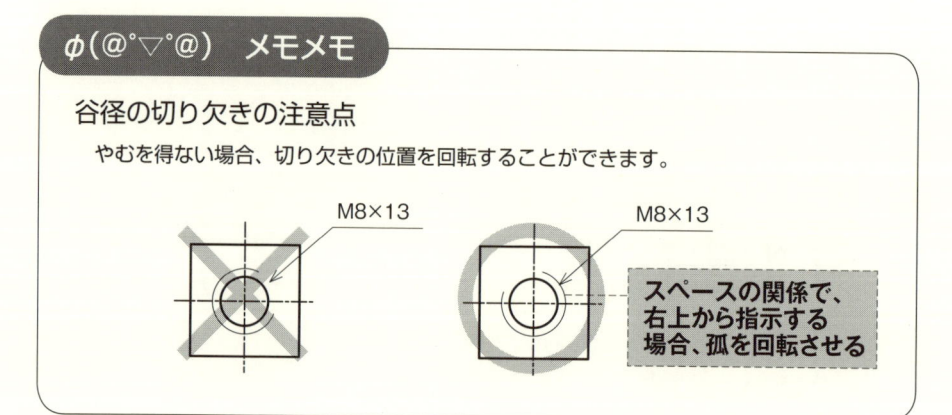

φ(@˚▽˚@)　メモメモ

谷径の切り欠きの注意点

やむを得ない場合、切り欠きの位置を回転することができます。

スペースの関係で、右上から指示する場合、孤を回転させる

6-1-4 Dimensioning of thread

Dimensions of the thread can be indicated as Figure 6-3.

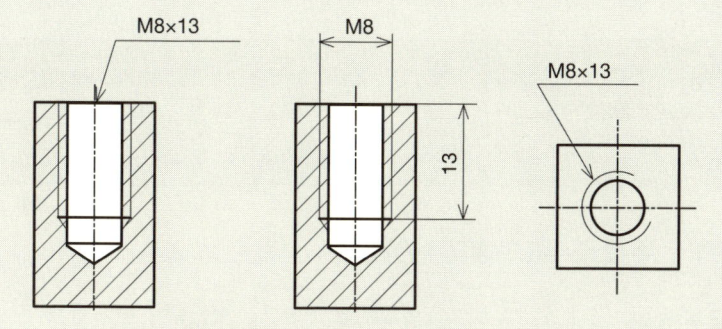

a) Typical dimensioning of thread

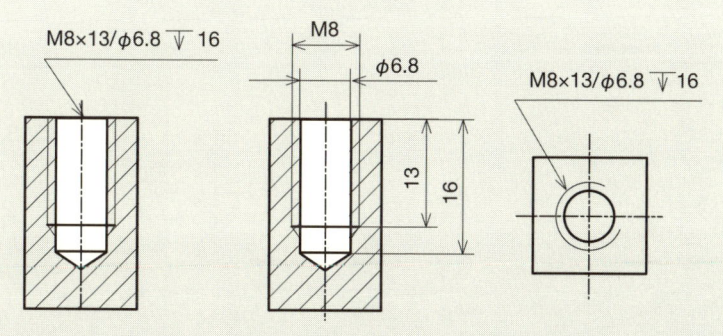

b) Dimensioning of thread with drill size and drill depth

Figure 6-3 Dimensioning of thread

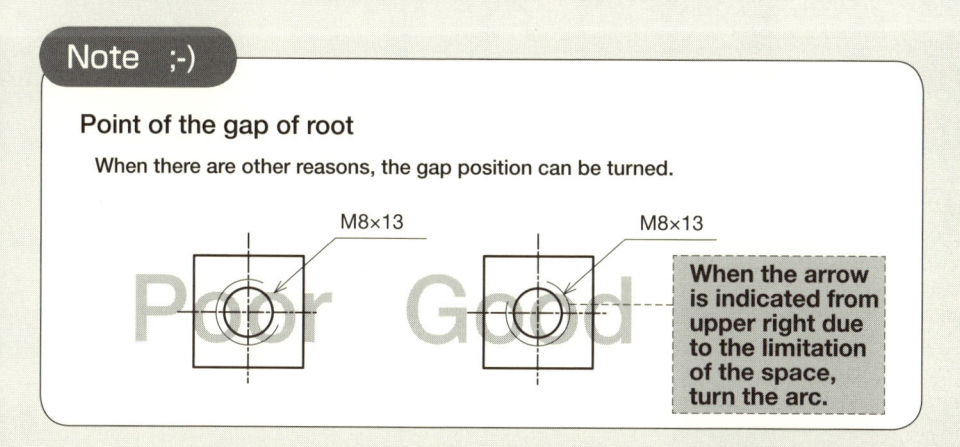

Note ;-)

Point of the gap of root

When there are other reasons, the gap position can be turned.

When the arrow is indicated from upper right due to the limitation of the space, turn the arc.

Chapter6 In what way should I use special symbols!

第6章 2 溶接記号の理解

> 溶接は、圧力や高い融点をもつ溶加材によって熱を加えて、2つ以上の金属部品間を永続的に固定する手法です。

6-2-1 溶接継手（つぎて）の種類

表6-3に示されるように、5つの基本的なタイプの溶接継手があります。

表6-3　基本的なタイプの溶接継手

突合せ継手	T継手	重ね継手
かど継手	へり継手	

へ〜！
溶接で2部品を接合するのに、いろんな継手があるんやね〜！

Chapter 6 - 2 Understanding weld symbols

Welding is a method of making permanent joints between two or more metal workpieces by applying heat, sometimes with pressure and sometimes with filler metal having a high melting point.

6-2-1 Type of weld joints

As shown in Table 6-3, there are five basic types of weld joints.

Table 6-3 Basic types of weld joints

Butt joint	T-joint	Lap joint
Corner joint	Edge joint	

Hmmm, there are various joints to join two work-pieces by welding, are there!

6-2-2 基本記号と補助記号

a) 溶接の基本記号を表6-4に示します。

表6-4　基本記号と位置の意味

溶接位置	すみ肉	プラグ スロット	スポット プロジェクション	シーム	スタッド
矢の側					
矢の反対側	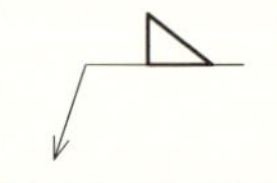	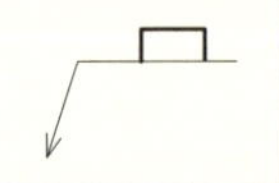			使用しない
両側	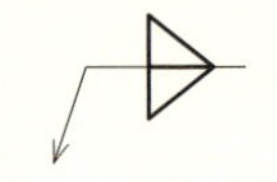	使用しない			使用しない

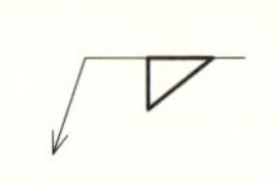

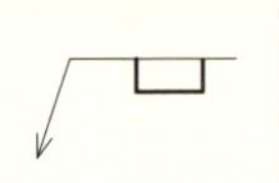

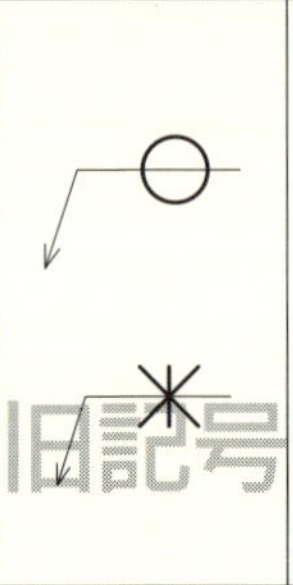

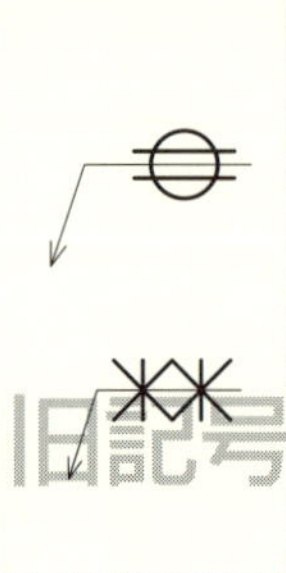

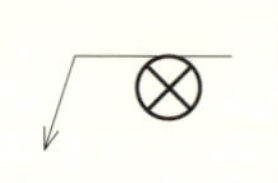

溶接位置	I形	V形	レ形	U形	J形
矢の側	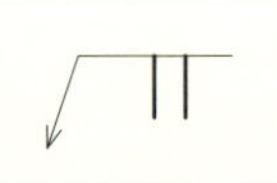	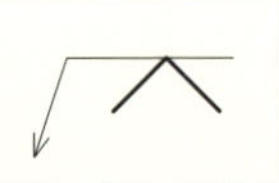	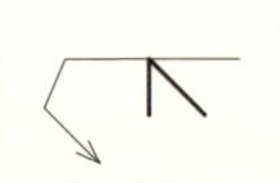	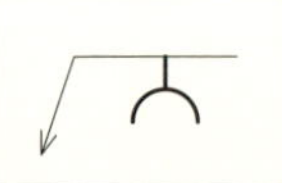	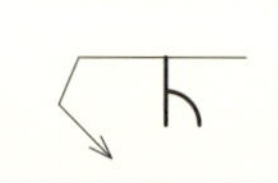
矢の反対側	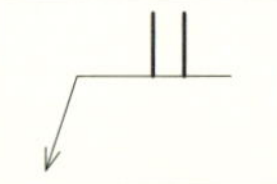	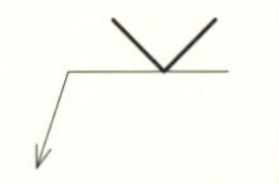	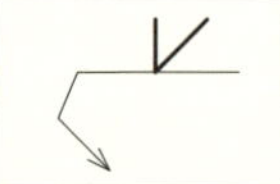	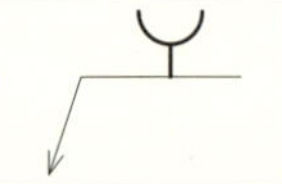	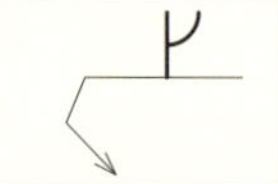
両側	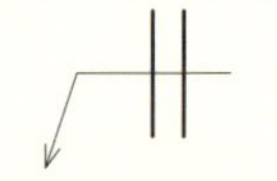	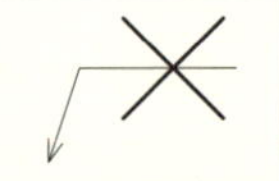	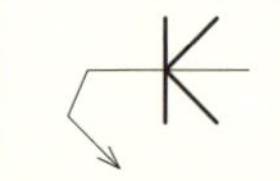	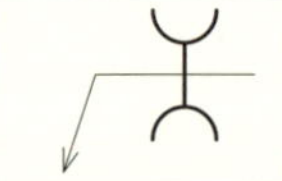	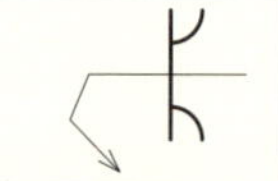

溶接位置	フレアV形	フレアレ形	へり	ビード	肉盛り
矢の側	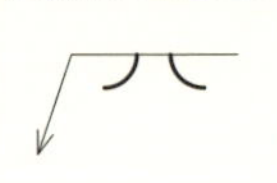	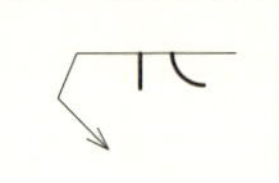	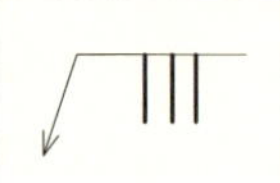	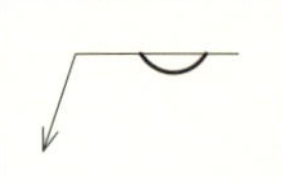	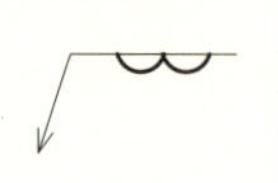
矢の反対側	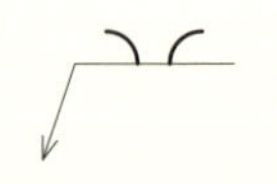	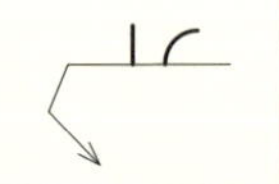	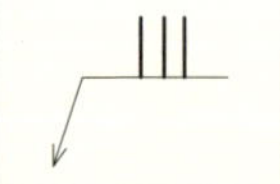		使用しない
両側	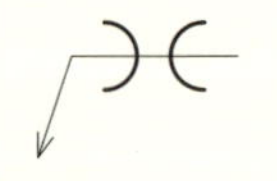	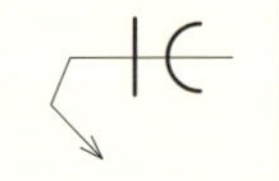	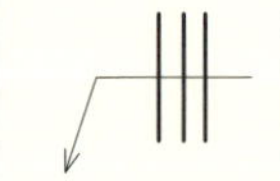	使用しない	使用しない

矢の側　　矢の反対側　　両側

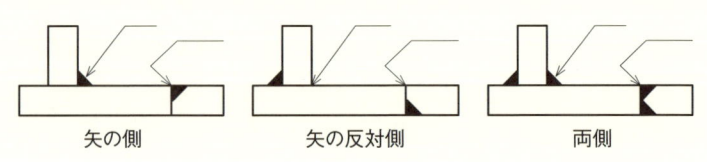

6-2-2 Basic symbols and supplementary symbols

a) Basic symbols for welding are shown in Table 6-4.

Table 6-4 Basic symbols and their location significance

Location	Filet	Plug or Slot	Spot or Projection	Seam	Stud
Arrow side	▽	⊓	○	⊖	⊗
Other side	△	⊔			Not used
Both side	▷	Not used			Not used

Location	Square	V	Bevel	U	J
Arrow side	‖	∧	⌐	⌒	⌐
Other side	‖	∨	⌙	⌣	⌙
Both side	‖	✕	K	⌬	⌬

Location	Flare-V	Flare-Bevel	Edge	Back	Surfacing
Arrow side	)(	)(	‖‖	⌒	⌒⌒
Other side	)(	)(	‖‖	⌒	Not used
Both side	)(	)(	‖‖	Not used	Not used

Arrow side   Other side   Both side

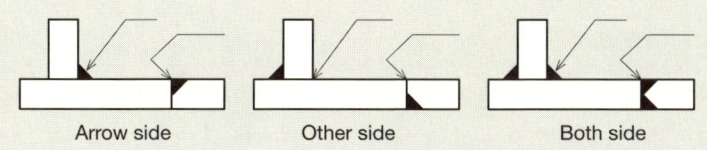

Chapter6 In what way should I use special symbols!

b) 開先溶接の記号と端部の溶接前形状を**表6-5**に示します。

表6-5 開先溶接記号と端部の事前形状

溶接の種類	図面	端部の事前形状	溶接状態
I形開先			
V形開先			
レ形開先			
V形フレア			
レ形フレア			
U形開先			
J形開先			

開先溶接記号は、溶接前の形状やから、めっちゃ簡単なんやで！

ひゃー素敵！めっちゃわかりやすいやんか！

b) Symbols of groove welding and edge preparation are shown in Table 6-5.

Table6-5 Symbols of groove welding and edge preparation

Type	Drawing	Edge preparation	Welding condition
Square groove			
V groove			
Bevel groove			
Flare-V groove			
Flare-bevel groove			
U groove			
J groove			

Symbols of groove welding are edge preparation before welding.
So it is very simple!

That's lovely! It's very easy to understand, isn't it!

Chapter6 In what way should I use special symbols!

c) 溶接の補助記号を**表6-6**、**表6-7**、**表6-8**に示します。

表6-6　補助記号(工程)と使用例

記号	工程	図面と要求形状
▲	現場溶接	溶接は現場で実施されます。
○	全周溶接	
⊓	裏当て	裏当て

表6-7　補助記号(仕上げ)

記号	C	G	M	P
仕上げ方法	チッピング	グラインダ	切削	研磨

表6-8　補助記号(表面形状)と使用例

記号	表面形状	図面	要求形状
—	平ら仕上げ	M	
⌒	凸形仕上げ		
⌣	へこみ仕上げ		
⌣⌣	止端を滑らかに仕上げ	G	

第6章　特殊な記号って、どない使うねん！

c) Supplementary symbols for welding are shown in **Table 6-6, 6-7, 6-8.**

Table 6-6 Supplementary symbols (Process) and examples

Symbol	Process	Drawing and shape of desire
⚑	Site weld	Weld to be made on site.
○	Weld all around	
⊐	Backing bar	Backing bar

Table 6-7 Supplementary symbols (Finish)

Symbol	C	G	M	P
Finish	Chipping	Grinding	Machining	Polishing

Table 6-8 Supplementary symbols (Contour) and examples

Symbol	Contour	Drawing	Shape of desire
—	Flat	M	
⌒	Convex		
⌣	Concave		
⏛	Smooth blend	G	

Chapter6 In what way should I use special symbols!

6-2-3 溶接記号要素の配置

溶接記号要素は、それぞれ決められた位置に記入されます。（図6-4）

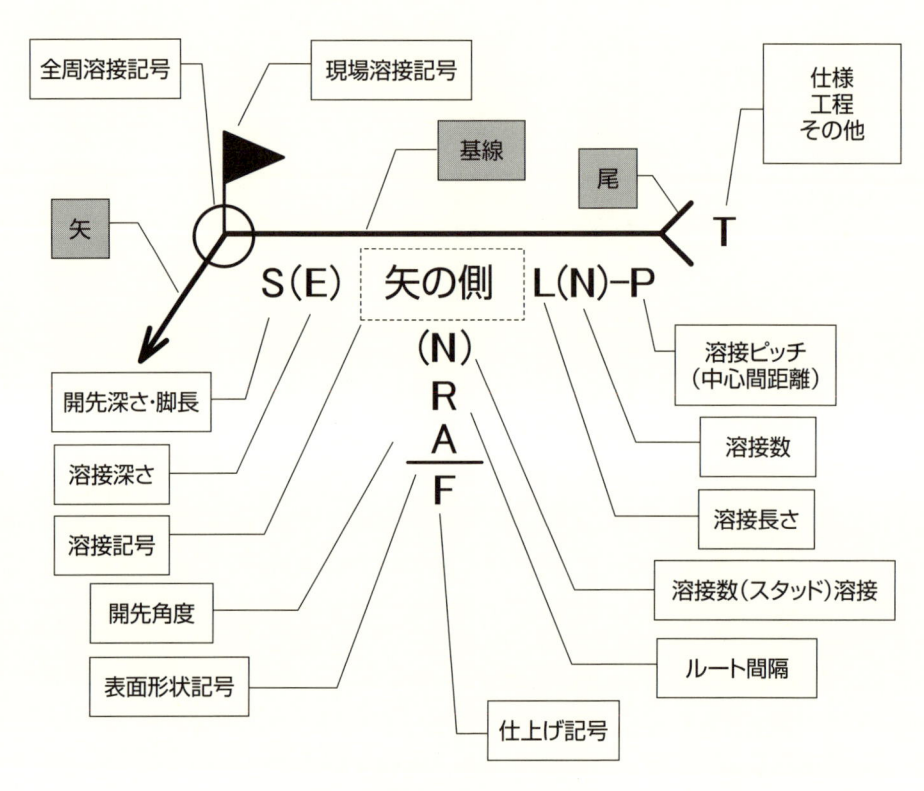

a) 矢の手前側を溶接する場合

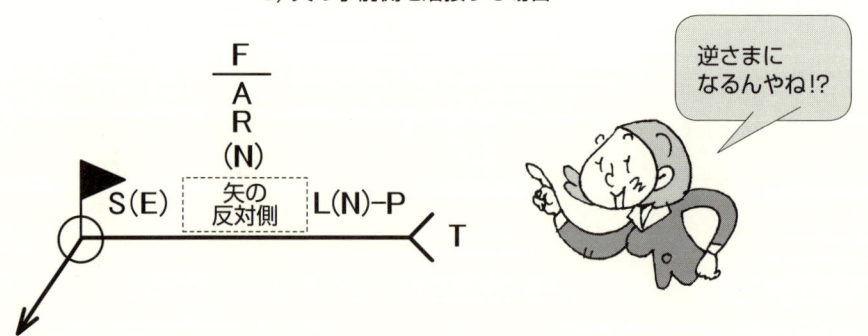

b) 矢の反対側を溶接する場合

図6-4　溶接記号要素

6-2-3 Position of elements of welding symbol

Elements of welding symbol are put on the position decided each. Refer to Figure 6-4.

Labels on diagram (a):
- Weld symbol for all around
- Field weld symbol
- Reference line
- Tail
- Specification, Process, or Other reference
- Arrow line
- S(E)
- Arrow side
- L(N)–P
- (N) R A F
- Depth of bevel or Leg size
- Weld depth
- Weld symbol
- Groove angle
- Contour symbol
- Finish symbol
- Pitch of welds (center to center spacing)
- Number of welds
- Length of weld
- Number of welds (Stud welding)
- Root opening

a) In case of arrow side welds

Diagram (b): S(E), other side, L(N)–P, T, F A R (N)

Becomes upside down, right!?

b) In case of other side welds

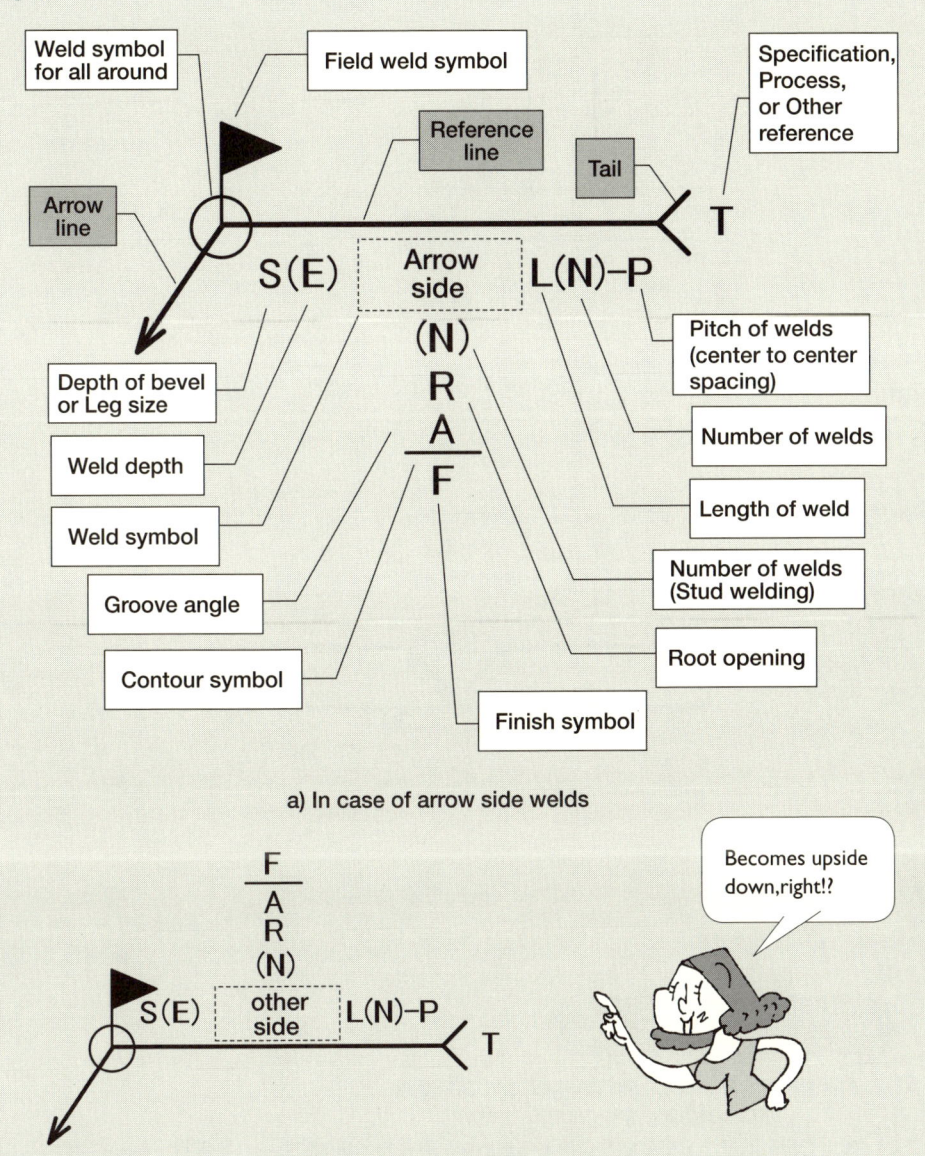

Figure 6-4 Elements of welding symbol

Chapter6 In what way should I use special symbols!

6-2-4　知っておくべき溶接用語

a) すみ肉溶接の場合

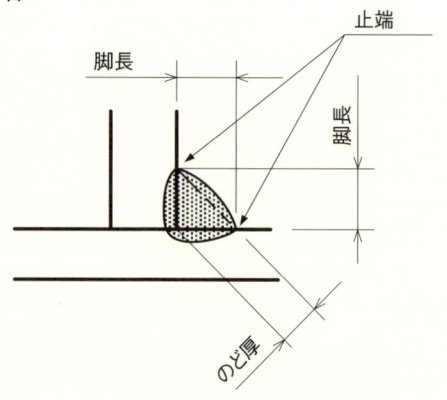

図6-5　すみ肉溶接の用語

b) 開先溶接の場合

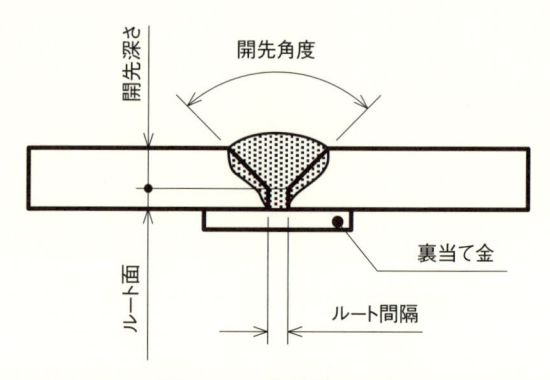

図6-6　開先溶接の用語

φ(@°▽°@)　メモメモ

のど厚…接合部の根元と隅肉溶接面の間の最短距離
ルート…傾斜や溝のない融合面の根元の部分
裏当て金…溶けた溶接金属を保持するために、継ぎ目の裏側に置かれる板材。裏当て金は恒久的または一時的の取り付けになります。

6-2-4 Weld term which you should know

a) In case of fillet welds

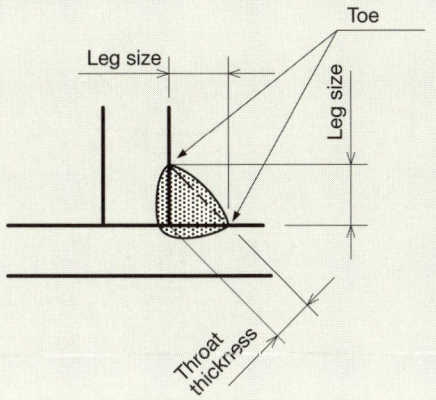

Figure 6-5 Term of fillet welds

b) In case of groove welds

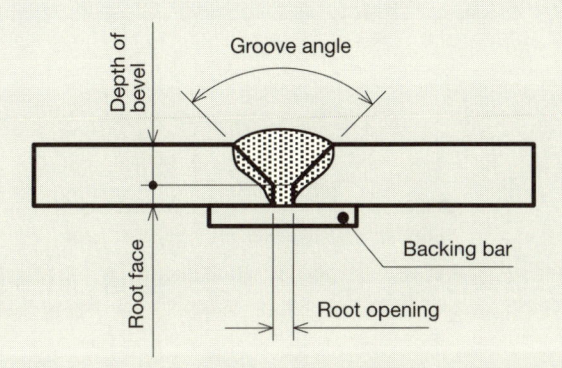

Figure 6-6 Term of groove welds

> **Note ;-)**
>
> Throat thickness : The shortest distance between the weld root and the face of a fillet weld.
> Root: The part of a fusion face at the root that is not beveled or grooved.
> Backing bar : Plate material placed against the back of the joint to retain molten weld metal. The backing bar may be either permanent or temporary.

6-2-5　溶接記号適用時の守るべきルール

a) 矢の側に溶接する場合、記号は基線の下側に記入します（**図6-7**）。

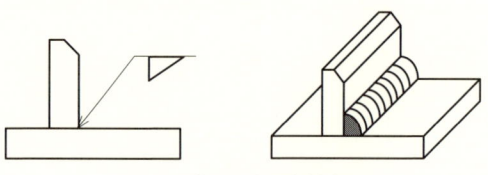

図6-7　矢の側の溶接指示

b) 矢の反対側に溶接する場合、記号は基線の上側に記入します（**図6-8**）。

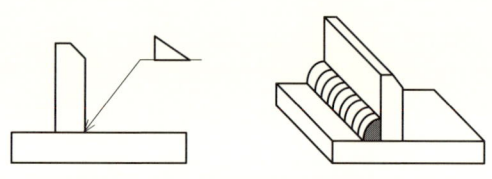

図6-8　矢の反対側の溶接指示

c) 矢の両側に溶接する場合、記号は基線の両側に記入します（**図6-9**）。

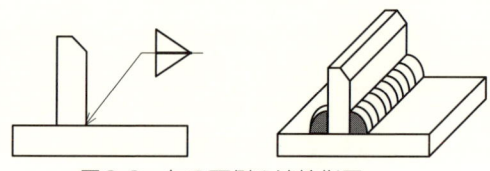

図6-9　矢の両側の溶接指示

d) 溶接自体の向きにかかわらず、溶接記号にある垂直線は、常に左側に描かれます（**図6-10**）。

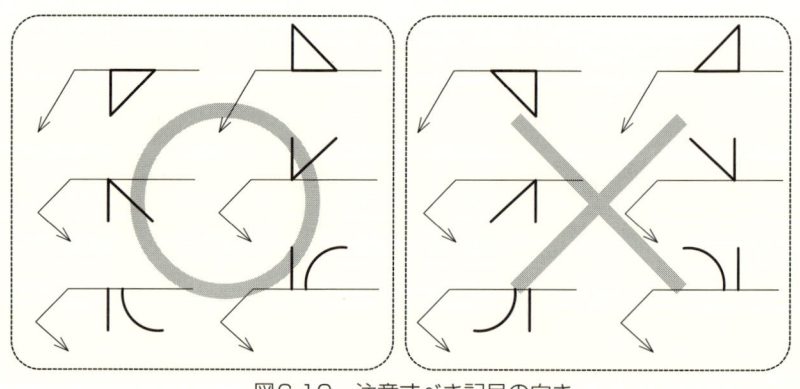

図6-10　注意すべき記号の向き

6-2-5 Rules to be observed when indicating weld symbols

a) When the welds are to be made on the arrow side, symbol is placed below the reference line. Refer to Figure 6-7.

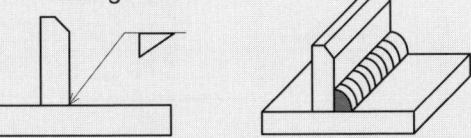

Figure6-7 Indication of arrow side welding

b) When the welds are to be made on the other side of the arrow, symbol is placed above the reference line. Refer to Figure 6-8.

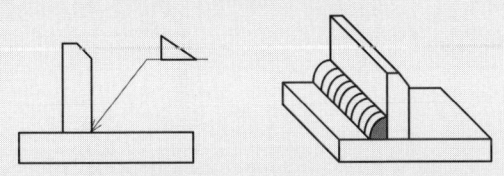

Figure6-8 Indication of other side welding

c) When the welds are to be made on both sides of the arrow, symbols are placed on both sides of the reference line. Refer to Figure 6-9.

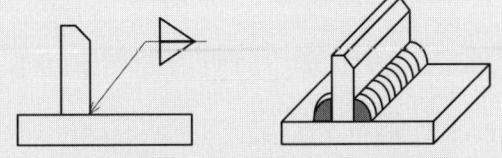

Figure6-9 Indication of both sides welding

d) Perpendicular line of welding symbol is always drawn on the left side, regardless of the orientation of the weld itself. Refer to Figure 6-10.

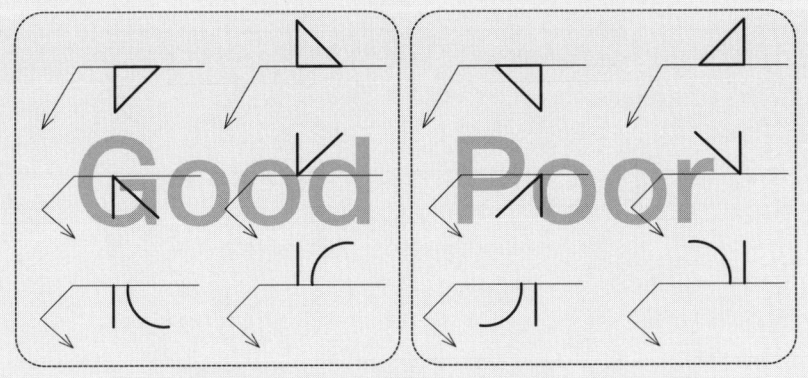

Figure 6-10 Direction of the symbol to be noted

e) すみ肉の脚長は記号の左側に記入します（**図6-11**）。

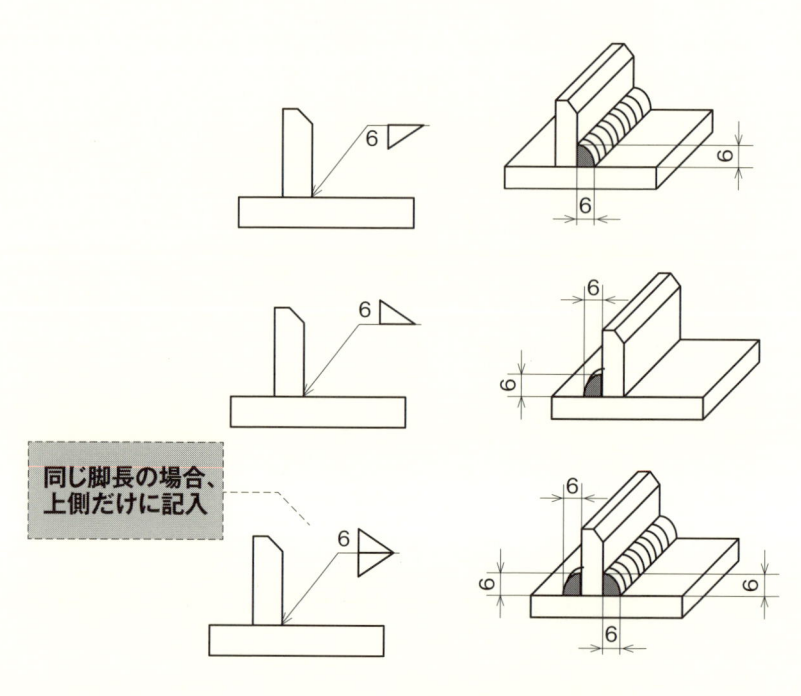

同じ脚長の場合、上側だけに記入

図6-11　すみ肉溶接の脚長の指示

f) 片側のみが面取りされる場合、矢は面取りされる部分を指します。この場合、矢の線は折り曲げなければいけません（**図6-12**）。

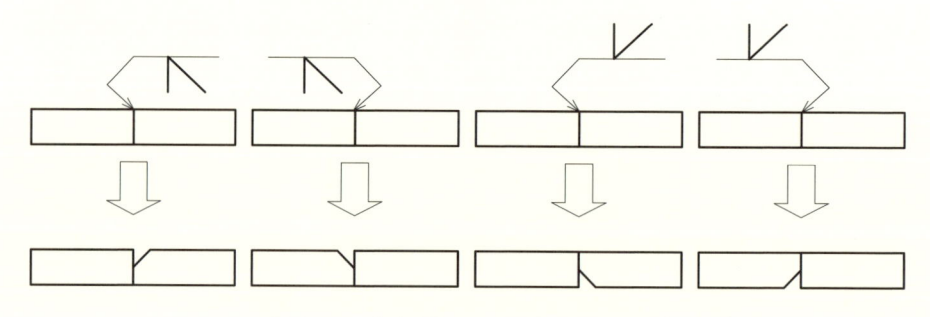

図6-12　矢を折らなければいけない例

e) The leg length of fillet welding is placed to the left side of the symbol. Refer to Figure 6-11.

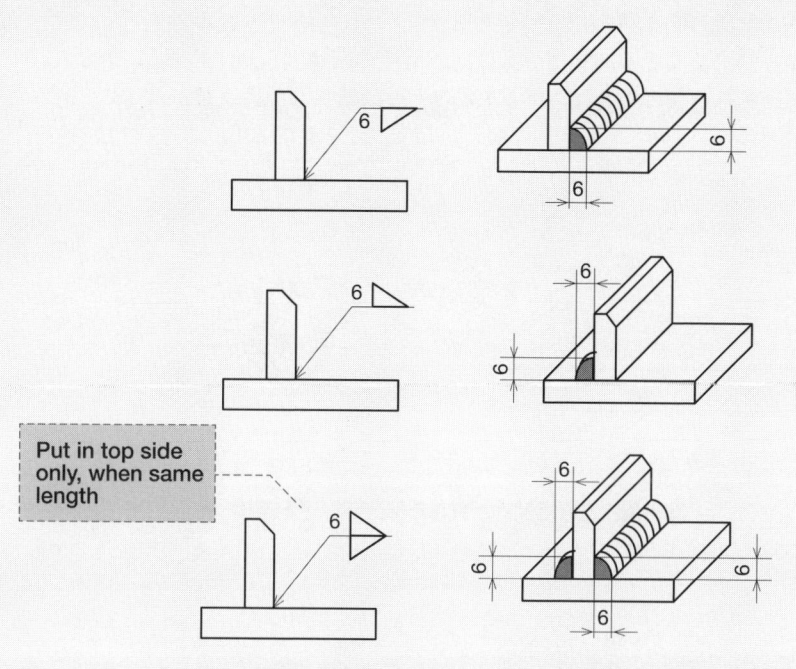

Put in top side only, when same length

Figure6-11 Indication of the leg length of filet welding

f) When only one side is chamfered, the arrow points toward the part that is to be chamfered. In this case, the arrow line has to bend. Refer to Figure 6-12.

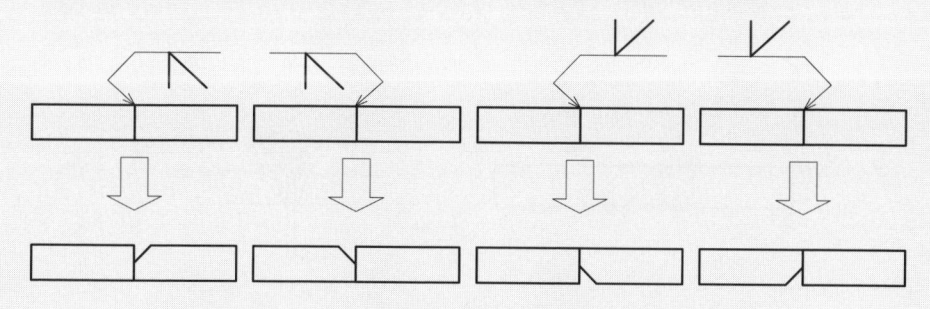

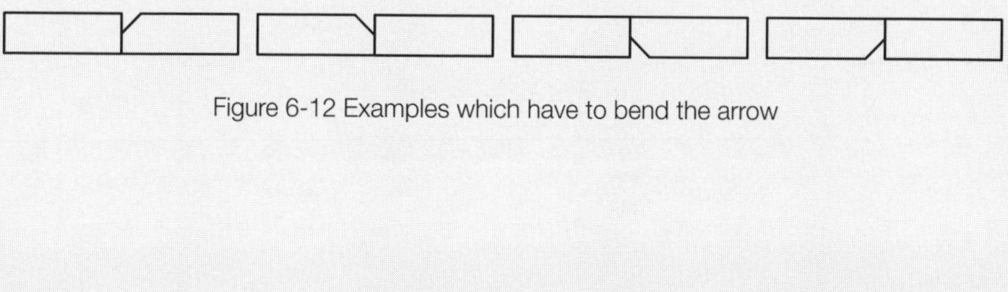

Figure 6-12 Examples which have to bend the arrow

g) 開先溶接の主寸法は、開先深さと溶接深さで表します。
　溶接深さは括弧をつけて開先深さに続けます。開先深さと溶接深さが異なる部分溶け込み溶接の例を図6-13 aに示します。
　開先深さと溶接深さが同じ部分溶け込み溶接の例を図6-13 bに示します。
　完全溶け込み溶接の場合は、図6-13cのように括弧をつけません。

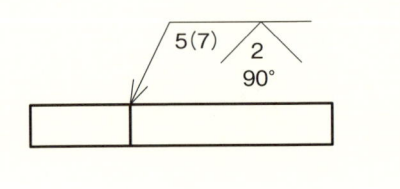

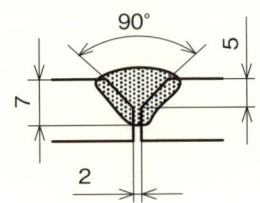

a) 開先深さと溶接深さが異なる部分溶け込み溶接

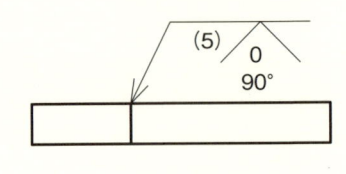

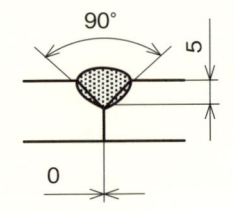

b) 開先深さと溶接深さが同じ部分溶け込み溶接

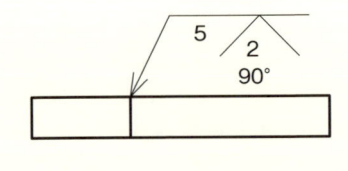

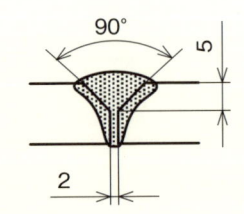

c) 完全溶け込み溶接

図6-13　開先溶接の指示

g) The main dimension of groove weld is indicated by the groove depth and the welding depth.

The welding depth attaches a parenthesis and continues it to the groove depth. The example of partial penetration welding from which the groove depth and the welding depth differ is shown in **Figure 6-13 a.**

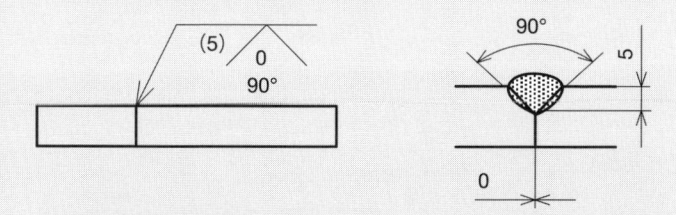

The example of partial penetration welding from which the groove depth and the welding depth are same is shown in Figure 6-13 b.

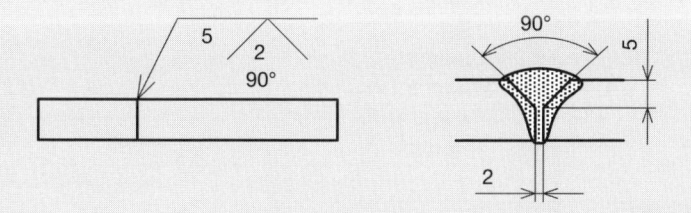

In full penetration welding, parenthesis is not put like Figure 6-13 c.

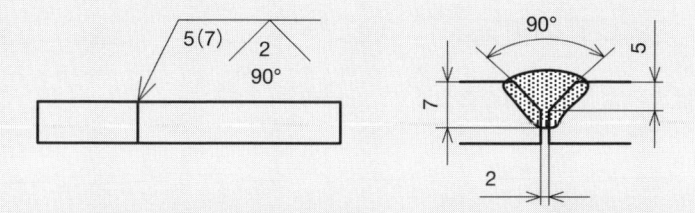

a) Partial penetration welding from which the groove depth and the welding depth differ

b) Partial penetration welding from which the groove depth and the welding depth are same

c) Full penetration weld

Figure 6-13 Indication of groove weld

第6章　3　幾何公差記号の理解

> 幾何公差は、形状の規制に関係し、大きさの規制に関係する寸法公差と異なります。
> 幾何公差は生産や検査工程に影響を与えます。

6-3-1　幾何特性の種類

幾何特性には次のような種類があります（**表6-9**）。

表6-9　幾何特性の種類

形体	公差の種類	特性	記号	適用 表面形体	適用 サイズ形体
単独形体（データムに関連しない）	形状公差	真直度	─	可	可
		平面度	▱	可	否
		真円度	○	可	否
		円筒度	⌭	可	否
		線の輪郭度	⌒	可	否
		面の輪郭度	⌓	可	否
関連形体（データムに関連する）	姿勢公差	平行度	∥	可	可
		直角度	⊥	可	可
		傾斜度	∠	可	可
		線の輪郭度	⌒	可	否
		面の輪郭度	⌓	可	否
	位置公差	同軸度	◎	否	可
		同心度	◎	否	可
		対称度	⹀	否	可
		位置度(*)	⌖	否	可
		線の輪郭度	⌒	可	否
		面の輪郭度	⌓	可	否
	振れ公差	円周振れ	↗	可	否
		全振れ	↗↗	可	否

* データムに関連しない場合もある

φ(@°▽°@)　メモメモ

表面形体とサイズ形体

表面形体とは、表面によって作られる形体で、大きさの概念をもちません。
サイズ形体とは、寸法によって形を作られる形体です。

| Chapter6 | 3 | # Understanding geometrical tolerance symbols |

Geometric tolerance concerns itself with shape control, unlike dimensional tolerance that concerns itself with size control.
Geometric tolerance influences the manufacturing and inspection process.

6-3-1 Types of geometric characteristics

The geometric characteristics have the following types. Refer to Table 6-9.

Table 6-9 Types of geometric characteristics

Features	Tolerance type	characteristic	symbols	Applied to	
				Feature-of-surface	Feature-of-size
For single features (not related to a DATUM)	Form	Straightness	—	Yes	Yes
		Flatness	▱		No
		Roundness	○		
		Cylindricity	⌭		
		Profile of a line	⌒		
		Profile of a surface	⌓		
For related features (related to a DATUM)	Orientation	Parallelism	//	Yes	Yes
		Perpendicularity	⊥		
		Angularity	∠		
		Profile of a line	⌒		No
		Profile of a surface	⌓		
	Location	Coaxiality	◎	No	Yes
		Concentricity			
		Symmetry	═		
		Position(*)	⊕		
		Profile of a line	⌒	Yes	No
		Profile of a surface	⌓		
	Run-out	Circular run-out	↗	Yes	No
		Total run-out	↗↗		

* May also no be related to a DATUM

Note ;-)

Feature-of-surface and Feature-of-size

Feature-of-surface is a shape made by the surface. It does not have a concept of size.
Feature-of-size is a shape made by the dimension.

6-3-2 データム

データムとは、位置や姿勢などを測定する際の基準線あるいは基準面です。もちろん、データムは加工や設計上の基準も意味します。

a) データム記号の指示

データム記号を図6-14に示します。

データム記号の三角形は、塗りつぶしたものと塗りつぶさないもののどちらでも構いません。

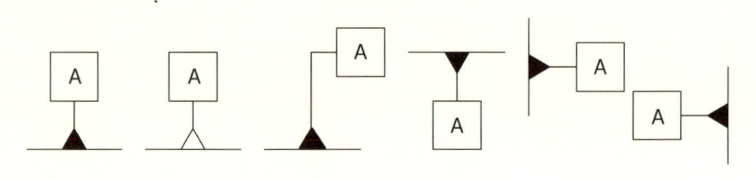

図6-14　データム記号

b) 基準となる中心軸へのデータム記号の指示

データムを中心軸に指示する場合、データム記号は次の例のうちの1つを使用して、図面に記入することができます（図6-15）。

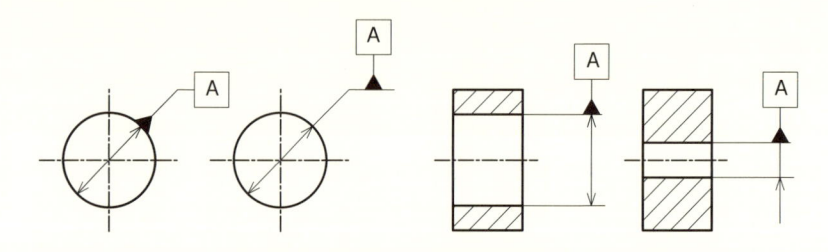

図6-15　基準となる中心軸への指示

> 寸法線に記号を付けることは、サイズ形体を意味するんやで！

6-3-2 DATUM system

DATUM (DE^TAMU in Japanese) means a base line or base surface when measuring position or orientation etc.

Of course DATUM means the base when processing or designing too.

a) Indication of DATUM symbol

DATUM symbols are shown in **Figure 6-14**.

In triangle of a DATUM symbol, filled or unfilled symbol can be used.

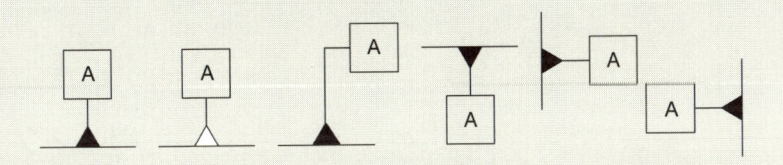

Figure 6-14 DATUM symbols

b) Placement of the DATUM symbol for datum center line

When the DATUM is indicated to a center line, the DATUM symbol can be placed on the drawing using one of the following examples. Refer to **Figure 6-15**.

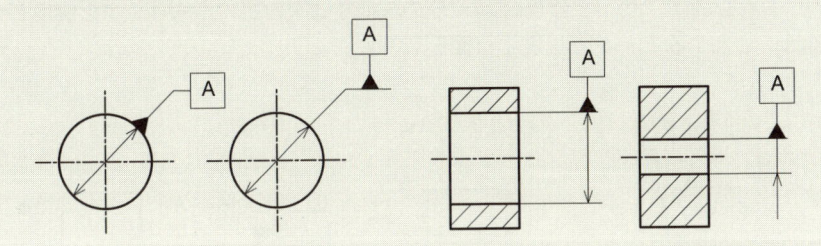

Figure 6-15 Indication to the datum center line

Attaching a symbol to a dimension line means a feature-of-size!

Chapter6 In what way should I use special symbols!

c) 基準となる中心平面へのデータム記号の指示
　データムを中心平面に指示する場合、データム記号は次の例のうちの1つを使用して、図面に記入することができます（**図6-16**）。

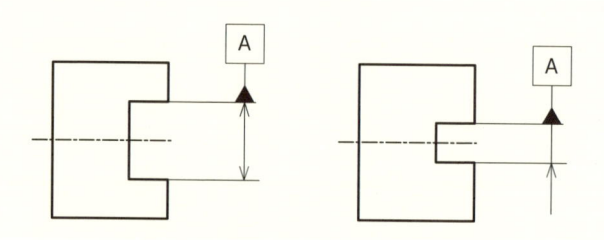

図6-16　基準となる中心平面への指示

> ということは、
> 上はサイズ形体で・・
> 下は表面形体なんや〜

d) 基準となる表面へのデータム記号の指示
　データムを表面に指示する場合、データム記号は次の例のうちの1つを使用して、図面に記入することができます（**図6-17**）。

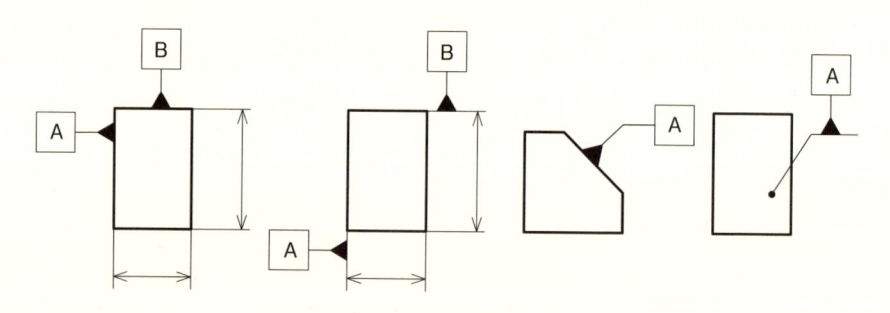

図6-17　基準となる表面への指示

c) Placement of the DATUM symbol for datum center plane

When the DATUM is indicated to a center plane, the DATUM symbol can be placed on the drawing using one of the following examples. Refer to Figure 6-16.

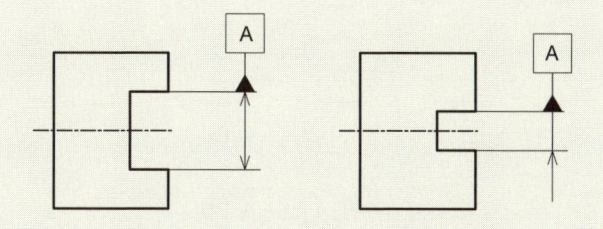

Figure 6-16 Indication to the datum center plane

> That is, the top figures are a feature-of-size...
> The bottom figures are a feature-of-surface...

d) Placement of the DATUM symbol for datum surface

When the DATUM is indicated to a surface, the DATUM symbol can be placed on the drawing using one of the following examples. Refer to Figure 6-17.

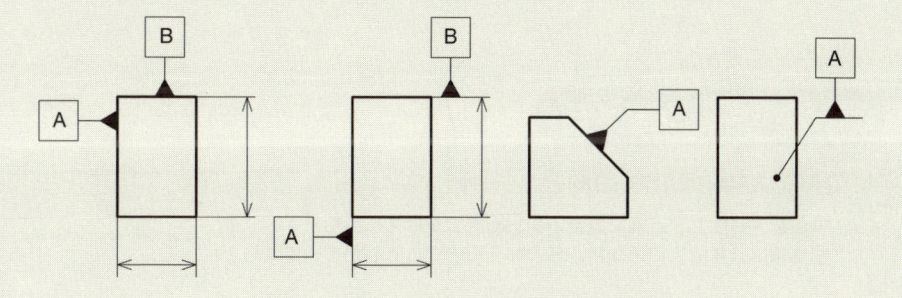

Figure 6-17 Indication to the datum surface

6-3-3　公差記入枠の指示

a) 公差記入枠

公差記入枠を図6-18に示します。

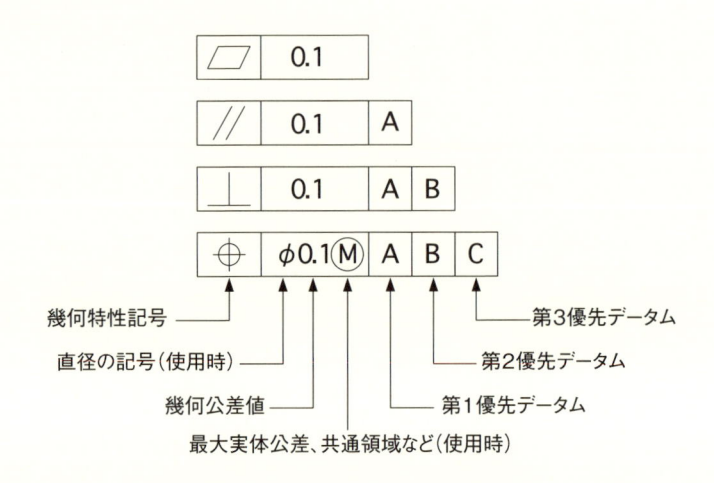

```
┌──┬─────┐
│ ▱ │ 0.1 │
└──┴─────┘

┌──┬─────┬───┐
│ ∥ │ 0.1 │ A │
└──┴─────┴───┘

┌──┬─────┬───┬───┐
│ ⊥ │ 0.1 │ A │ B │
└──┴─────┴───┴───┘

┌──┬────────┬───┬───┬───┐
│ ⊕ │ φ0.1Ⓜ │ A │ B │ C │
└──┴────────┴───┴───┴───┘
```

- 幾何特性記号
- 直径の記号（使用時）
- 幾何公差値
- 最大実体公差、共通領域など（使用時）
- 第1優先データム
- 第2優先データム
- 第3優先データム

図6-18　公差記入枠と要素の順序

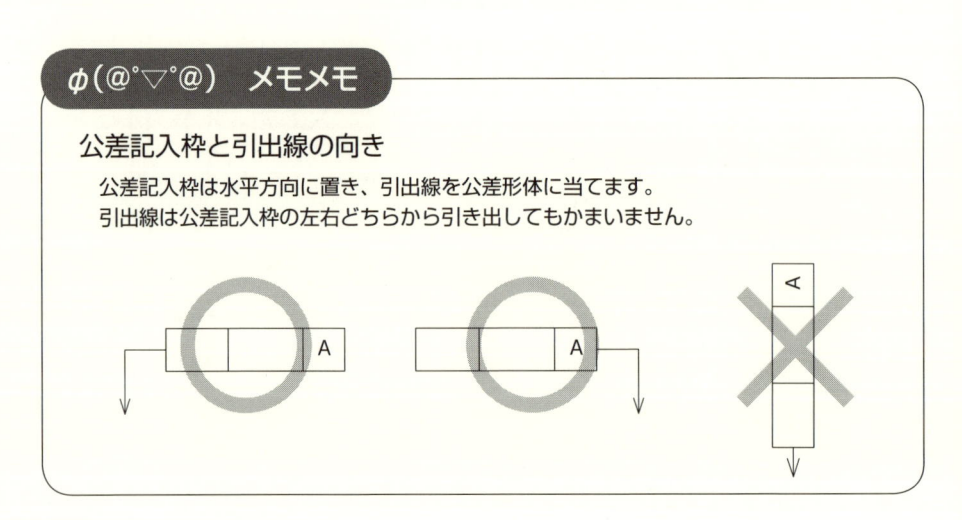

φ(@°▽°@)　メモメモ

公差記入枠と引出線の向き

公差記入枠は水平方向に置き、引出線を公差形体に当てます。
引出線は公差記入枠の左右どちらから引き出してもかまいません。

6-3-3 Indication of tolerance frame

a) Tolerance frame

Tolerance frames are shown in Figure 6-18.

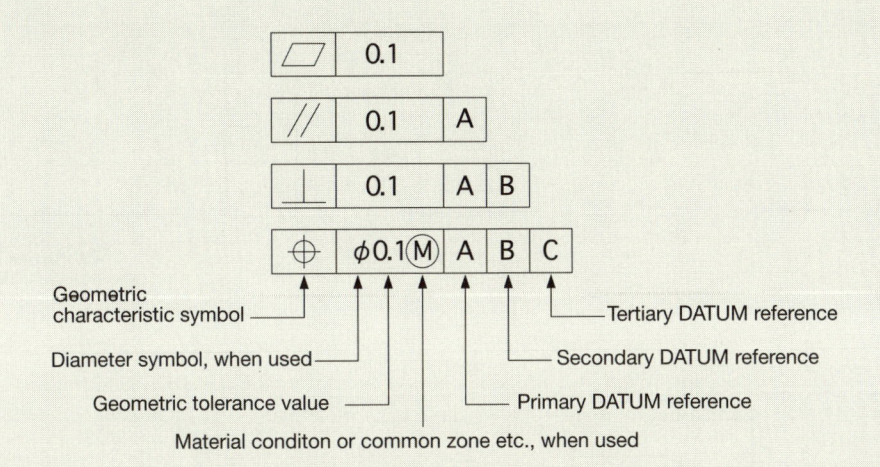

⌭	0.1

∥	0.1	A

⊥	0.1	A	B

⊕	⌀0.1Ⓜ	A	B	C

- Geometric characteristic symbol
- Diameter symbol, when used
- Geometric tolerance value
- Material condition or common zone etc., when used
- Primary DATUM reference
- Secondary DATUM reference
- Tertiary DATUM reference

Figure 6-18 Tolerance frame with order of elements

Note ;-)

Direction of tolerance frame and leader line

Tolerance frame should be placed horizontally, and a leader line is indicated to a tolerance feature.

A leader line can be drawn to either of both sides.

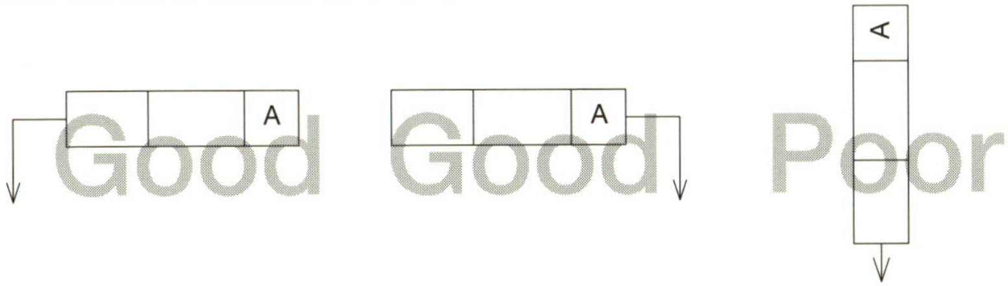

Good   Good   Poor

b) 中心軸あるいは中心平面への幾何公差の指示

　幾何公差を中心軸あるいは中心平面に指示する場合、引出線の矢は次の例のうちの1つを使用して、図面に記入します（**図6-19**）。

a) 中心軸への指示

b) 中心平面への指示

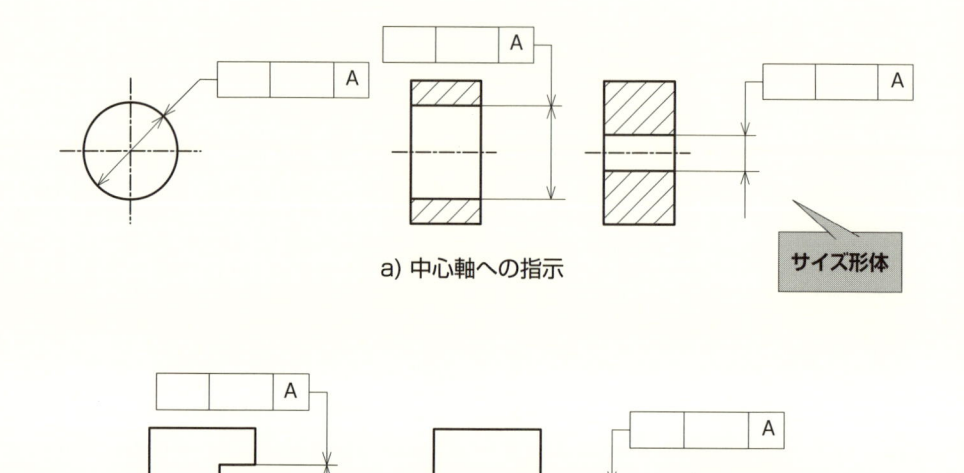

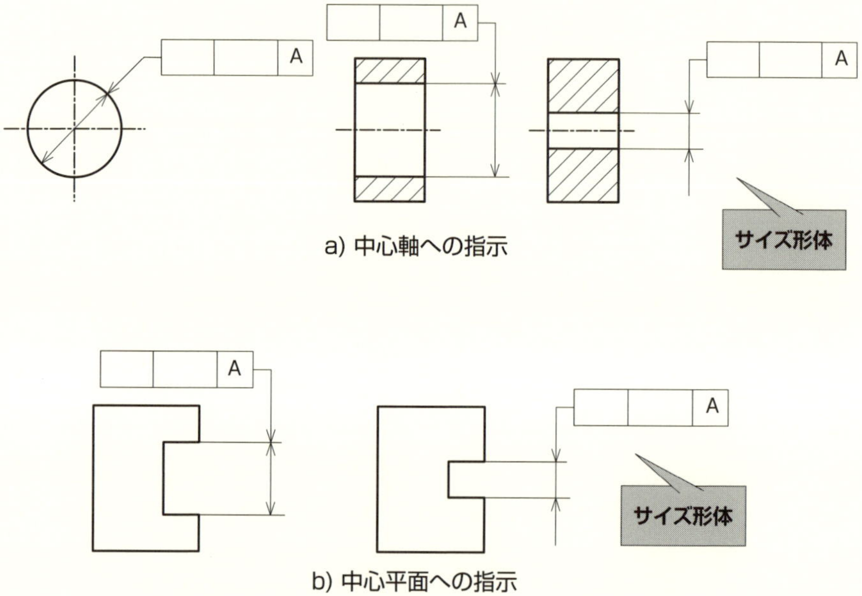

図6-19　中心軸あるいは中心平面への幾何特性指示

c) 表面への幾何公差の指示

　データムを表面に指示する場合、引出線の矢は次の例のうちの1つを使用して、図面に記入します（**図6-20**）。

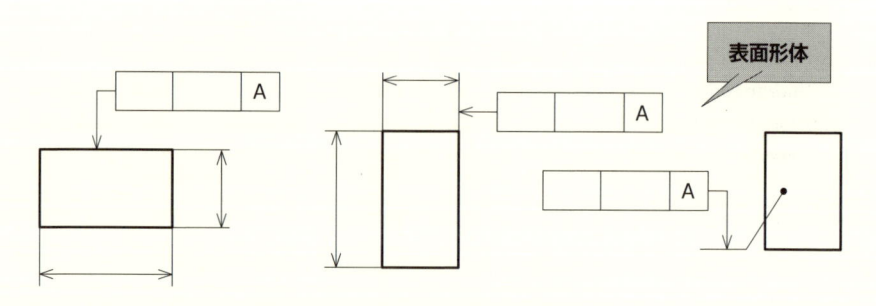

図6-20　表面への幾何特性指示

b) Indication of the geometrical tolerance for center line or center plane

When the geometrical tolerance is indicated to a axis or a center plane, arrow of the leader line is placed on the drawing using one of the following examples. Refer to Figure 6-19.

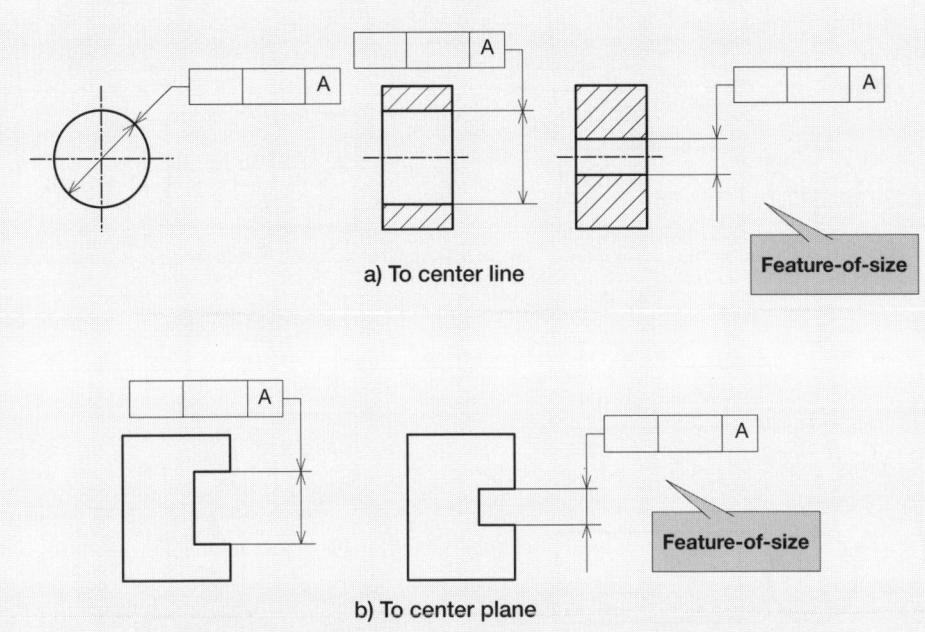

a) To center line

b) To center plane

Figure 6-19 Indication of geometric characteristic to center line or center plane

c) Indication of the geometrical tolerance for surface

When the geometrical tolerance is indicated to a surface, arrow of the leader line can be placed on the drawing using one of the following examples. Refer to Figure 6-20.

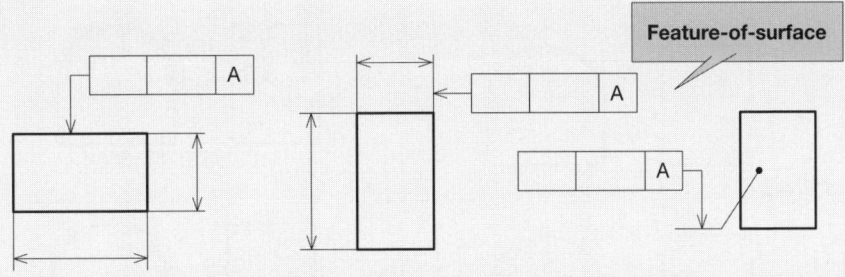

Figure 6-20 Indication of Geometric characteristic to surface

6-3-4 幾何公差指示例

a) 形状公差

表6-10 形状公差の指示例とその解釈

代表的な図面例	公差域
真直度 ─ φ0.1	φ0.1
平面度 ⌭ 0.05	0.05
真円度 ○ 0.1	0.1 / 切口の線として計測する
円筒度 ⌯ 0.1	0.1
線の輪郭度 ⌒ 0.1 (R20)	0.1 / 切口の線として計測する
面の輪郭度 ⌓ 0.1 (R20)	0.1

6-3-4 Examples of geometrical tolerance indication

a) Form tolerance

Table 6-10 Example of indication and explanation of form tolerance.

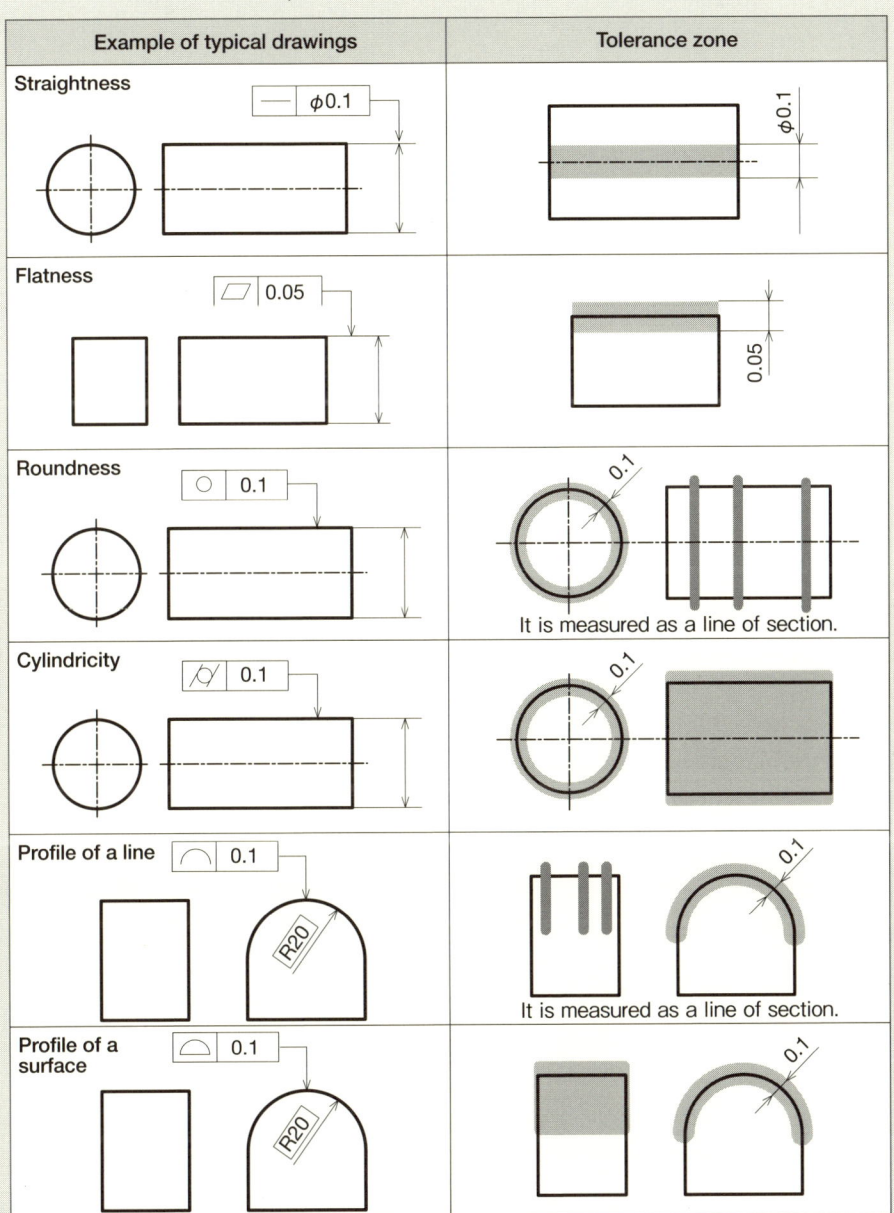

Example of typical drawings	Tolerance zone
Straightness — ⌀0.1	⌀0.1
Flatness ▱ 0.05	0.05
Roundness ○ 0.1	0.1 — It is measured as a line of section.
Cylindricity ⌭ 0.1	0.1
Profile of a line ⌒ 0.1 (R20)	0.1 — It is measured as a line of section.
Profile of a surface ⌓ 0.1 (R20)	0.1

b) 姿勢公差

図6-11　姿勢公差の指示例とその解釈

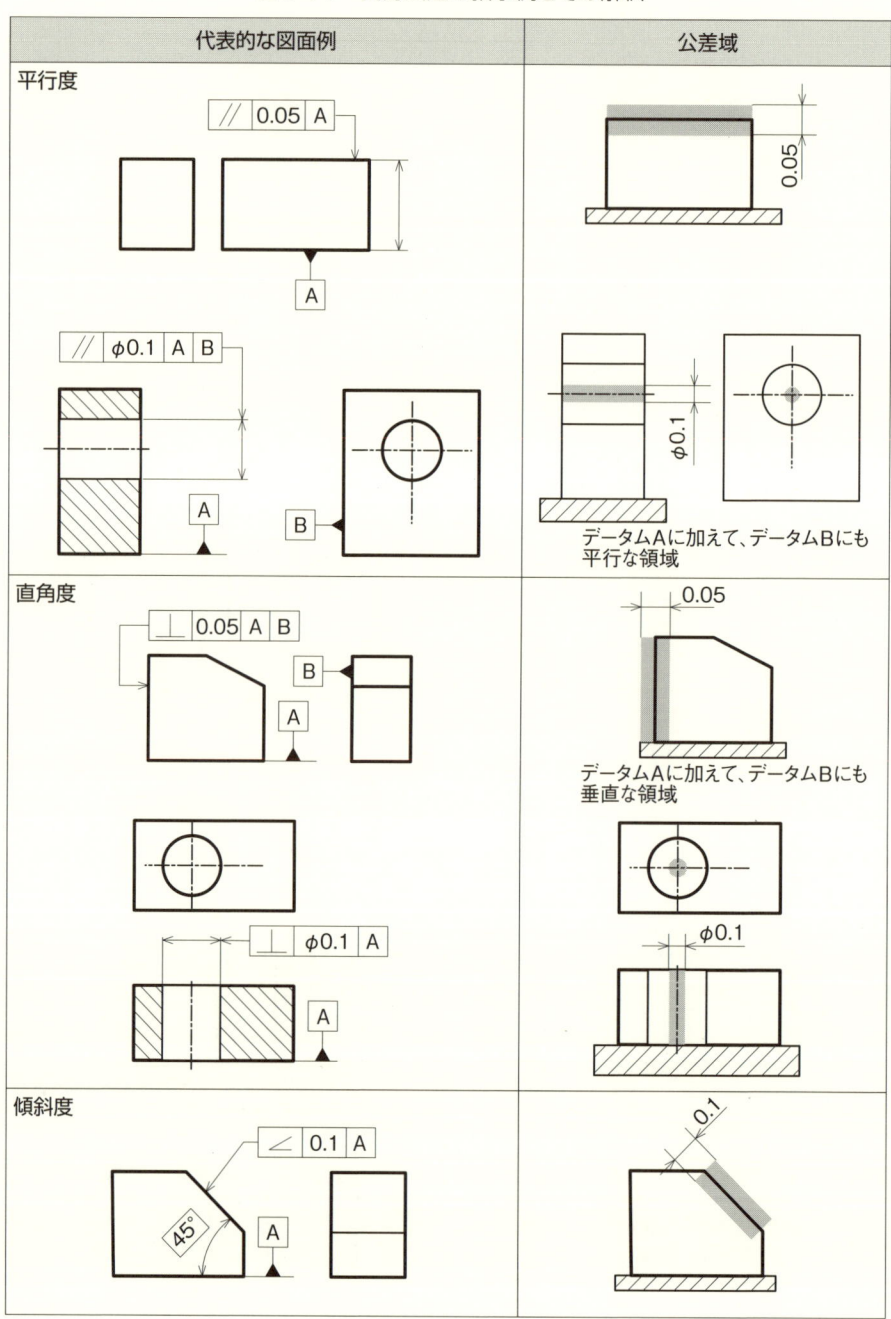

代表的な図面例	公差域
平行度	
直角度	
傾斜度	

データムAに加えて、データムBにも平行な領域

データムAに加えて、データムBにも垂直な領域

b) Orientation tolerance

Table 6-11 Example of indication and explanation of orientation tolerance.

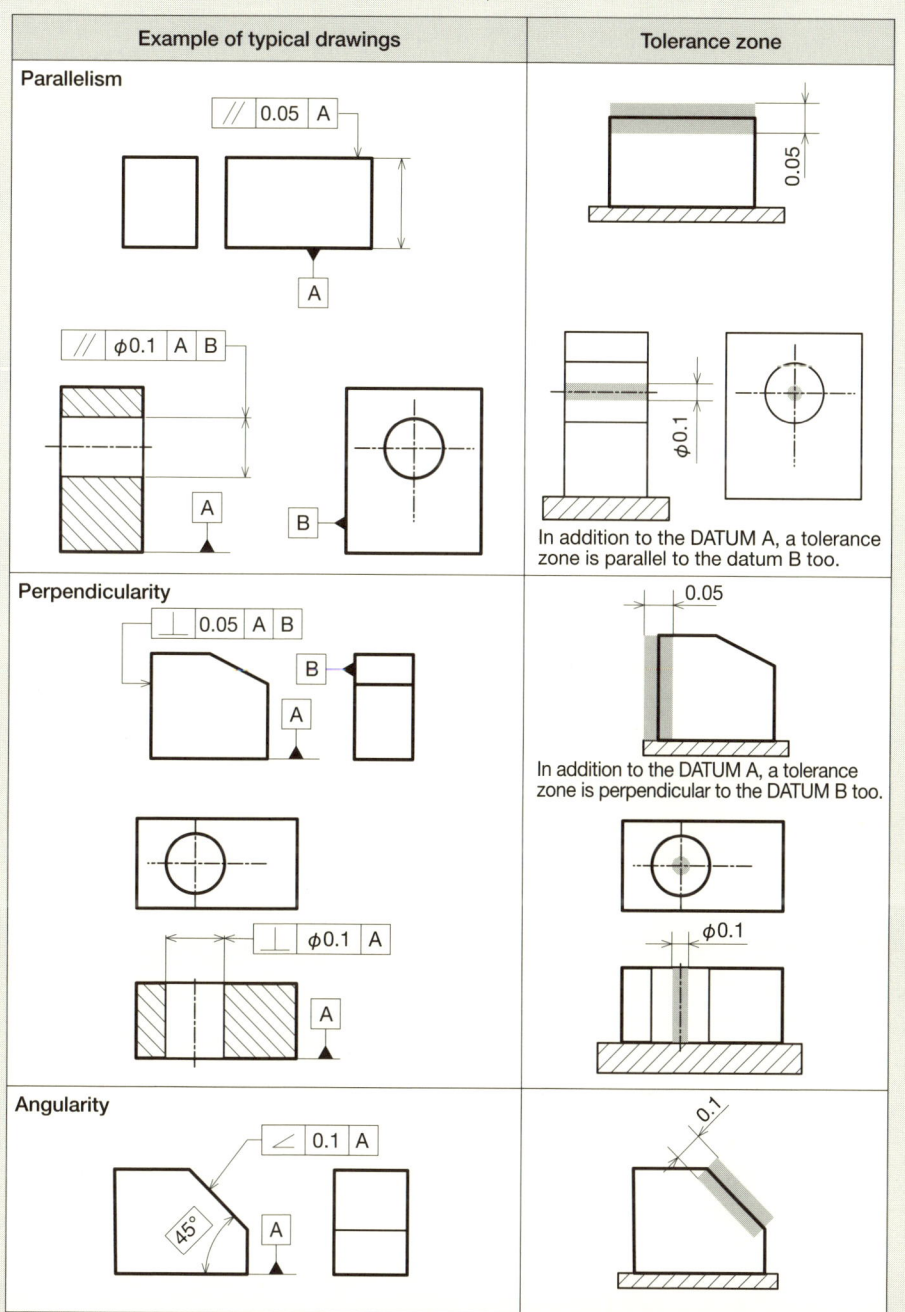

Example of typical drawings	Tolerance zone
Parallelism (// 0.05 A)	
(// φ0.1 A B)	In addition to the DATUM A, a tolerance zone is parallel to the datum B too.
Perpendicularity (⊥ 0.05 A B)	In addition to the DATUM A, a tolerance zone is perpendicular to the DATUM B too.
(⊥ φ0.1 A)	φ0.1
Angularity (∠ 0.1 A), 45°	0.1

Chapter6 In what way should I use special symbols!

c) 位置公差

図6-12 位置公差の指示例とその解釈

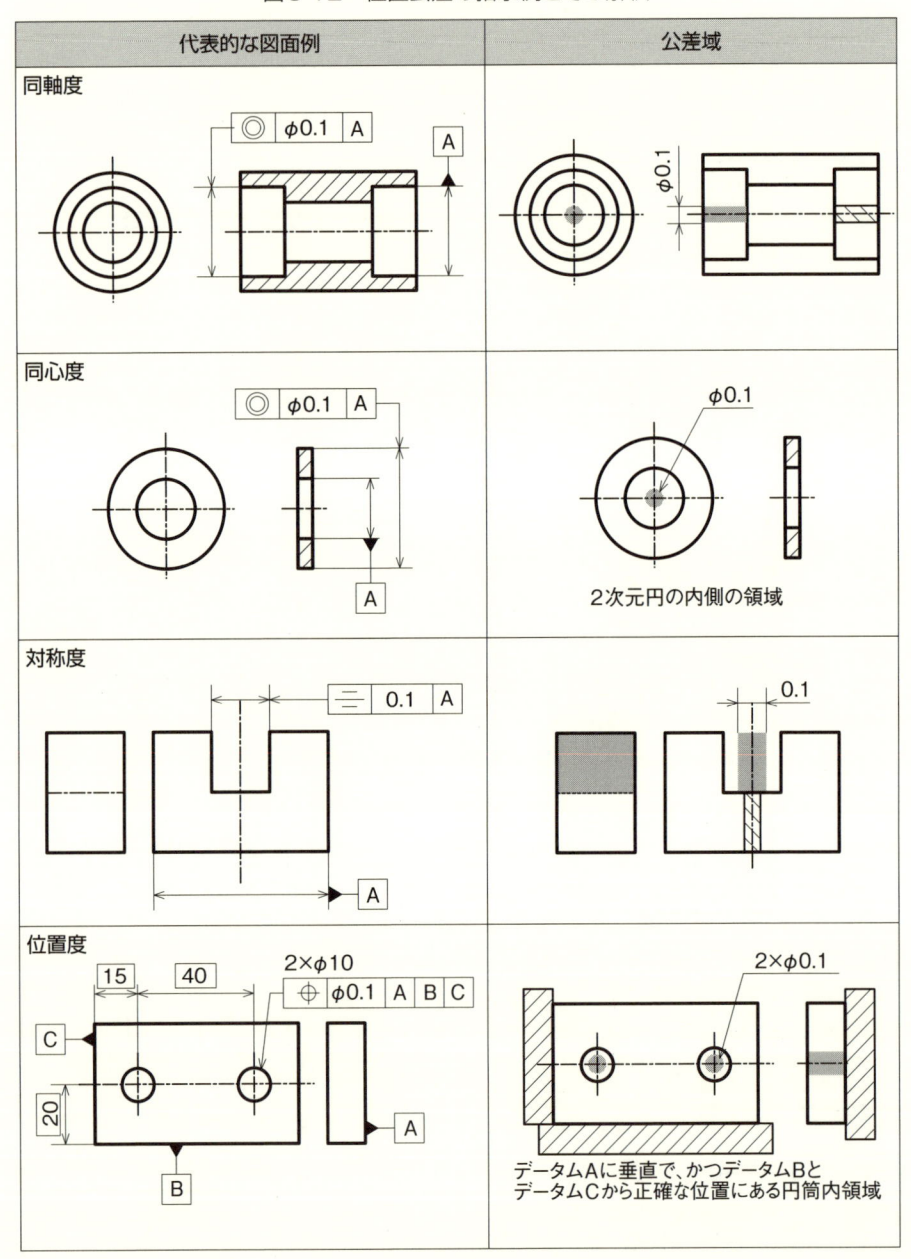

代表的な図面例	公差域
同軸度	
同心度	2次元円の内側の領域
対称度	
位置度	データムAに垂直で、かつデータムBとデータムCから正確な位置にある円筒内領域

c) Location tolerance

Table 6-12 Example of indication and explanation of location tolerance.

Example of typical drawings	Tolerance zone
Coaxiality ⌀0.1 A	
Concentricity ⌀0.1 A	⌀0.1 Tolerance zone is an inside of two-dimensional circle
Symmetry 0.1 A	0.1
Position 2×⌀10, ⌀0.1 A B C, 15, 40, 20	2×⌀0.1 Tolerance zone is inside of a cylinder which is perpendicular to the DATUM A and in an exact position from the DATUM B and C.

Chapter6 In what way should I use special symbols!

d) 振れ公差

図6-13 振れ公差の指示例とその解釈

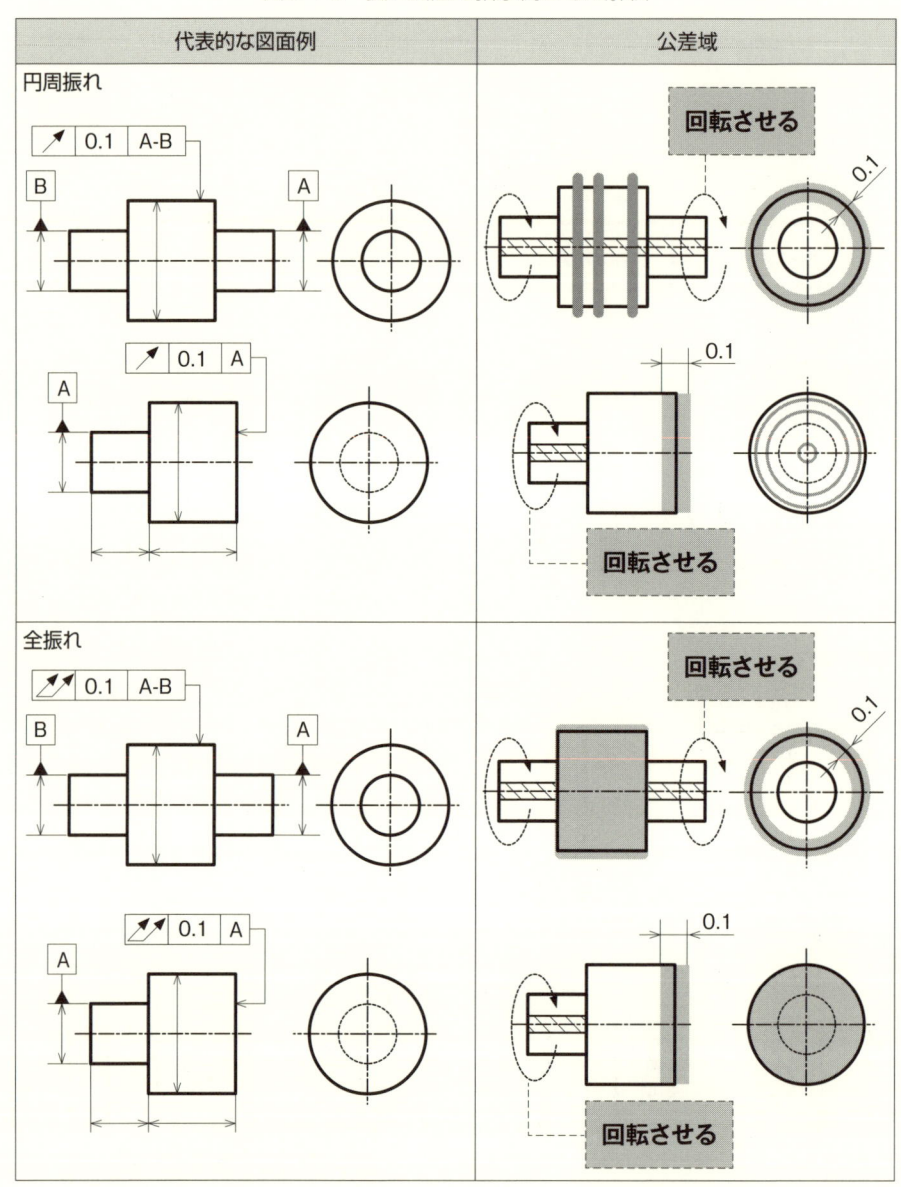

代表的な図面例	公差域
円周振れ	
全振れ	

d) Run-out tolerance

Table 6-13 Example of indication and explanation of run-out tolerance

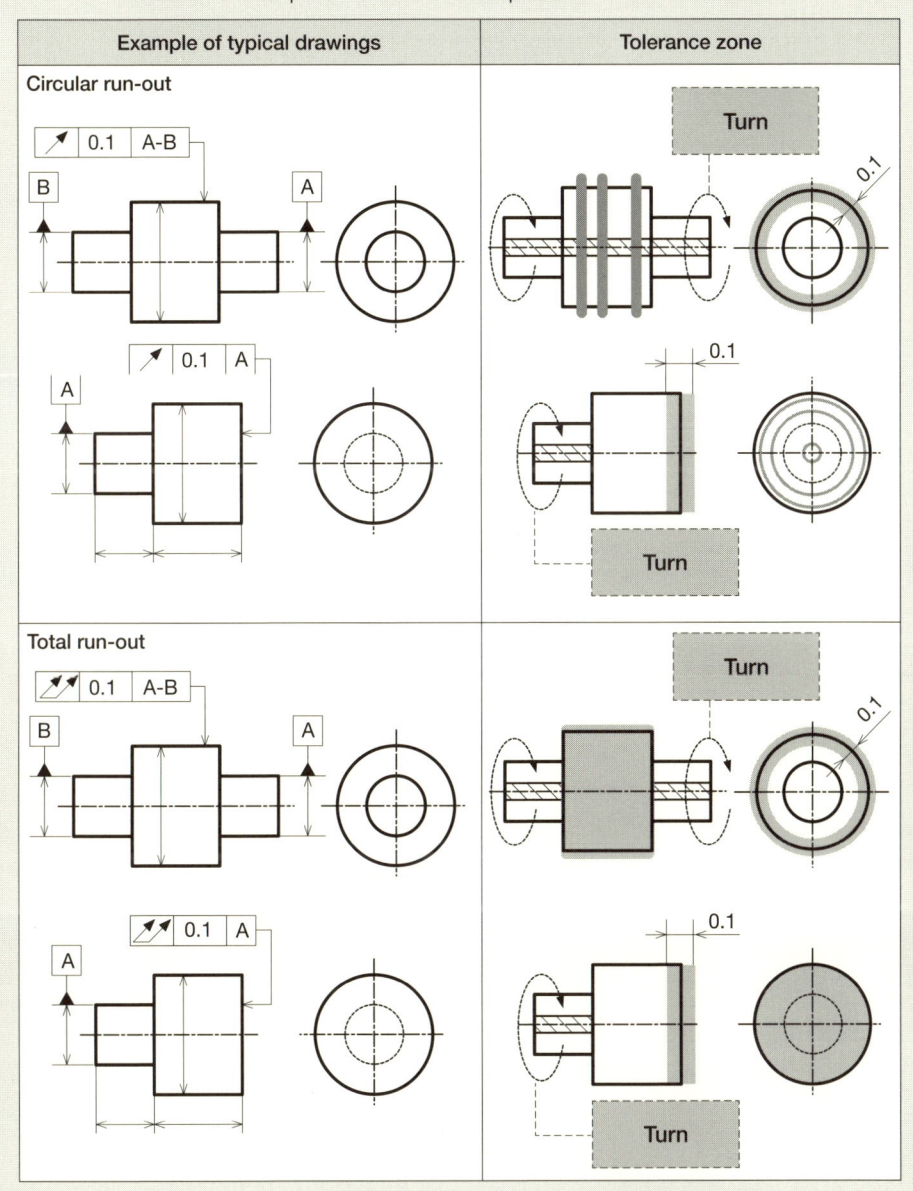

Example of typical drawings	Tolerance zone
Circular run-out	
Total run-out	

6-3-5　その他のテクニック

a) 全周記号

円を引出線に付けると、その投影図のすべての面が適用されます。ただし、投影図の表裏には適用されませんので、注意してください（**図6-21**）。

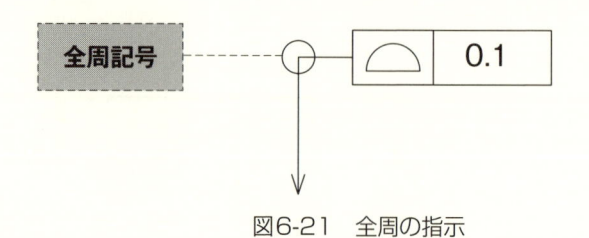

図6-21　全周の指示

b) 共通領域

いくつかの分割された形状に一つの公差域を適用することができます。この場合、公差記入枠の中に、「共通領域」の略語として記号「CZ」を加えます（**図6-22**）。

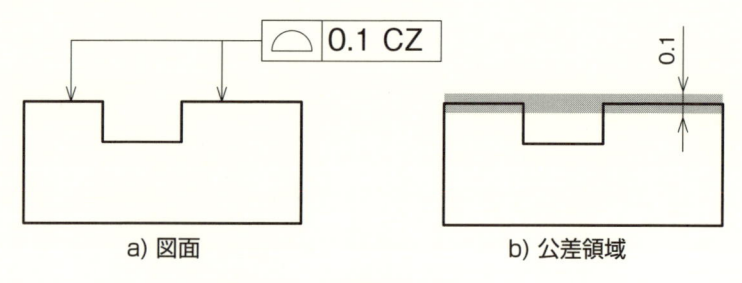

a) 図面　　　　b) 公差領域

図6-22　共通領域の指示

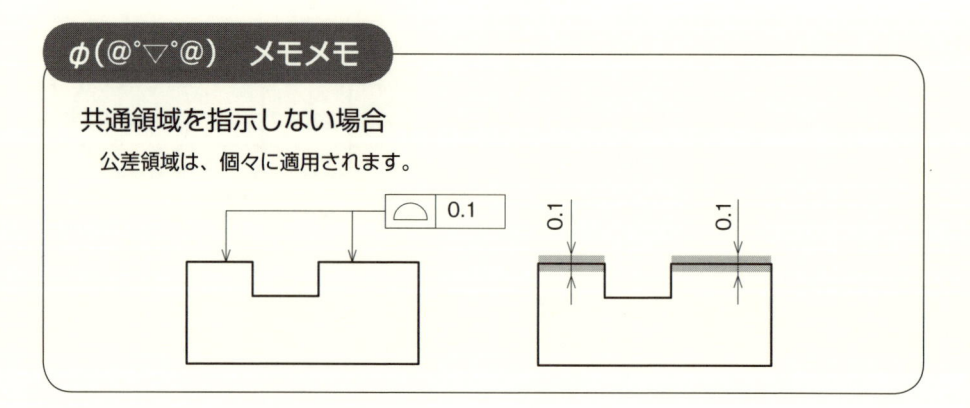

φ(@˚▽˚@)　メモメモ

共通領域を指示しない場合

公差領域は、個々に適用されます。

第6章　特殊な記号って、どない使うねん！

6-3-5 Other techniques

a) Symbol for all around

When the leader line with circle applies to all surface in the view. Take care since the back-side and front-side of the view do not apply. Refer to Figure 6-21.

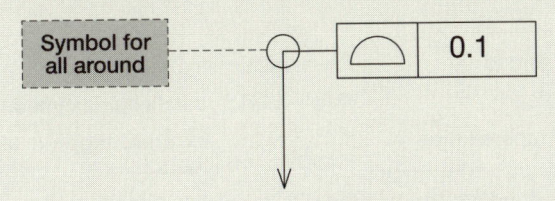

Figure 6-21 Indication for all around

b) Common zone

It is possible to apply a single tolerance zone to several separate shapes.
In this case, in tolerance frame the symbol "CZ" is added as an abbreviation for "common zone". Refer to Figure 6-22.

a) Drawing b) Tolerance zone

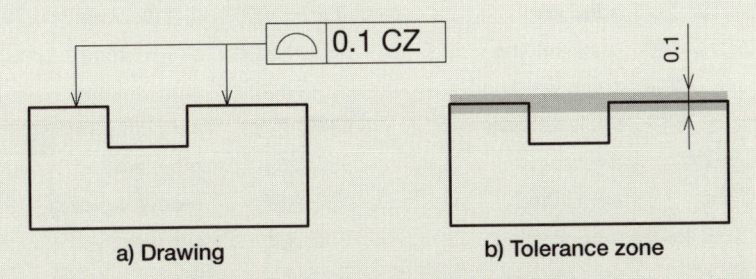

Figure 6-22 Indication of common zone

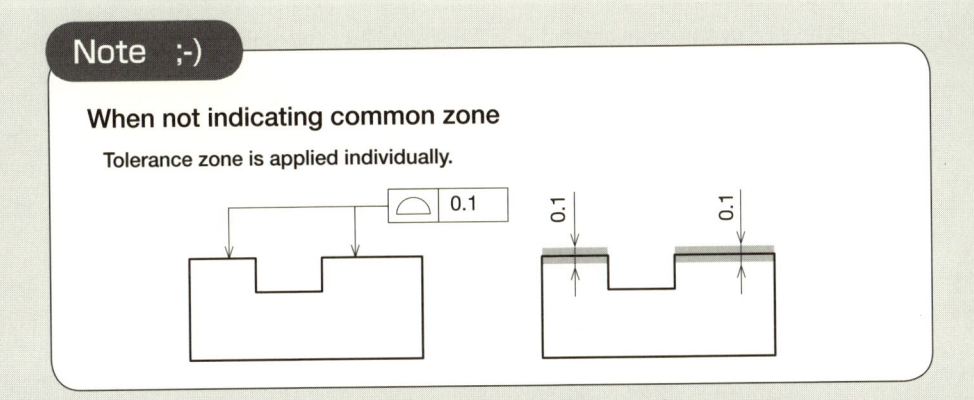

Note ;-)

When not indicating common zone

Tolerance zone is applied individually.

用語集　glossary

本章で使用した単語です。他の用語も使うことができます。

The list of words used for this chapter is shown. You can use another words, too.

ねじ	(screw) thread	メートルねじ	metric thread
台形ねじ	acme thread	管用平行ねじ	parallel pipe thread
管用テーパめねじ	internal tapered pipe thread	管用テーパおねじ	external tapered pipe thread
並目ねじ	coarse pitch thread	細目ねじ	fine pitch thread
呼び長さ	nominal length	ねじ長さ	thread length
外径（山の頂）	crest	谷径（谷底）	root
下穴深さ	drill depth	溶接	weld
突合せ継手	butt joint	T継手	T-joint
重ね継手	lap joint	かど継手	corner joint
へり継手	edge joint	すみ肉溶接	filet welding
プラグ溶接	plug welding	スロット溶接	slot welding
シーム溶接	seam welding	スタッド溶接	stud welding
ビード溶接	back welding	肉盛り溶接	surfacing welding
矢の側	arrow side	矢の反対側	other side
両側	both side	開先溶接	groove welding
ルート間隔	root opening	現場溶接	site weld
裏当て	backing bar	止端	toe
幾何公差	geometrical tolerance	データム	DATUM
表面形体	feature-of-surface	サイズ形体	feature-of-size
幾何特性	geometrical character	公差記入枠	tolerance frame
第一優先データム	primary DATUM reference	第二優先データム	secondary DATUM reference
第三優先データム	tertiary DATUM reference	中心平面	center plane
形状公差	form tolerance	姿勢公差	orientation tolerance
位置公差	location tolerance	振れ公差	run-out tolerance
全周記号	symbol for all around	共通領域	common zone

第7章

図面って、どない描くねん！

Chapter 7
In what way should I make drawings!

本章で示す寸法や公差などは、記入例を表したものです。
これらは、機能や加工を保証するものではありません。
ご了承下さい。
　Dimensions and tolerance etc. Shown in this chapter are the dimensioning example.
　These dimensioning may guarantee neither function nor processing.
　I appreciate your agreement.

この章を学習すると、次の知識を得ることができます。
1）寸法記入の論理的思考（治具の例）
2）寸法記入の論理的思考（ドリルプレスバイスの例）

After studying this chapter, you will be able to get knowledge as follows:
-1. Logical mind of the dimensions indication (Example of Jig)
-2. Logical mind of the dimensions indication
　　(Example of Drilling-press vice)

| 第7章
Chapter7 | 1 | 寸法指示の論理的思考(治具の例)
Logical mind of the dimensions indication (Example of Jig) |

具体的な組品を使って基準となる形体や機能を探し、寸法記入を考えてみましょう。

Using real assembly products, let's learn the dimensioning with looking for datum features and functions.

まず、丸棒を固定する治具の組立図から、構造を理解しましょう。

For a start, let's understand structure from the assembly drawing of the jig which fixes a rod.

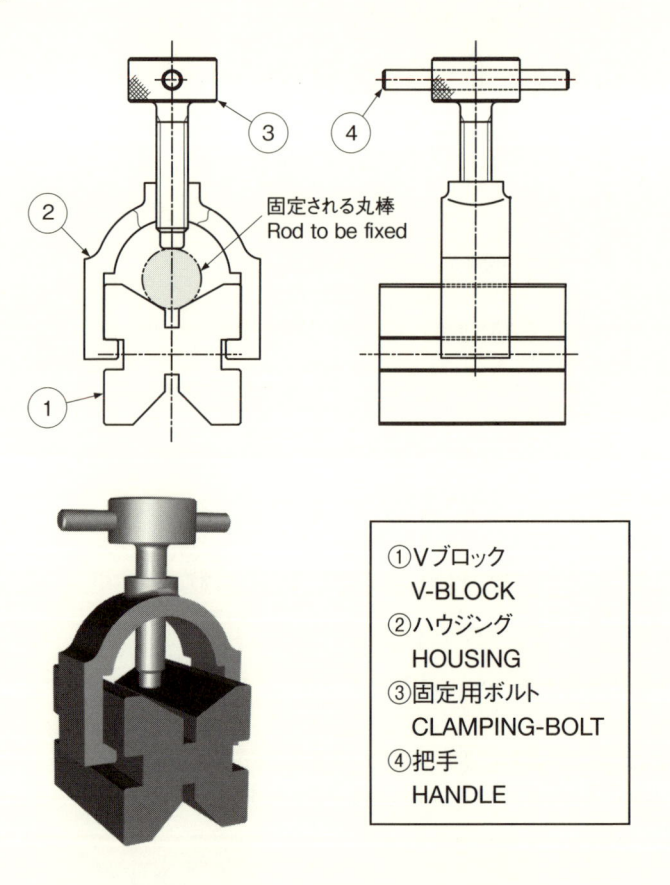

固定される丸棒
Rod to be fixed

① Vブロック
　 V-BLOCK
② ハウジング
　 HOUSING
③ 固定用ボルト
　 CLAMPING-BOLT
④ 把手
　 HANDLE

治具の組立図　Assembling drawing of jig

第7章　図面って、どない描くねん！

① Vブロックの図面作成
Drawing of V-block

Vブロックは丸棒を保持するために使用される要素です。
V-block is a device used to hold a rod.

想定される材質 Assumed material
材質:FC200 Material:FC200
表面処理:なし Finishing:None

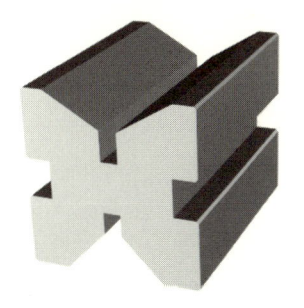

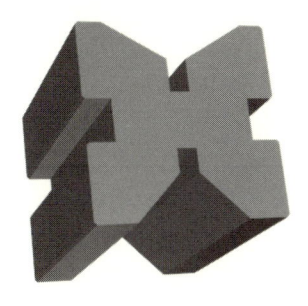

Vブロックの3次元モデル　3D-model of V-block

a) 投影図の準備

a) Views preparation

　正面図は上下左右がほぼ対称形状であるため、中心線を記入します。

　The front view is almost vertical and horizontal symmetry, so center lines are placed.

　Vブロックは、2つの投影図で表すことができます。

　V-block can be expressed by two views.

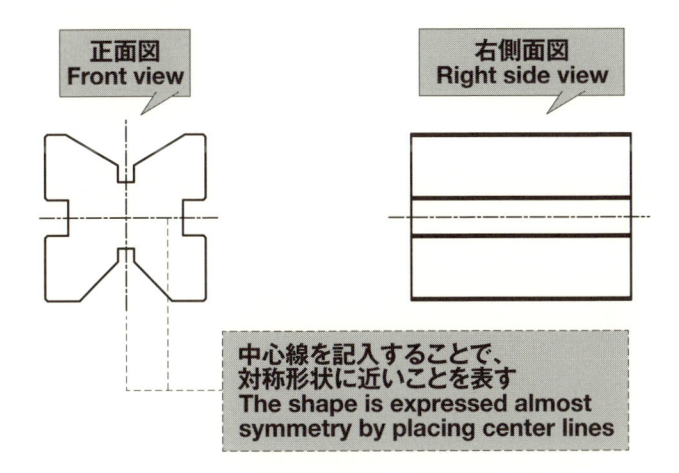

正面図 / Front view

右側面図 / Right side view

中心線を記入することで、対称形状に近いことを表す
The shape is expressed almost symmetry by placing center lines

Vブロックの投影図　Views of V-block

Chapter7　In what way should I make drawings!

b) Vブロックの機能をばらしてみましょう。
b) Let's break down the function of V-block.

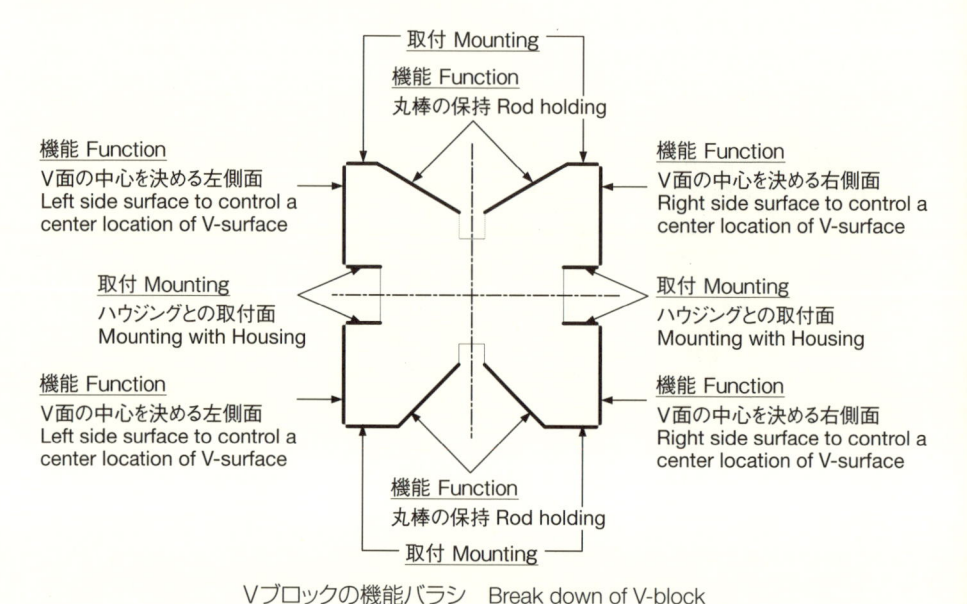

取付 Mounting

機能 Function
丸棒の保持 Rod holding

機能 Function
V面の中心を決める左側面
Left side surface to control a center location of V-surface

機能 Function
V面の中心を決める右側面
Right side surface to control a center location of V-surface

取付 Mounting
ハウジングとの取付面
Mounting with Housing

取付 Mounting
ハウジングとの取付面
Mounting with Housing

機能 Function
V面の中心を決める左側面
Left side surface to control a center location of V-surface

機能 Function
V面の中心を決める右側面
Right side surface to control a center location of V-surface

機能 Function
丸棒の保持 Rod holding

取付 Mounting

Vブロックの機能バラシ　Break down of V-block

c) 最初に、機能に影響する重要な寸法を指示します。
c) First, priority dimensions which influence functions are indicated.

　横幅70mmの寸法公差は、ハウジングとの関係から検討します。
　The dimensional tolerance of 70mm width is considered from relation with Housing.

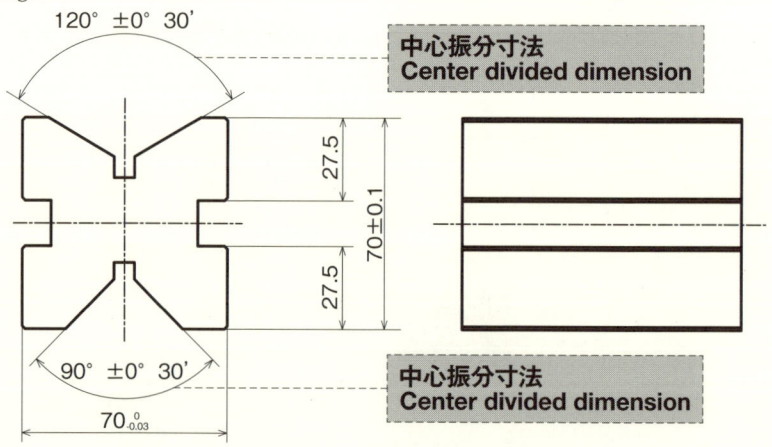

120° ±0° 30'

27.5

27.5

70±0.1

90° ±0° 30'

$70_{-0.03}^{0}$

中心振分寸法
Center divided dimension

中心振分寸法
Center divided dimension

重要寸法の指示　Indication of priority dimensions

d) V面に関連する寸法をまとめて指示します。
d) Dimensions related to V-surfaces are indicated collectively.

　　溝幅7mmは上下共通なので、1か所に指示してもう一方は省略します。
　The 7mm groove width is common to the upper and lower sides. So dimension is placed to one side, and the other side is omitted.

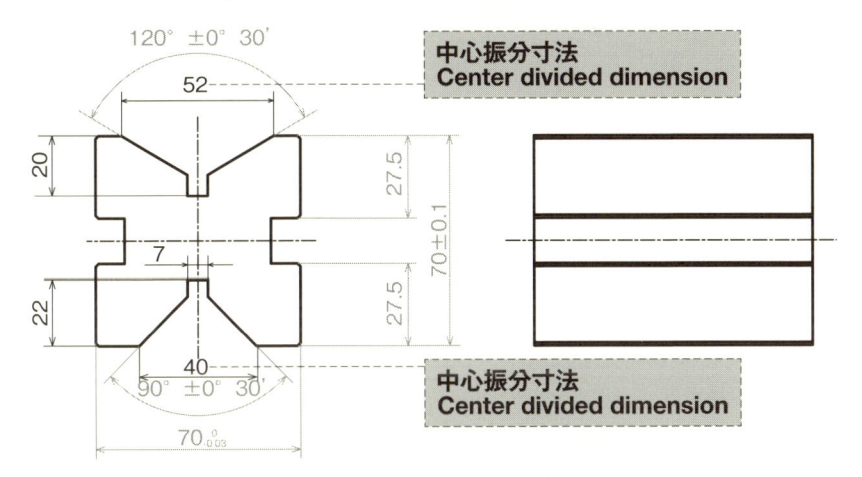

関連寸法の指示（V面）　Indication of related dimensions (V-surface)

e) 両横の溝に関連する寸法をまとめて指示します。
e) Dimensions related to both side grooves are indicated collectively.

　　加工に配慮して、溝幅15mmを参考寸法で示します。
　For convenience of the processing, groove width 15mm is indicated as reference dimension.

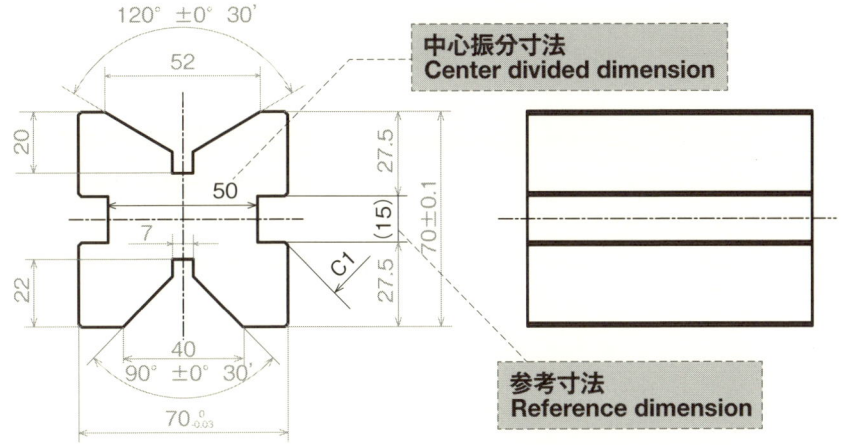

関連寸法の指示（側面の溝）　Indication of related dimensions (side grooves)

Chapter7　In what way should I make drawings!

f) その他、足りない寸法を指示します。
f) Other required dimensions are indicated.

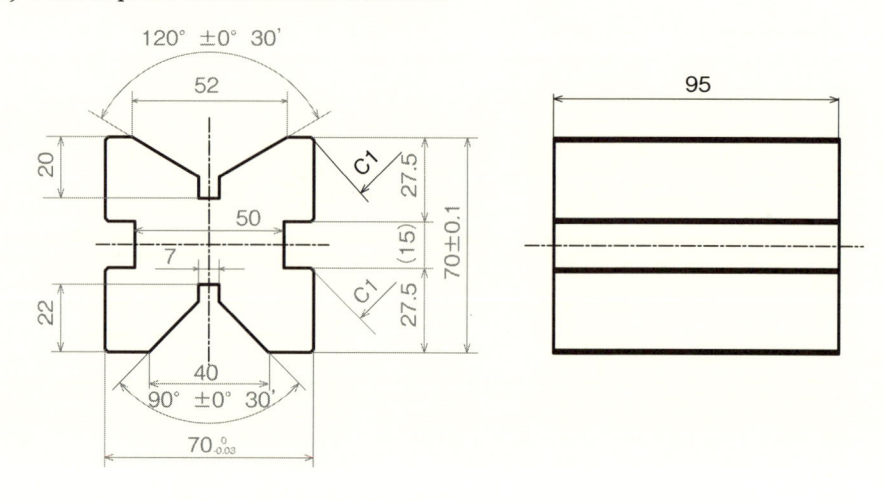

足りない寸法の指示　Indication of other required dimensions

g) さらに、面の肌記号を記入します。
g) In addition, surface texture symbols are placed.

良好な面の肌は、重要寸法と一致していることがわかるでしょう。
You can understand that good surface texture accord with priority dimensions.

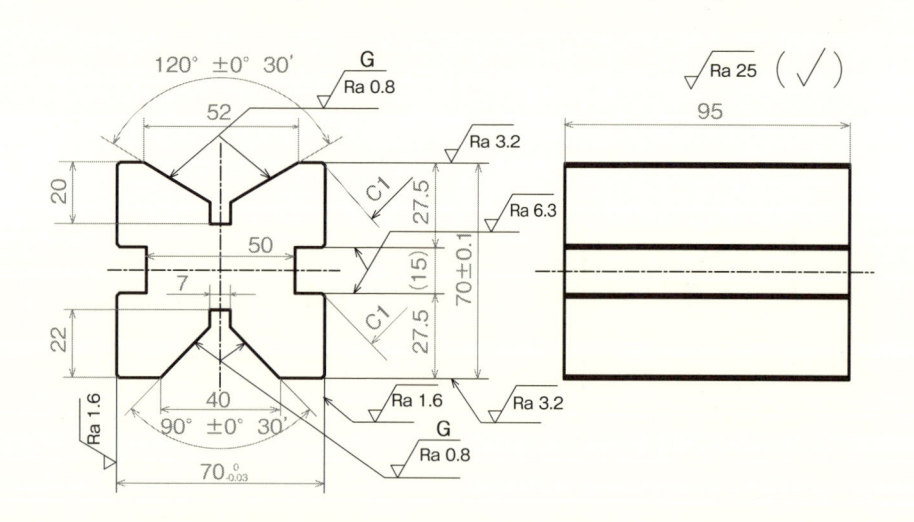

面の肌記号の指示　Indication of indicating surface texture symbols

丸棒を使った平行度の指示
Indication of parallelism using a rod

丸棒を使った平行度の指示
Indication of parallelism using a rod

計測に関する注記
Note about measurement

丸棒(ROD)φ35×115　√Ra 25　(√)
// φ0.05 A

95

丸棒(ROD)φ20×115
// φ0.05 B

注記　丸棒を準備して、平行度を計測すること
NOTE Prepare cylindrical rods and measure parallelism.

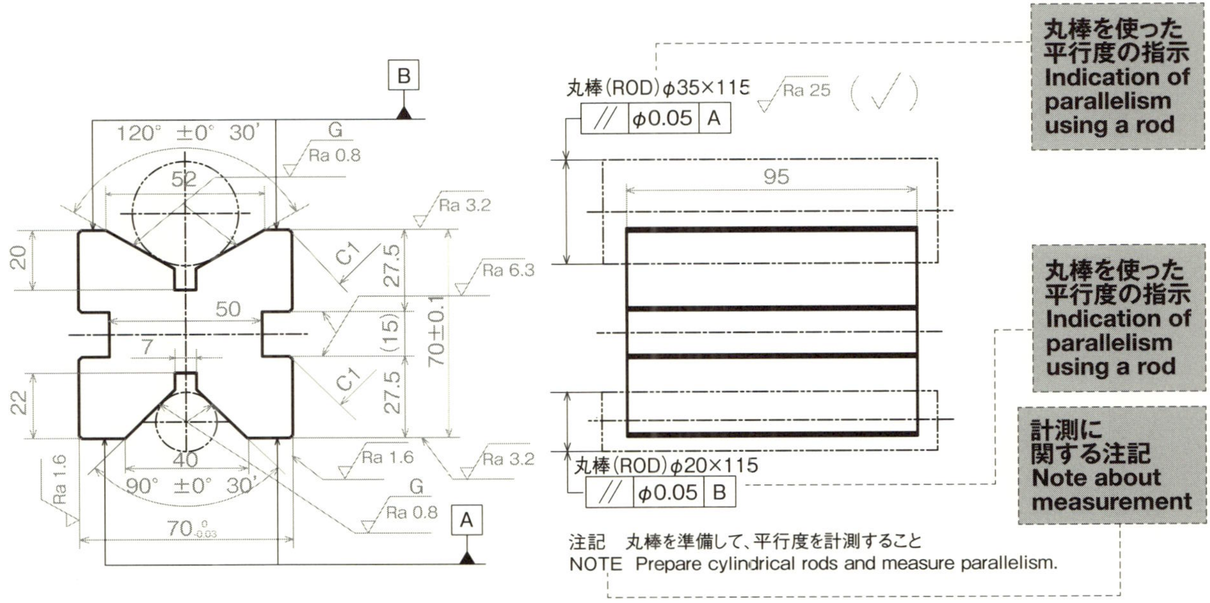

幾何公差の指示　Indication of geometrical tolerance

h) 必要に応じて幾何公差を記入します。
h) Geometrical tolerances are placed as necessary.

V面が取り付け面と平行であることを平行度で指示します。平行度は検査に配慮して、ロッドを使います。
It is indicated by parallelism that V-surface is parallel to a mounting surface. For convenience of the measurement, parallelism uses a rod.

Chapter7　In what way should I make drawings!

②ハウジングの図面作成
Drawing of Housing

ハウジングは固定ボルトで丸棒を締め付けるために使われる要素です。
Housing is a device which is used to tighten a rod with Clamping bolt.

ハウジングは鋳造後、切削加工するものとします。
Housing shall be made by cutting after casting.

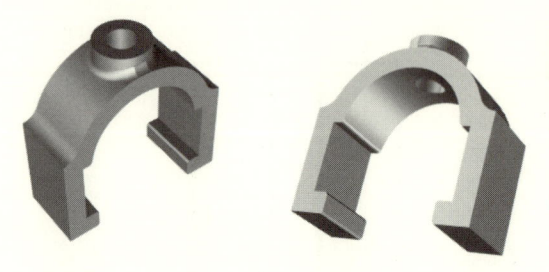

想定される材質
Assumed material

材質:FC200
Material:FC200

表面処理:Ep-Fe/Zn8/CM2
Finishing:Ep-Fe/Zn8/CM2

ハウジングの3次元モデル　3D-model of Housing

a) 投影図の準備
a) Views preparation

　　正面図は左右対称形状であるため、中心線を記入します。
　　The front view is horizontal symmetry, so center lines are placed.
　　ハウジングは、2つの投影図で表すことができます。
　　Housing can be expressed by two views.

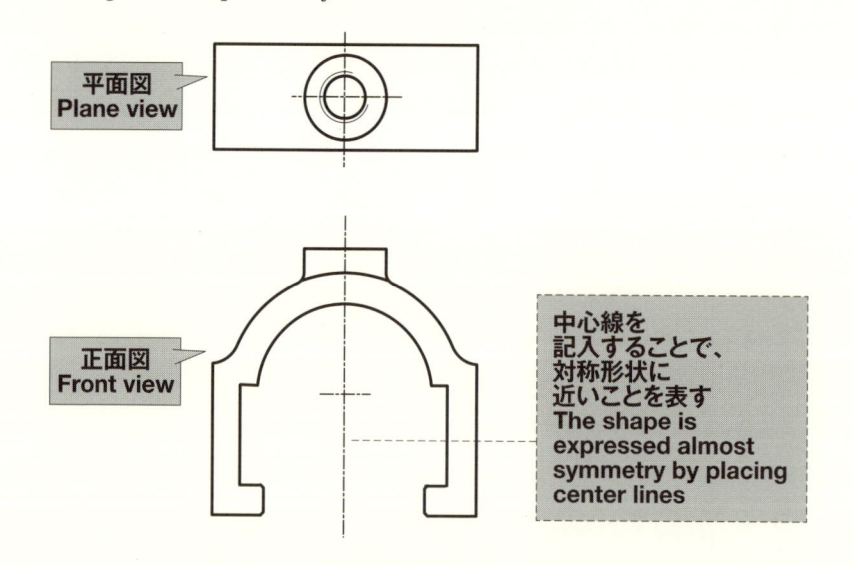

平面図
Plane view

正面図
Front view

中心線を記入することで、対称形状に近いことを表す
The shape is expressed almost symmetry by placing center lines

ハウジングの投影図　Views of Housing

b) ハウジングの機能をばらしてみましょう。
b) Let's break down Housing.

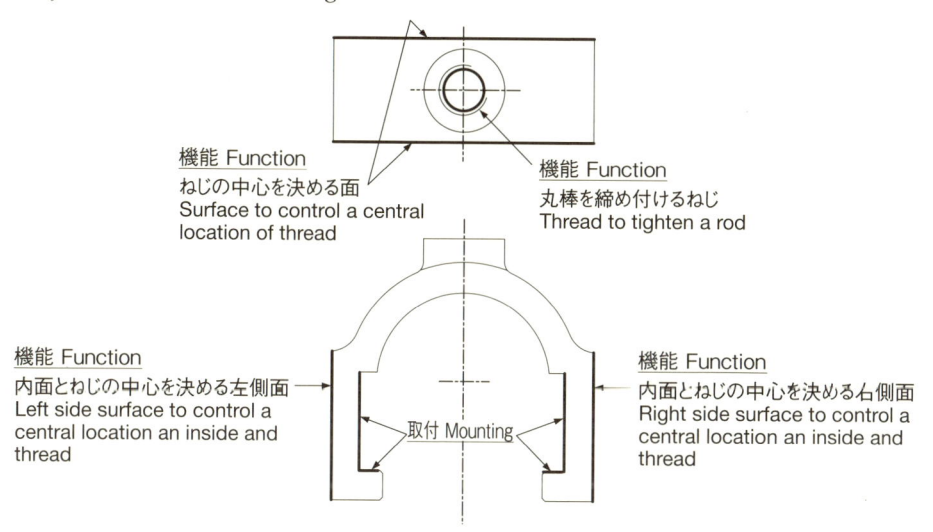

機能 Function
ねじの中心を決める面
Surface to control a central location of thread

機能 Function
丸棒を締め付けるねじ
Thread to tighten a rod

機能 Function
内面とねじの中心を決める左側面
Left side surface to control a central location an inside and thread

取付 Mounting

機能 Function
内面とねじの中心を決める右側面
Right side surface to control a central location an inside and thread

ハウジングの機能バラシ　Break down of Housing

c) 最初に、機能に影響する重要な寸法を指示します。
c) First, priority dimensions which influence functions are indicated.

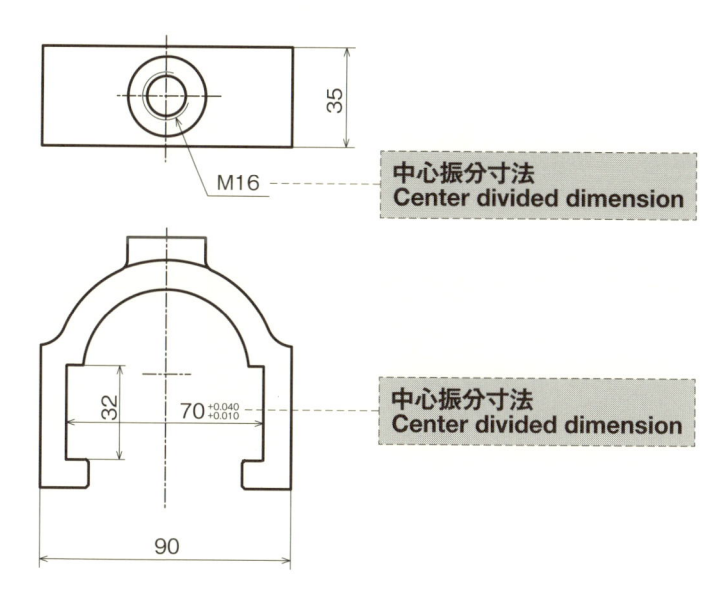

35

M16

中心振分寸法
Center divided dimension

32

70 $^{+0.040}_{+0.010}$

90

中心振分寸法
Center divided dimension

重要寸法の指示　Indication of priority dimensions

Chapter7　In what way should I make drawings!　247

d) 内側形体に関連する寸法をまとめて指示します。
d) Dimensions related to inside-feature are indicated collectively.

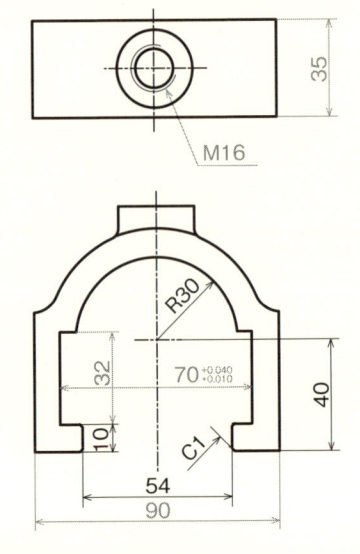

関連寸法の記入例（内側形体）　Indication of related dimensions (Inside-feature)

e) 外側形体に関連する寸法をまとめて指示します。
e) Dimensions related to outside-feature are indicated collectively.

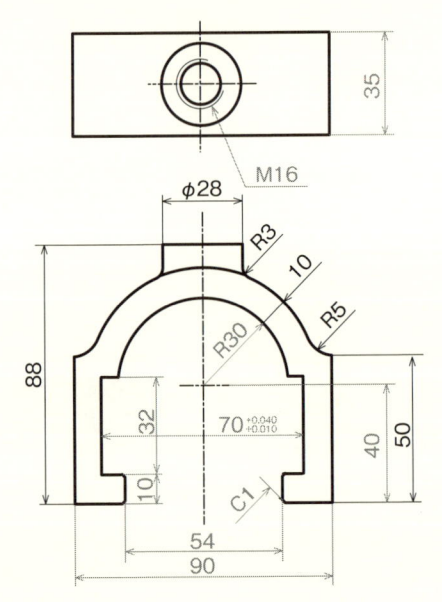

関連寸法の指示（外側形体）　Indication of related dimensions (Outside-feature)

f) さらに、面の肌記号を指示します。
f) In addition, surface texture symbols are indicated.

良好な面の肌は、重要寸法と一致していることがわかるでしょう。
You can understand that good surface texture accord with priority dimensions.

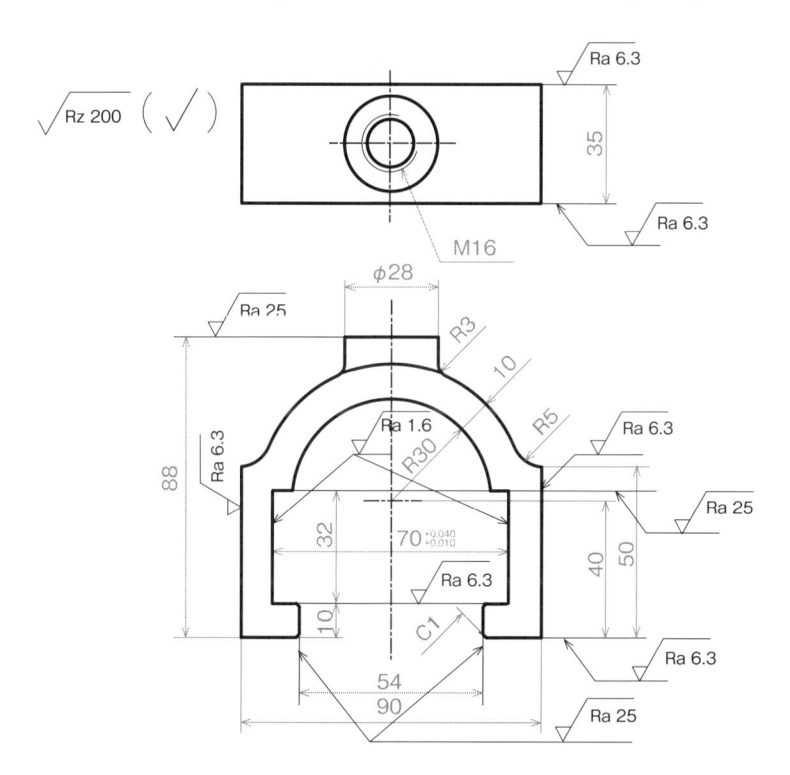

面の肌記号の指示　Indication of surface texture symbols

③固定用ボルトの図面作成
Drawing of Clamping-bolt

固定用ボルトは、ねじの締結力を利用して丸棒を固定するために使用される要素です。

Clamping-bolt is a device which is used to fix a rod by fastening.

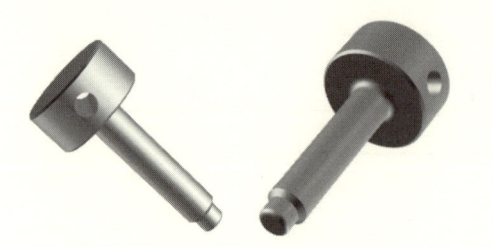

想定される材質 / Assumed material

材質:SGD400-D
Material:SGD400-D

表面処理:Ep-Fe/Zn8/CM2
Finishing:Ep-Fe/Zn8/CM2

固定用ボルトの３次元モデル　3D-model of Clamping-bolt

a) 投影図の準備
a) Views preparation

　本品は円筒形状であるため、正面図は水平とし加工部を右に向けて記入します。
　Because Clamping-bolt is cylindrical, front view is placed to horizontally and turns a processing part to the right.

　固定ボルトは、穴を除いて旋盤で加工できる単純な形状のため、一つの投影図で表すことができます。
　Clamping-bolt can be expressed by only one view. Because this is simple shape to be able to process with lathe except a hole.

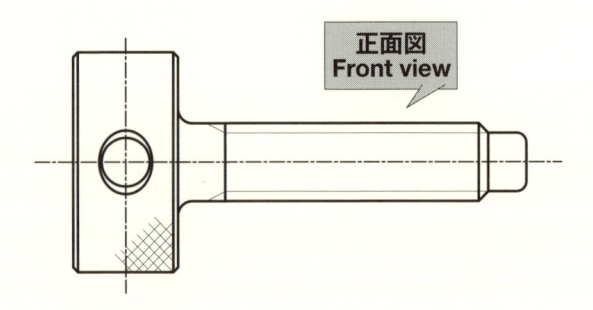

正面図
Front view

固定ボルトの投影図　Views of Clamping-bolt

b) 固定ボルトの機能をばらしてみましょう。
b) Let's break down Clamping-bolt.

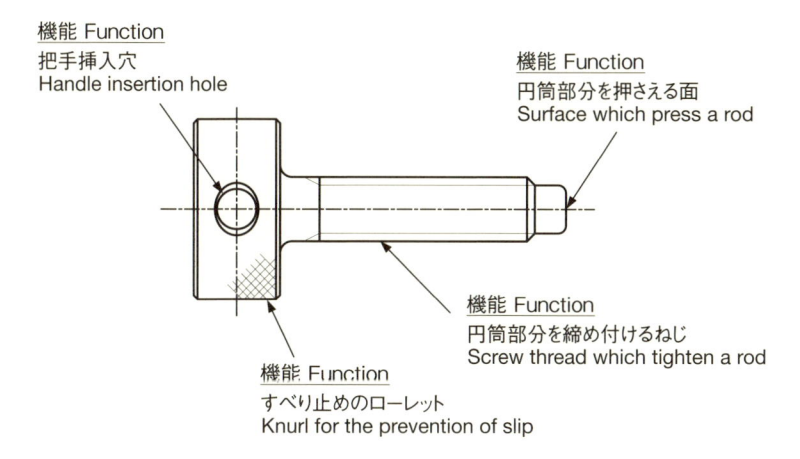

機能 Function
把手挿入穴
Handle insertion hole

機能 Function
円筒部分を押さえる面
Surface which press a rod

機能 Function
円筒部分を締め付けるねじ
Screw thread which tighten a rod

機能 Function
すべり止めのローレット
Knurl for the prevention of slip

固定ボルトの機能バラシ　Breakdown of Clamping-bolt

c) 最初に、機能に影響する重要な寸法を指示します。
c) First, priority dimensions which influence functions are indicated

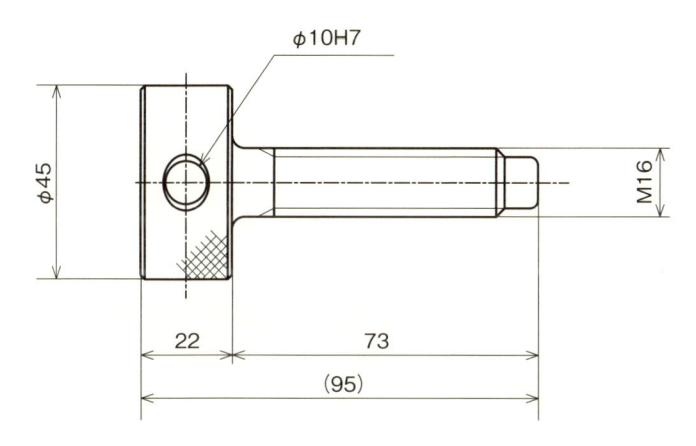

φ10H7
φ45
M16
22
73
(95)

重要寸法の指示　Indication of priority dimensions

d) その他、足りない寸法を指示します。
d) Other required dimensions are indicated.

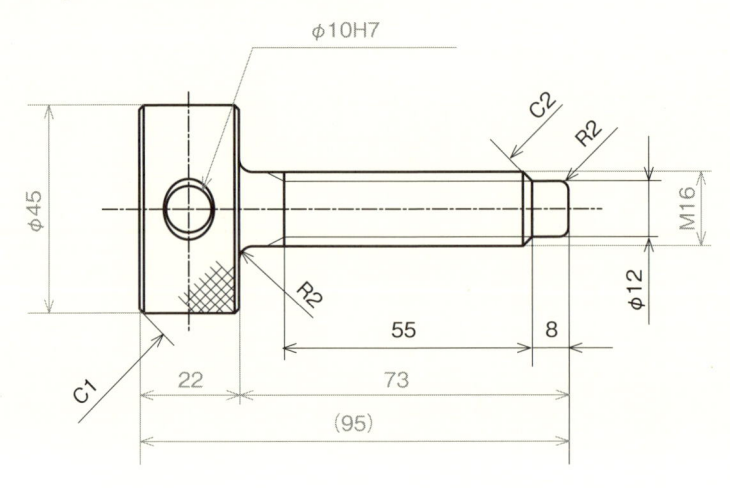

足りない寸法の指示　Indication of other required dimensions

e) 加工上の要求事項を指示します。
e) The requirements of processing are indicated.

　丸棒に傷をつけないよう、固定ボルトの端面はセンター穴を要求しません。
　Center drill hole is not required in end of Clamping-bolt so that scratch may not be given to a rod.

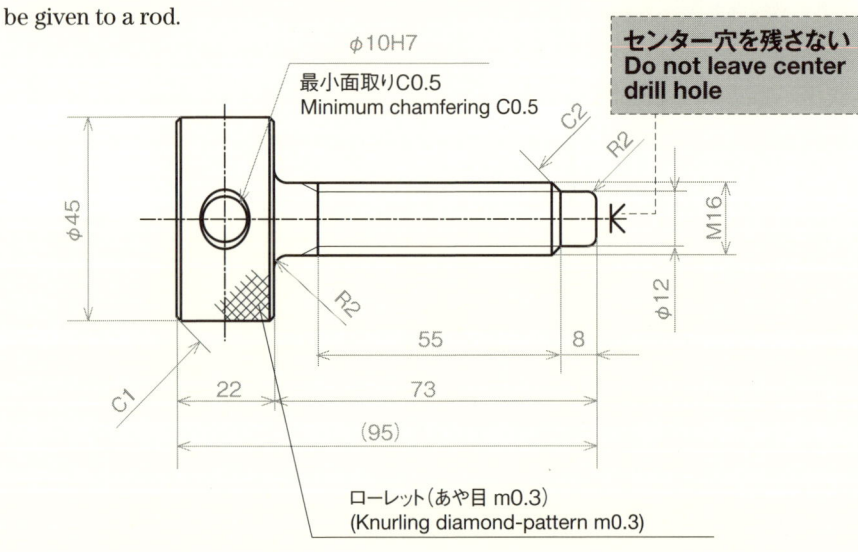

最小面取りC0.5
Minimum chamfering C0.5

センター穴を残さない
Do not leave center drill hole

ローレット（あや目 m0.3）
(Knurling diamond-pattern m0.3)

加工上の要求事項の指示　Indication of requirements of processing

f) さらに、面の肌記号を指示します。
f) In addition, surface texture symbols are indicated.

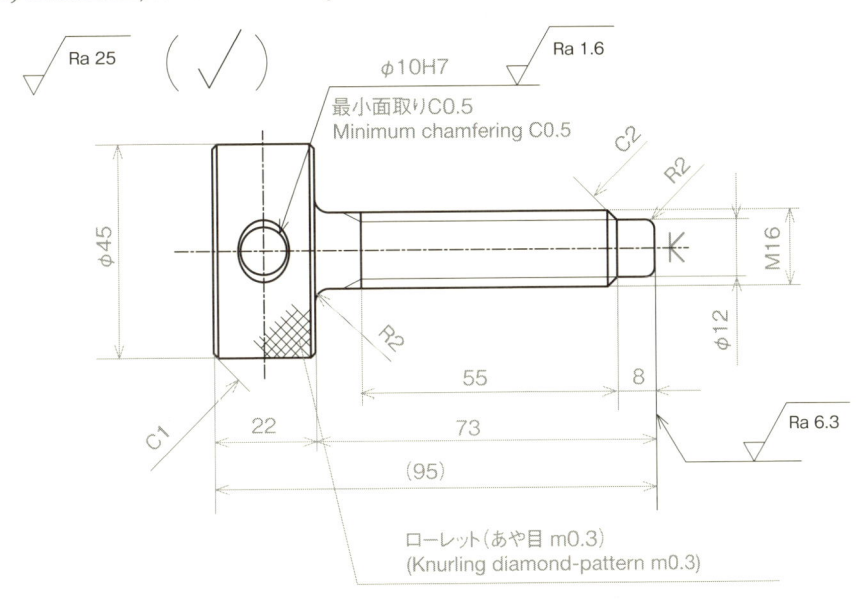

面の肌記号の指示　Indication of surface texture symbols

φ(@°▽°@) メモメモ　Note ;-)

円弧上の面取り形状の注意点
Point of the chamfering shape on arc

平面にある面取りの場合、穴の口元は全周を均一に面取り加工できます。しかし、円弧上にある穴の口元は全周を均一に面取りできません。

In chamfering on a flat surface, the hole-edge is chamfered uniformly. But In chamfering on an arc, the hole-edge is not chamfered uniformly.

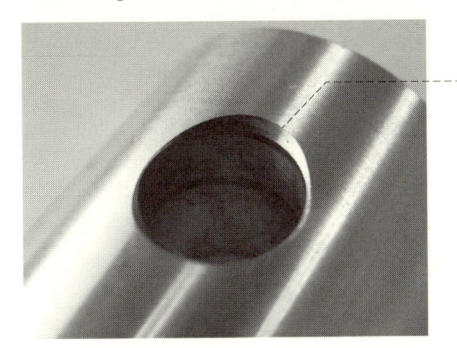

弧の頂点ほど面取り量が多い
There are many amounts of chamfering at peak of an arc.

④ 把手の図面作成
Drawing of Handle

把手は固定ボルトの回転力を増やすために使用される要素です。
Handle is a device which is used to increase torque of Clamping-bolt.
把手は固定ボルトから取外しできるよう、隙間ばめとします。
Handle is designed as a clearance fit to remove from Clamping-bolt.

想定される材質
Assumed material

材質:SGD400-D9
Material:SGD400-D9

表面処理:Ep-Fe/Zn8/CM2
Finishing:Ep-Fe/Zn8/CM2

把手の3次元モデル　3D-model of Handle

a) 投影図の準備
a) Views preparation

本品は円筒形状であるため、正面図は水平に記入します。
Handle is cylindrical, so front view is placed to horizontally.

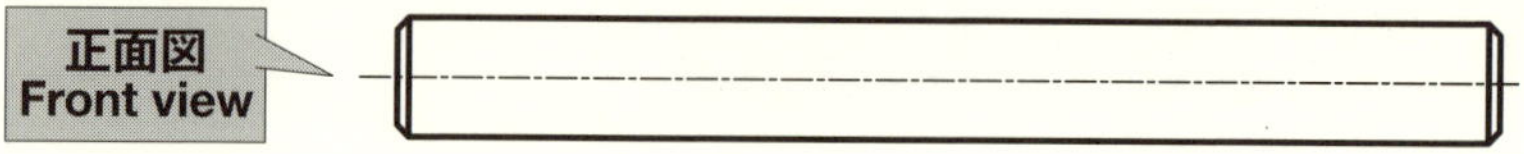

正面図
Front view

把手の投影図　Views of Handle

b) 把手の機能をばらしてみましょう。
b) Let's break down Handle.

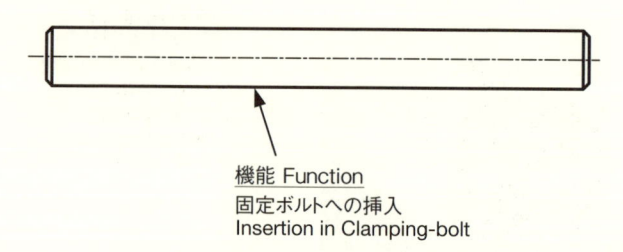

機能 Function
固定ボルトへの挿入
Insertion in Clamping-bolt

把手の機能バラシ　Break down of Handle

c) 単純形状なので、すべての寸法を指示します。
c) This is simple shape, all dimensions are indicated.

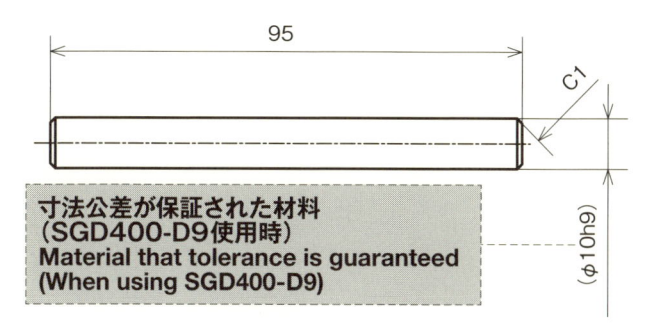

寸法公差が保証された材料
（SGD400-D9使用時）
Material that tolerance is guaranteed
(When using SGD400-D9)

全ての寸法の指示　Indication of all dimensions

d) さらに、面の肌記号を指示します。
d) In addition, surface texture symbols are indicated.

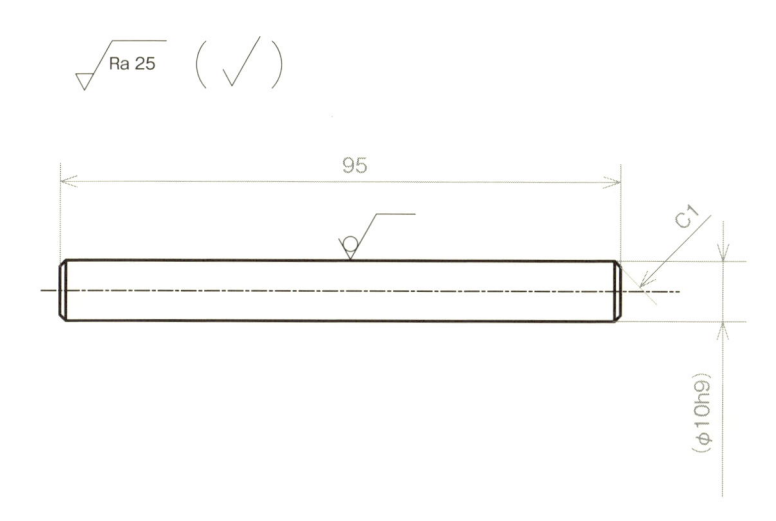

Ra 25　(✓)

面の肌記号の指示　Indication of surface texture symbols

第7章 Chapter7 — 2

寸法指示の論理的思考(ドリルプレスバイスの例)
Logical mind of the dimensions indication (Example of Drilling-press vice)

ドリルプレスバイスは、工作物を保持したり締め付けたりするために使用されるねじ装置です。

Drilling-press vice is a mechanical screw device that is used for holding or clamping a work-piece.

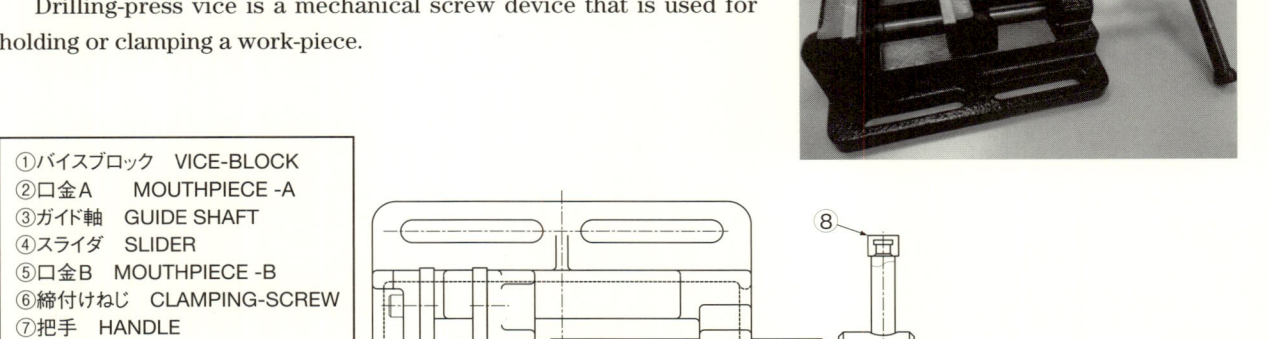

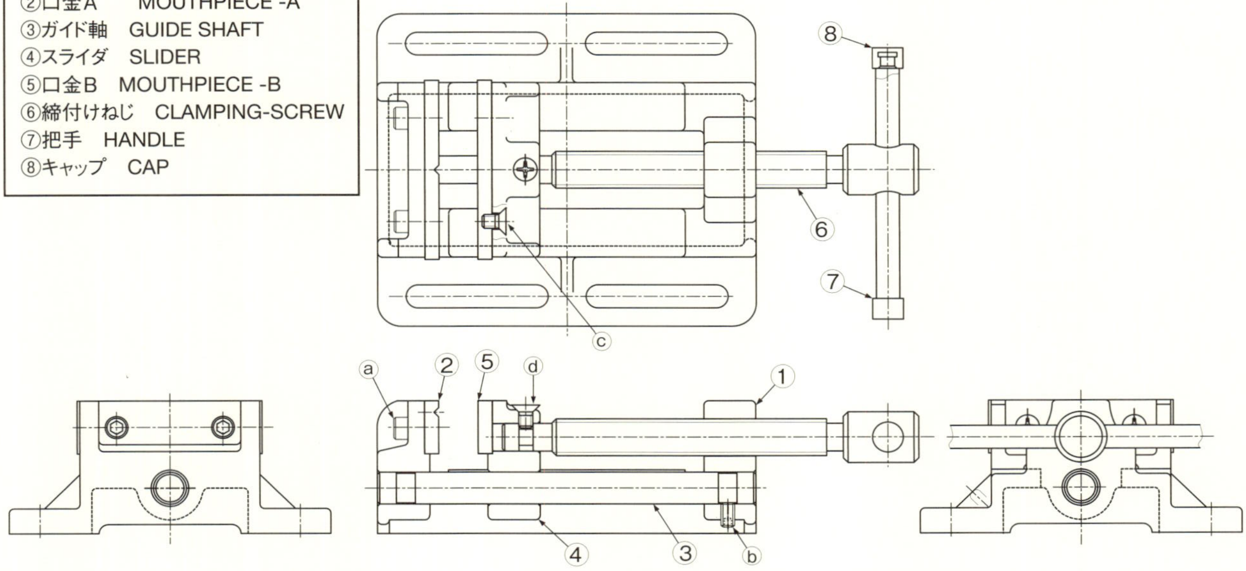

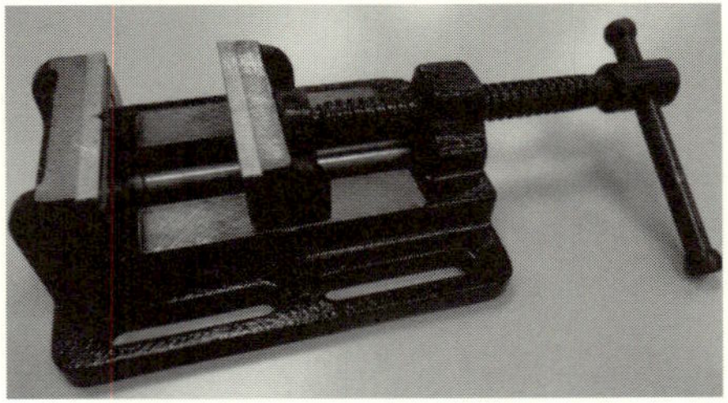

ドリルプレスバイスの組立図　Assembly drawing of Drilling-press vice

① バイスブロック　VICE-BLOCK
② 口金A　MOUTHPIECE-A
③ ガイド軸　GUIDE SHAFT
④ スライダ　SLIDER
⑤ 口金B　MOUTHPIECE-B
⑥ 締付けねじ　CLAMPING-SCREW
⑦ 把手　HANDLE
⑧ キャップ　CAP

第7章　図面って、どない描くねん!

① バイスブロックの図面作成
Drawing of Vice-block

a) 投影図の準備と機能バラシ
a) Views preparation and break down

バイスブロックは鋳造後、切削加工するものとします。
Vice-block shall be made by cutting after casting.

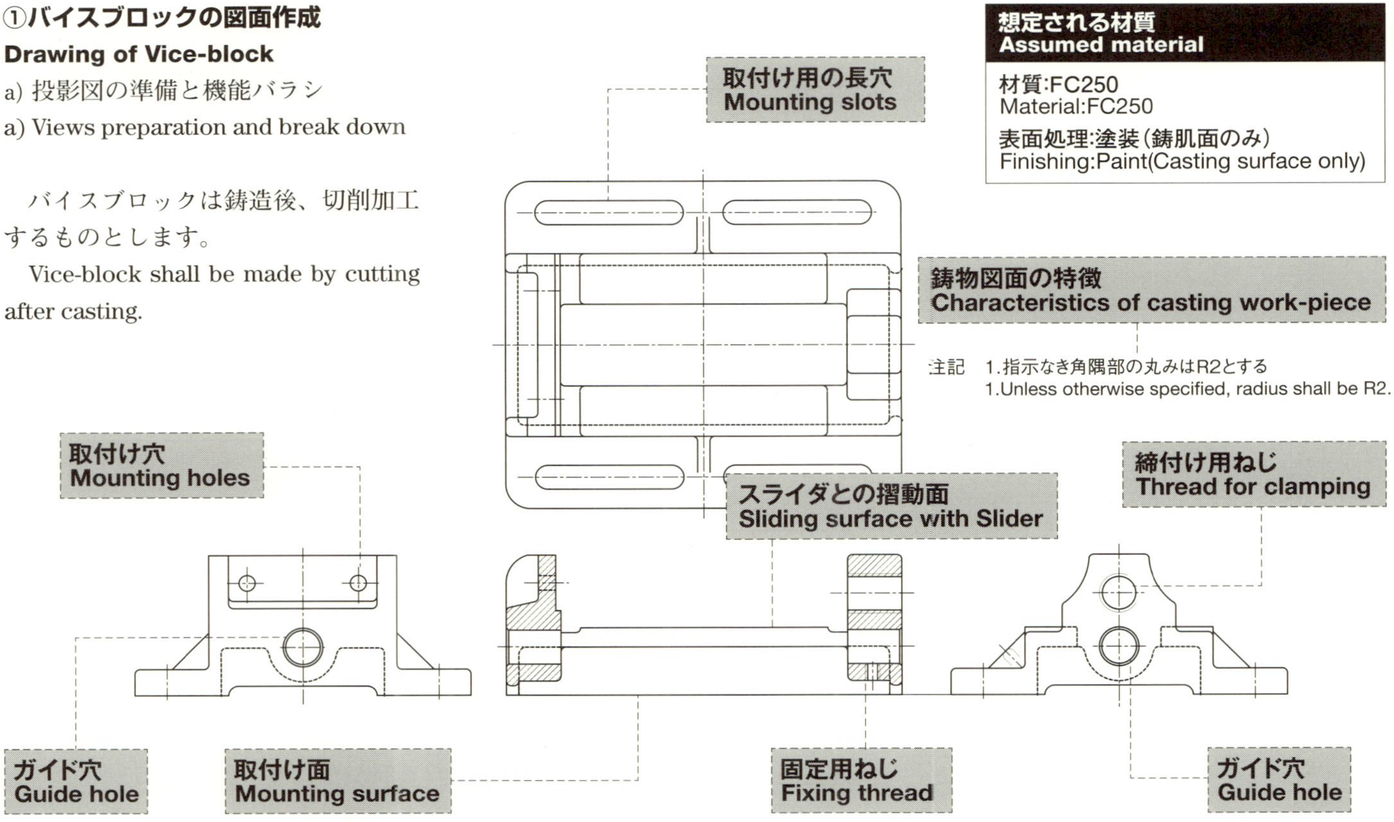

想定される材質
Assumed material

材質:FC250
Material:FC250
表面処理:塗装(鋳肌面のみ)
Finishing:Paint(Casting surface only)

鋳物図面の特徴
Characteristics of casting work-piece

注記 1.指示なき角隅部の丸みはR2とする
1.Unless otherwise specified, radius shall be R2.

取付け用の長穴
Mounting slots

スライダとの摺動面
Sliding surface with Slider

締付け用ねじ
Thread for clamping

ガイド穴
Guide hole

固定用ねじ
Fixing thread

取付け面
Mounting surface

取付け穴
Mounting holes

ガイド穴
Guide hole

バイスブロックの投影図　View of Vice-block

Chapter7　In what way should I make drawings!　257

b) 最初に、機能に影響する重要寸法を正面図に指示します。
b) First, priority dimensions which influence functions are indicated to front view.

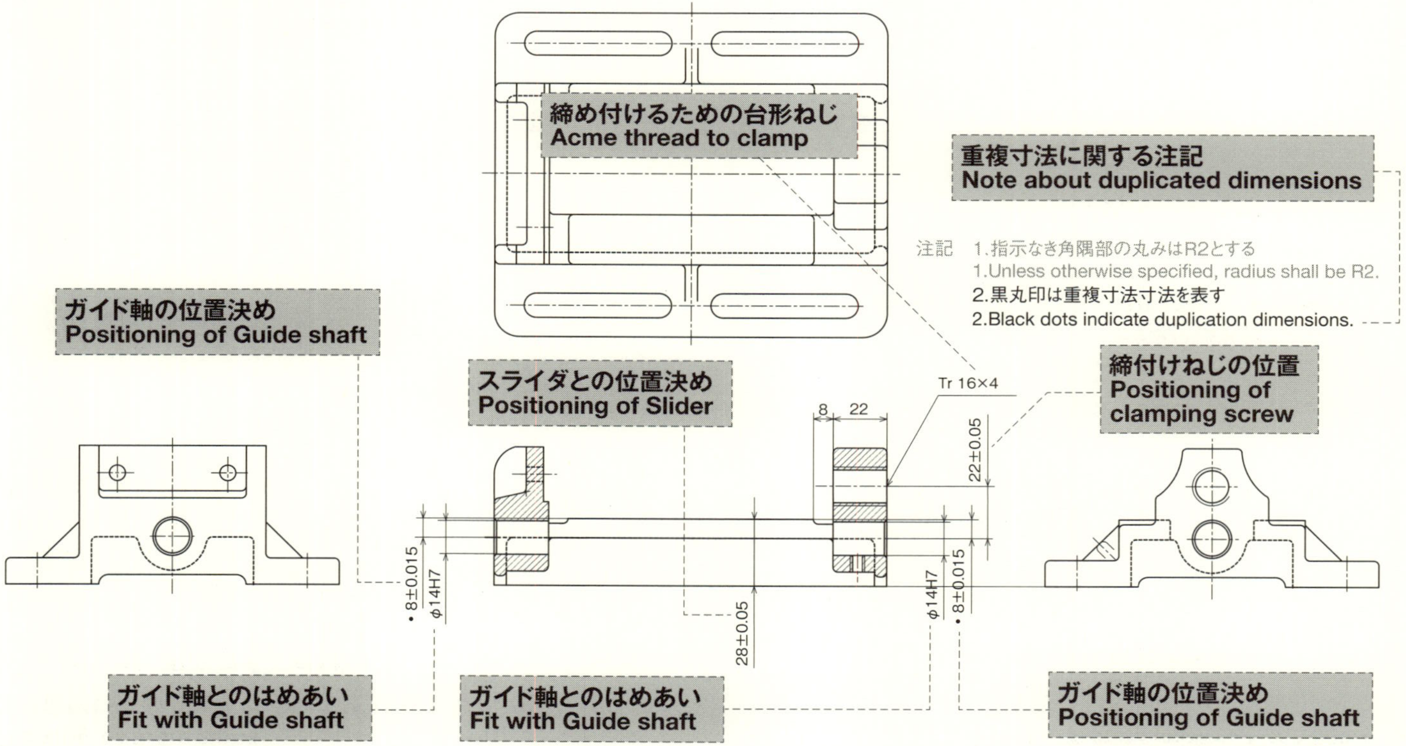

重複寸法に関する注記
Note about duplicated dimensions

注記
1. 指示なき角隅部の丸みはR2とする
1. Unless otherwise specified, radius shall be R2.
2. 黒丸印は重複寸法を表す
2. Black dots indicate duplication dimensions.

締付けねじの位置
Positioning of clamping screw

ガイド軸の位置決め
Positioning of Guide shaft

Tr 16×4
22±0.05
8±0.015
φ14H7
22
8
28±0.05

締め付けるための台形ねじ
Acme thread to clamp

スライダとの位置決め
Positioning of Slider

ガイド軸とのはめあい
Fit with Guide shaft

φ14H7
8±0.015

ガイド軸の位置決め
Positioning of Guide shaft

ガイド軸とのはめあい
Fit with Guide shaft

重要寸法の指示　Indication of priority dimensions

バイスブロックの高さ
Height of Vice-block

R25の中心高さ
Height of R25 center

注記
1. 指示なき角隅部の丸みはR2とする
1. Unless otherwise specified, radius shall be R2.
2. 黒丸印は重複寸法を表す
2. Black dots indicate duplication dimensions.

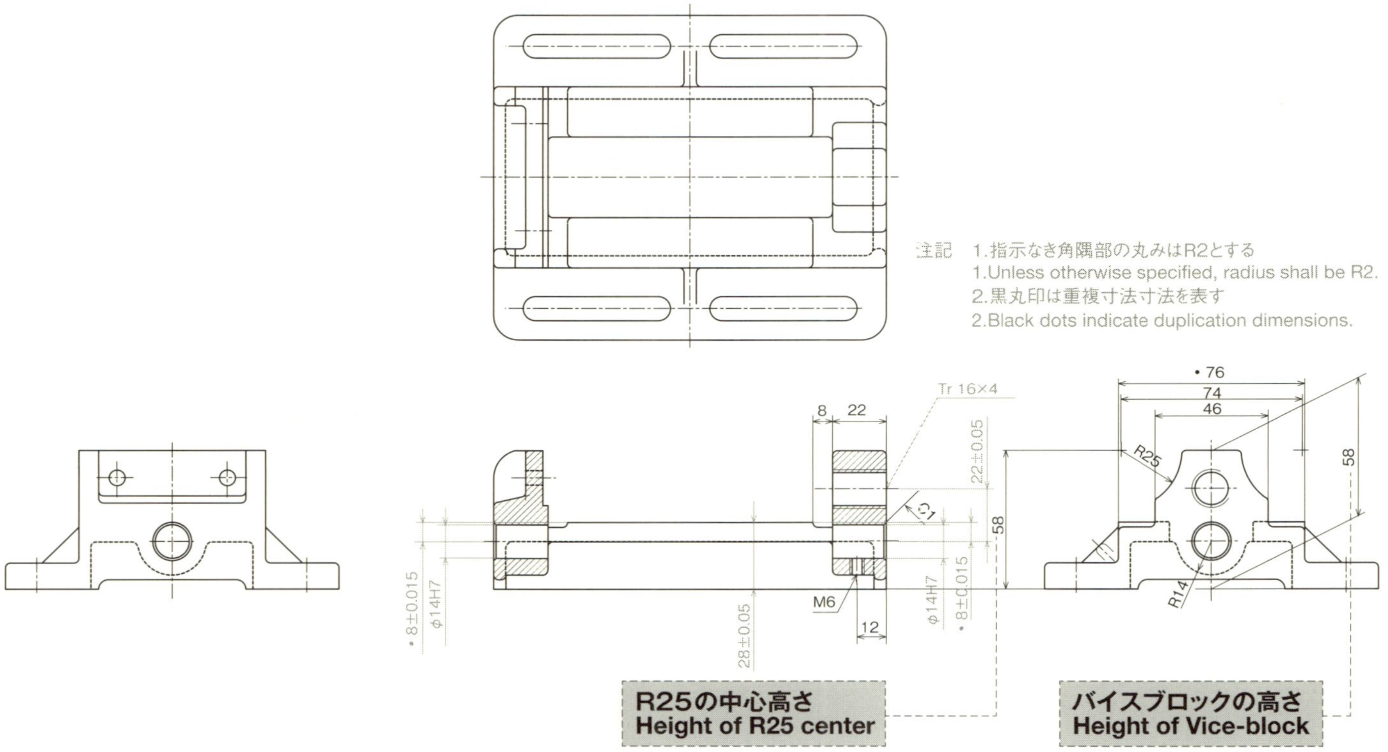

右側形状への寸法指示　Indication of right-side dimensions

c) 右側面に関連する寸法をまとめて指示します。
c) Dimensions related to right-side are indicated collectively.

Chapter7　In what way should I make drawings!

259

d) 左側面に関連する寸法をまとめて指示します。
d) Dimensions related to left-side are indicated collectively.

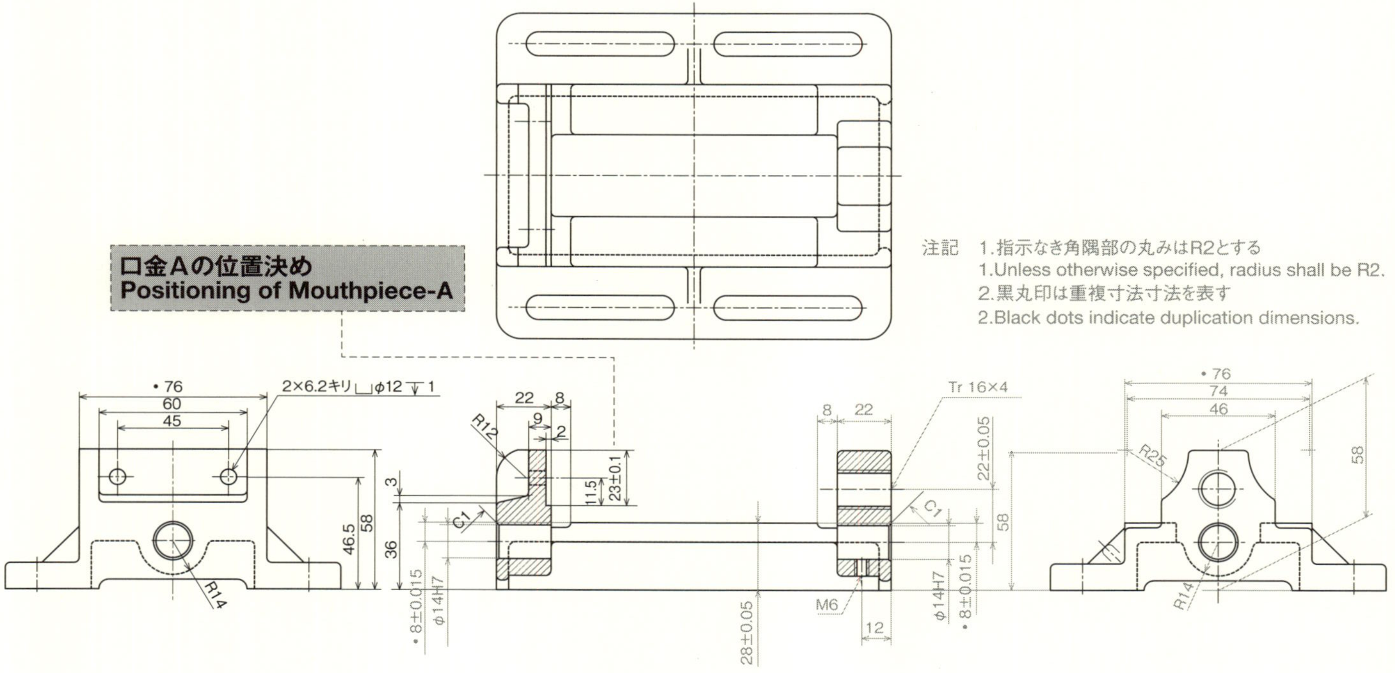

口金Aの位置決め
Positioning of Mouthpiece-A

注記 1. 指示なき角隅部の丸みはR2とする
1. Unless otherwise specified, radius shall be R2.
2. 黒丸印は重複寸法寸法を表す
2. Black dots indicate duplication dimensions.

左側形状への寸法指示　Indication of left-side dimensions

リブの厚み
Thickness of rib

注記
1. 指示なき角隅部の丸みはR2とする
1. Unless otherwise specified, radius shall be R2.
2. 黒丸印は重複寸法を表す
2. Black dots indicate duplication dimensions.

e) 取付け部に関連する寸法をまとめて指示します。
e) Dimensions related to mounting part are indicated collectively.

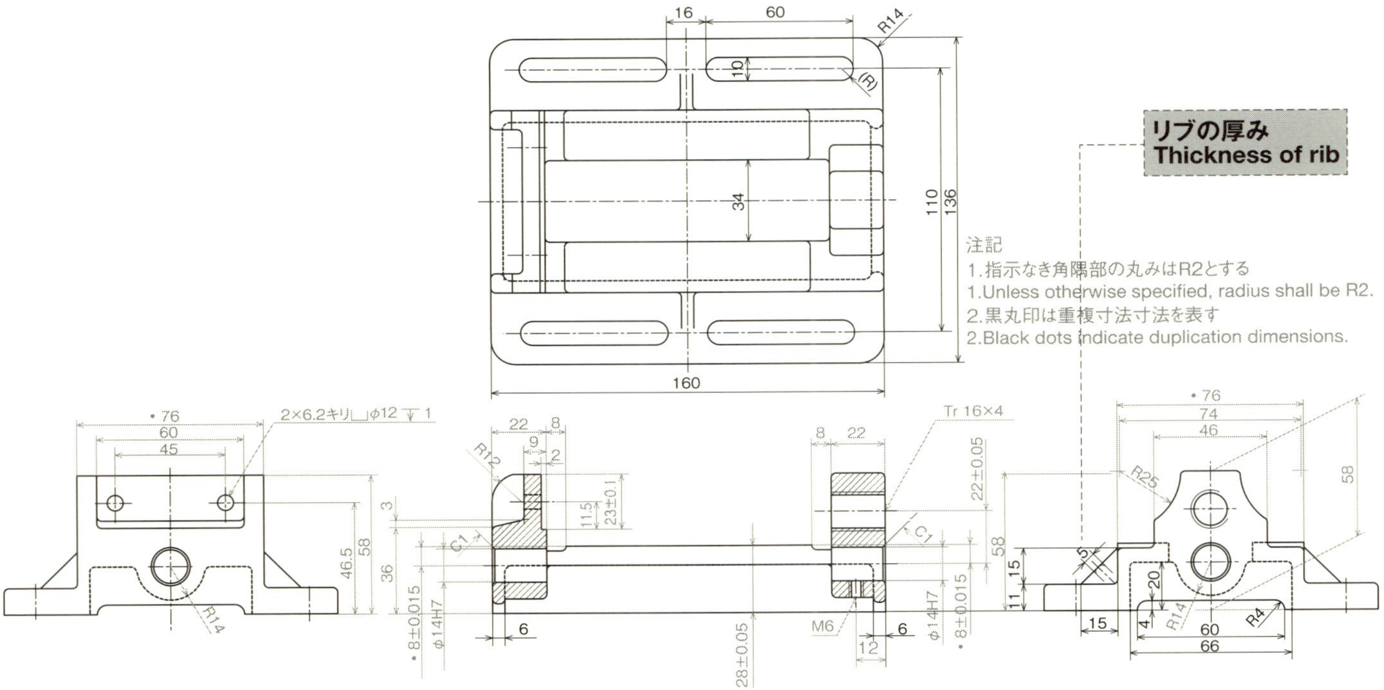

取付け部の寸法の指示　Indication of mounting part dimensions

Chapter7　In what way should I make drawings!

f) 面の肌記号と幾何公差を指示します。
f) In addition, surface texture symbols and geometrical tolerance are indicated.

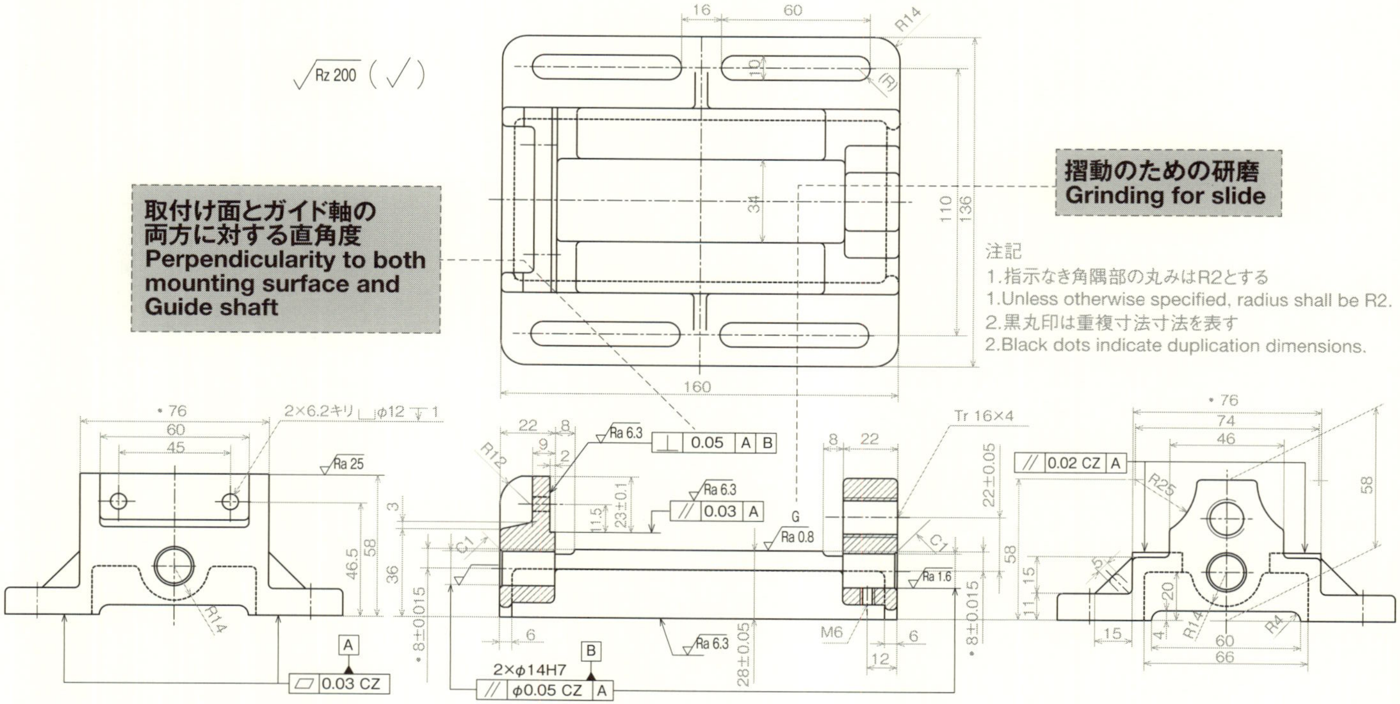

摺動のための研磨
Grinding for slide

注記
1. 指示なき角隅部の丸みはR2とする
1. Unless otherwise specified, radius shall be R2.
2. 黒丸印は重複寸法を表す
2. Black dots indicate duplication dimensions.

取付け面とガイド軸の両方に対する直角度
Perpendicularity to both mounting surface and Guide shaft

面の肌記号と幾何公差の指示例　Indication of surface texture symbols and geometrical tolerance

②口金Aの図面作成
Drawing of Mouthpiece-A

a) 投影図の準備と機能バラシ
a) Views preparation and break down

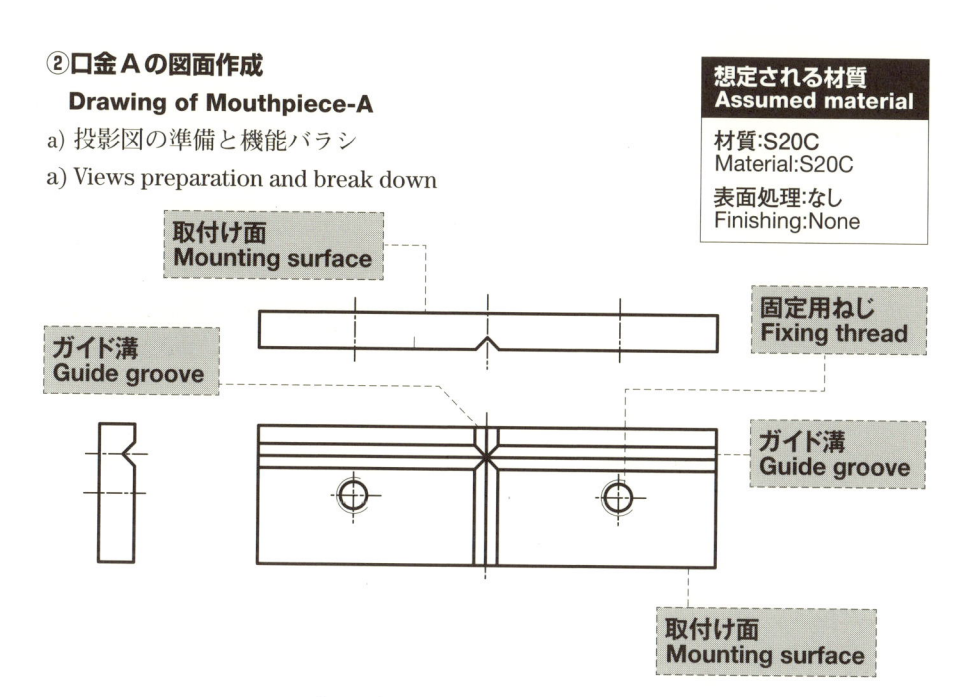

想定される材質 Assumed material
材質:S20C Material:S20C
表面処理:なし Finishing:None

取付け面 Mounting surface
ガイド溝 Guide groove
固定用ねじ Fixing thread
ガイド溝 Guide groove
取付け面 Mounting surface

口金Aの投影図　Views of Mouthpiece-A

b) 寸法と面の肌記号を指示します。
b) Dimensions and surface texture symbols are indicated.

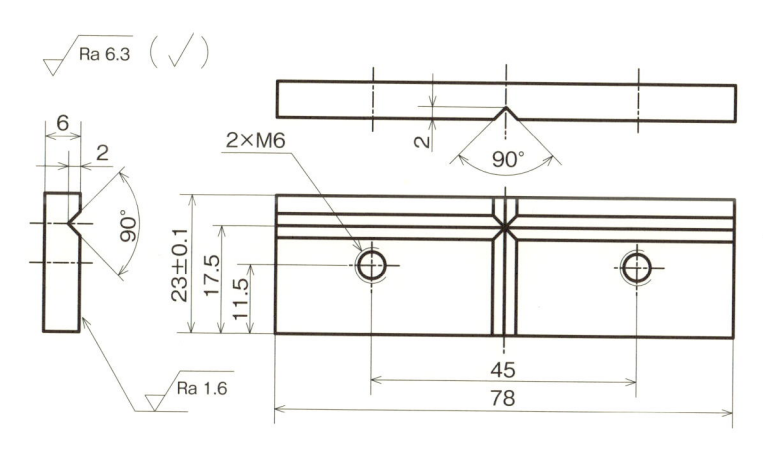

寸法と面の肌記号の指示　Indication of dimensions and surface texture symbols

Chapter7　In what way should I make drawings!

③ガイド軸の図面作成
Drawing of Guide shaft

a) 投影図の準備と機能バラシ

a) Views preparation and break down

本品は円筒形状であるため、正面図は水平とします。

Guide shaft is cylindrical shape. So, front view is arranged to horizontally.

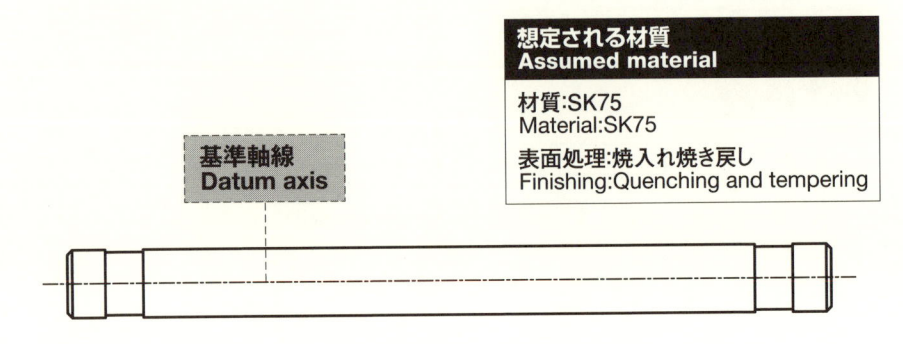

想定される材質
Assumed material

材質:SK75
Material:SK75

表面処理:焼入れ焼き戻し
Finishing:Quenching and tempering

基準軸線
Datum axis

ガイド軸の投影図　View of Guide shaft

b) 寸法と面の肌記号を指示します。

b) Dimensions and surface texture symbols are indicated.

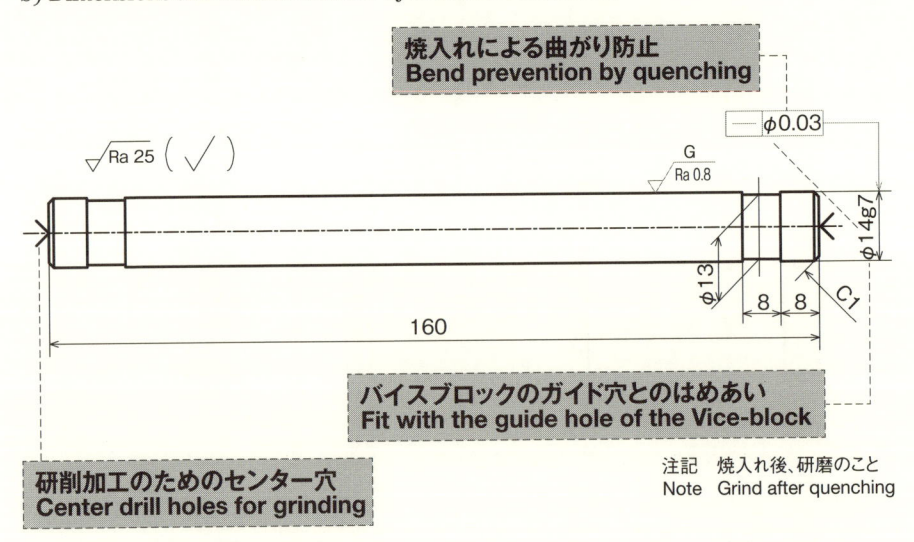

焼入れによる曲がり防止
Bend prevention by quenching

φ0.03

Ra 25

G Ra 0.8

φ13　φ14g7

8　8　C1

160

バイスブロックのガイド穴とのはめあい
Fit with the guide hole of the Vice-block

研削加工のためのセンター穴
Center drill holes for grinding

注記　焼入れ後、研磨のこと
Note　Grind after quenching

寸法と面の肌記号の指示　Indication of dimensions and surface texture symbols

④ スライダの図面作成
Drawing of Slider

a) 投影図の準備と機能バラシ
a) Views preparation and break down

スライダは鋳造後、切削加工するものとします。
Slider shall be made by cutting after casting.

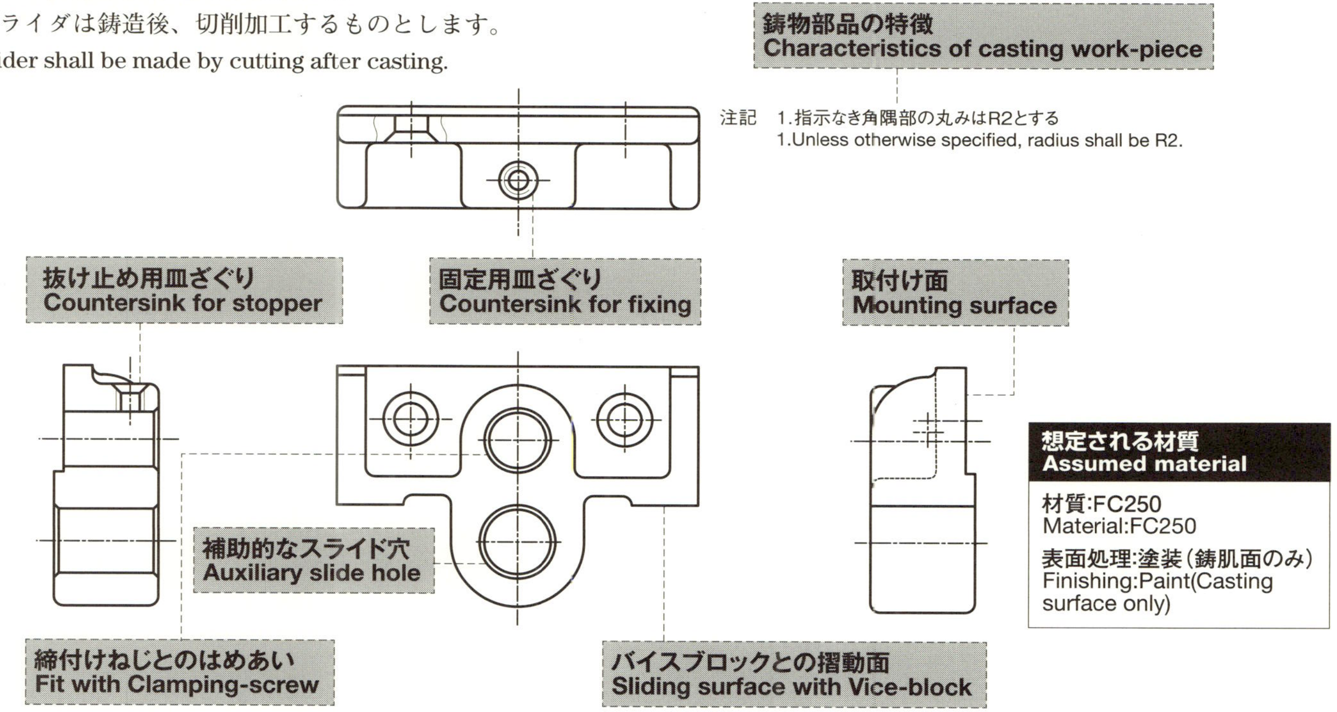

鋳物部品の特徴
Characteristics of casting work-piece

注記 1.指示なき角隅部の丸みはR2とする
1.Unless otherwise specified, radius shall be R2.

想定される材質
Assumed material

材質:FC250
Material:FC250
表面処理:塗装（鋳肌面のみ）
Finishing:Paint(Casting surface only)

取付け面
Mounting surface

固定用皿ざぐり
Countersink for fixing

バイスブロックとの摺動面
Sliding surface with Vice-block

抜け止め用皿ざぐり
Countersink for stopper

補助的なスライド穴
Auxiliary slide hole

締付けねじとのはめあい
Fit with Clamping-screw

スライダの投影図 Views of Slider

Chapter7 In what way should I make drawings!

皿穴のための位置決め
Positioning for countersink

注記
1. 指示なき角隅部の丸みはR2とする
1. Unless otherwise specified, radius shall be R2.

摺動のための研磨
Grinding for slide

ガイド軸とのすきまばめ
Clearance fit with Guide shaft

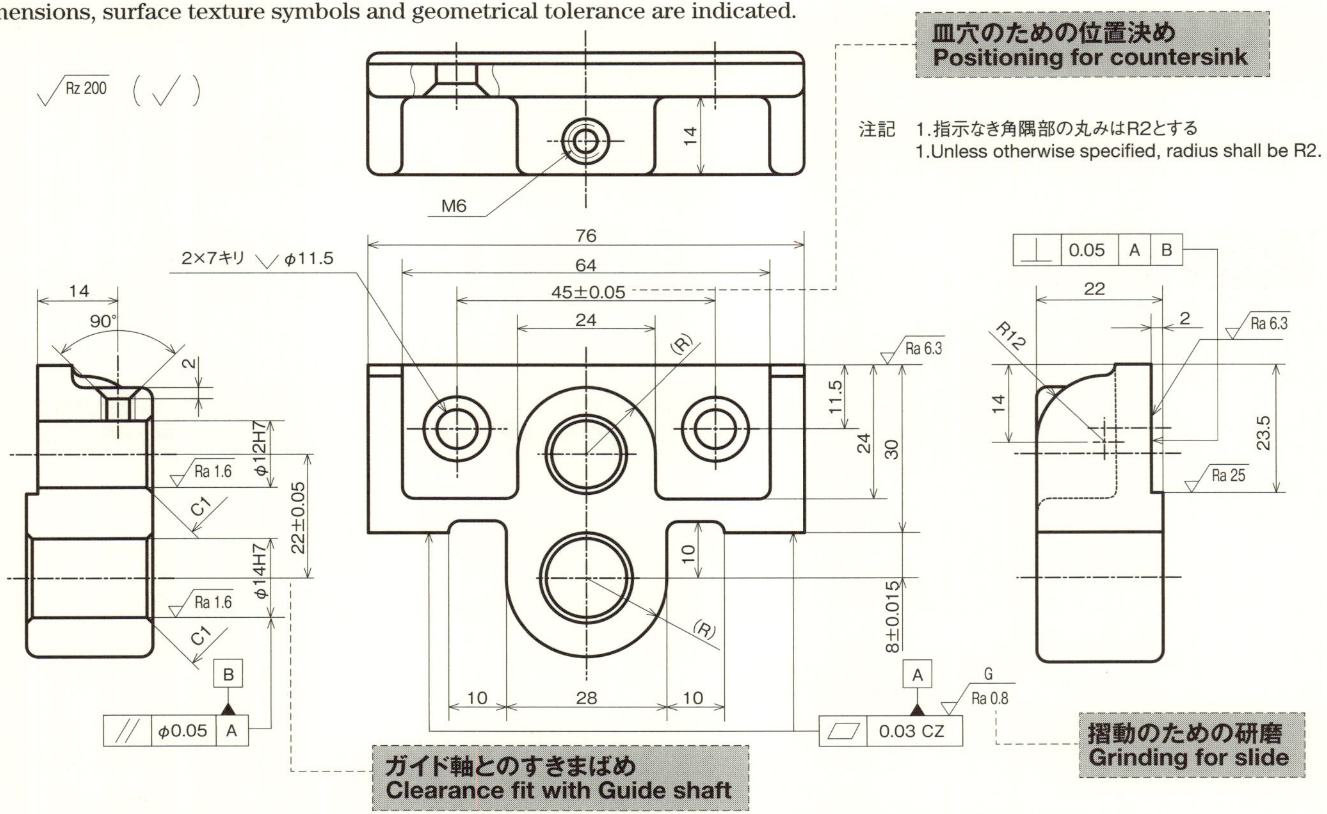

寸法と面の肌記号、幾何公差の指示　Indication of dimensions and surface texture and geometrical tolerance symbols

b) 寸法と面の肌記号、幾何公差を指示します。
b) Dimensions, surface texture symbols and geometrical tolerance are indicated.

266　第7章　図面って、どない描くねん！

⑤ 口金Ｂの図面作成
Drawing of Mouthpiece-B

a) 投影図の準備と機能バラシ

a) Views preparation and break down

　　平板のため、板厚を投影図の中に記入することで、この部品は正面図だけで表すことができます。しかし、表面粗さを指示するために２つの投影図で表します。

　　This work-piece is flat plate, so view can be expressed by only front view with thickness. But two views are expressed to indicate the surface texture.

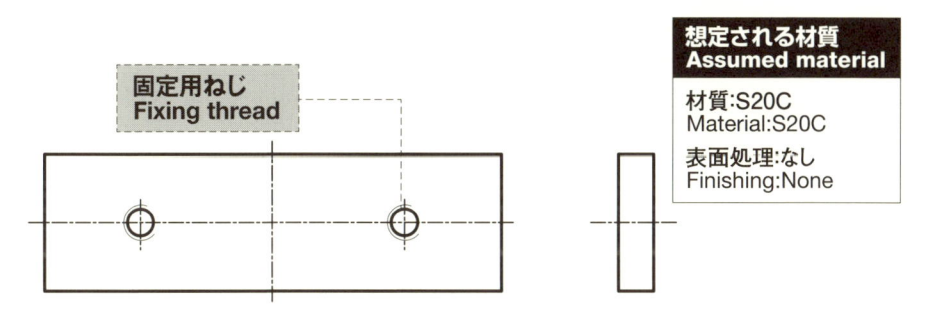

固定用ねじ / Fixing thread

想定される材質 / Assumed material
材質:S20C
Material:S20C
表面処理:なし
Finishing:None

口金Ｂの投影図　View of Mouthpiece-B

b) 寸法と面の肌記号を指示します。

b) Dimensions and surface texture symbols are indicated.

　　表裏どちらでも組めるよう、面の肌記号は両面に指示します。

　　The surface textures are indicated to both sides so that either of the back and front side can be assembled.

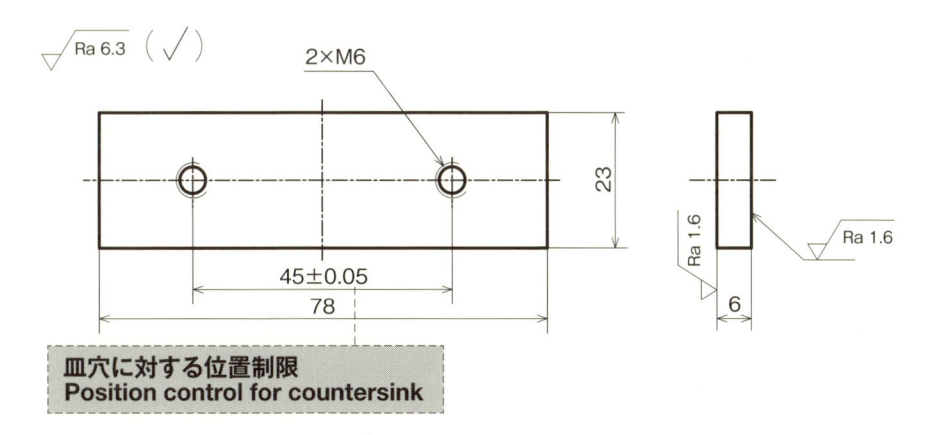

√Ra 6.3 (√)　　2×M6

23

45±0.05
78

Ra 1.6　　Ra 1.6
6

皿穴に対する位置制限 / Position control for countersink

寸法と面の肌記号の記入例　Indication of dimensions and surface texture symbols

Chapter7　In what way should I make drawings!

⑥ 締付けねじの図面作成
Drawing of Clamping-screw

a) 投影図の準備
a) Views preparation

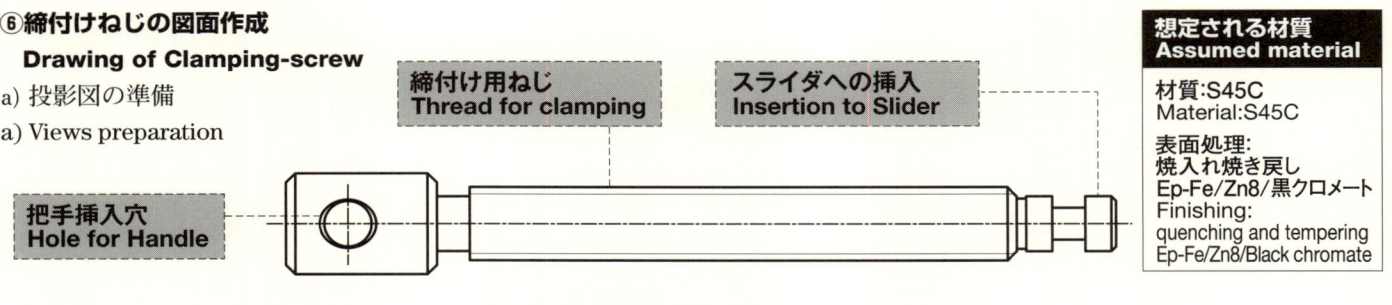

想定される材質 / Assumed material
材質：S45C
Material: S45C
表面処理：
焼入れ焼き戻し
Ep-Fe/Zn8/黒クロメート
Finishing:
quenching and tempering
Ep-Fe/Zn8/Black chromate

把手挿入穴 / Hole for Handle

締付け用ねじ / Thread for clamping

スライダへの挿入 / Insertion to Slider

締付けねじの投影図　View of Clamping-screw

b) 寸法と面の肌記号を指示します。
b) Dimensions and surface texture symbols are indicated.

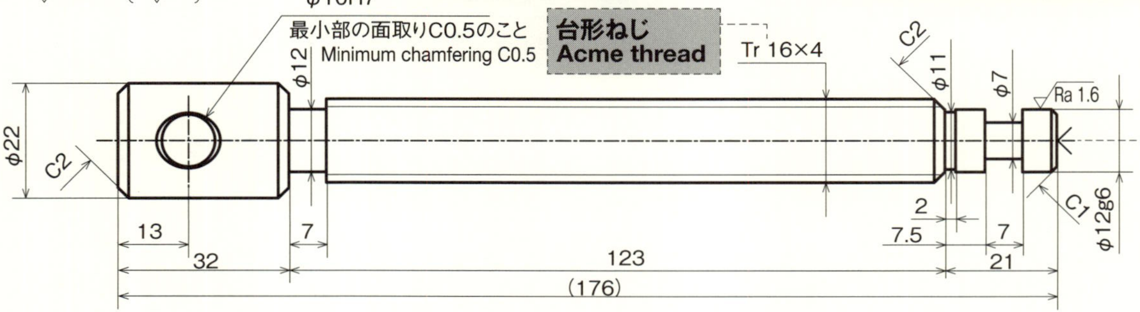

√Ra 25　(√)

φ10H7

最小部の面取りC0.5のこと
Minimum chamfering C0.5

把手とすきまばめ / Clearance fit with Handle

台形ねじ / Acme thread
Tr 16×4

加工時の振れ防止 / Run-out prevention at processing

Ra 1.6
φ12, C1
φ7
φ11
C2
7, 21
2
7.5
123
(176)
7
32
13
φ12
φ22
C2

寸法と面の肌記号の指示　Indication of dimensions and surface texture symbols

第7章　図面って、どない描くねん！

⑦ 把手の図面作成
Drawing of Handle

a) 投影図の準備と機能バラシ
a) Views preparation and break down

想定される材質
Assumed material
材質:SGD400-D
Material:SGD400-D
表面処理:Ep-Fe/Zn8/黒クロメート
Finishing:Ep-Fe/Zn8/Black chromate

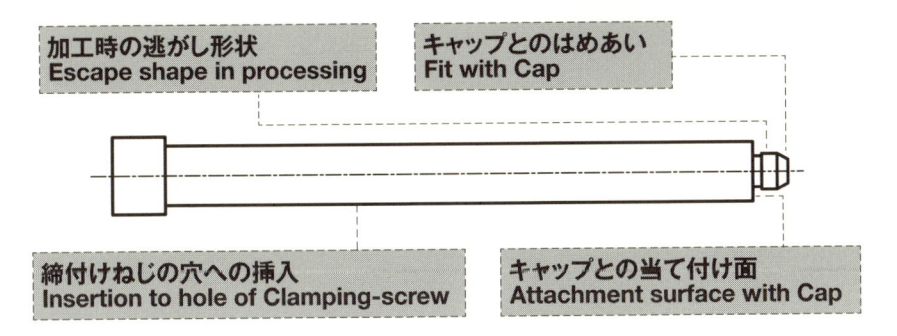

加工時の逃がし形状 / Escape shape in processing
キャップとのはめあい / Fit with Cap
締付けねじの穴への挿入 / Insertion to hole of Clamping-screw
キャップとの当て付け面 / Attachment surface with Cap

ハンドルの投影図　View of Handle

b) 寸法と面の肌記号を指示します。
b) Dimensions and surface texture symbols are indicated.

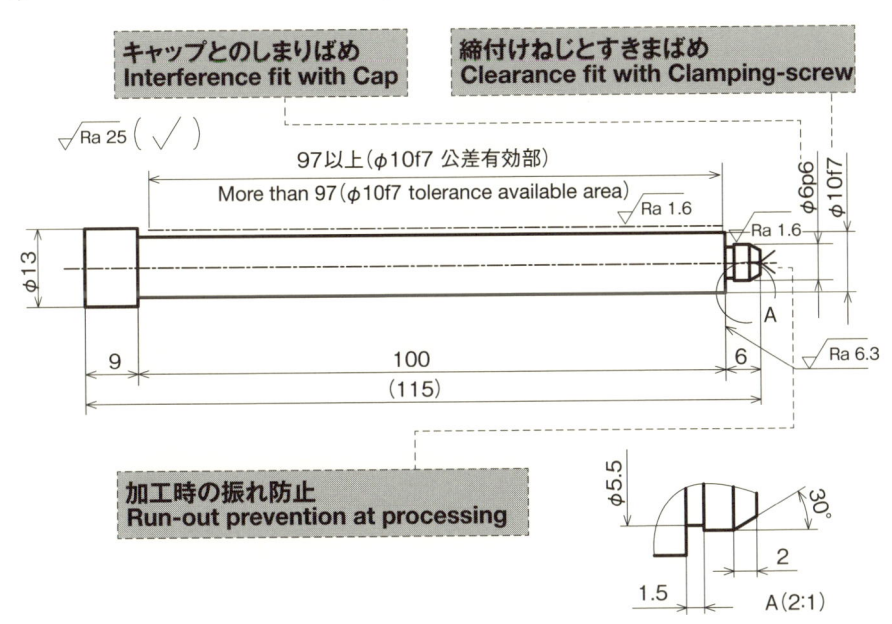

キャップとのしまりばめ / Interference fit with Cap
締付けねじとすきまばめ / Clearance fit with Clamping-screw

√Ra 25 (√)

97以上(φ10f7 公差有効部)
More than 97(φ10f7 tolerance available area)
Ra 1.6
φ6p6
φ10f7
Ra 1.6
φ13
A
Ra 6.3
9　100　6
(115)

加工時の振れ防止 / Run-out prevention at processing

φ5.5
30°
2
1.5
A(2:1)

寸法と面の肌記号の指示　Indication of dimensions and surface texture symbols

Chapter7　In what way should I make drawings!

⑧ キャップの図面作成
Drawing of Cap

a) 投影図の準備
a) Views preparation

想定される材質
Assumed material

材質:SGD400-D
Material:SGD400-D

表面処理:Ep-Fe/Zn8/黒クロメート
Finishing:Ep-Fe/Zn8/Black chromate

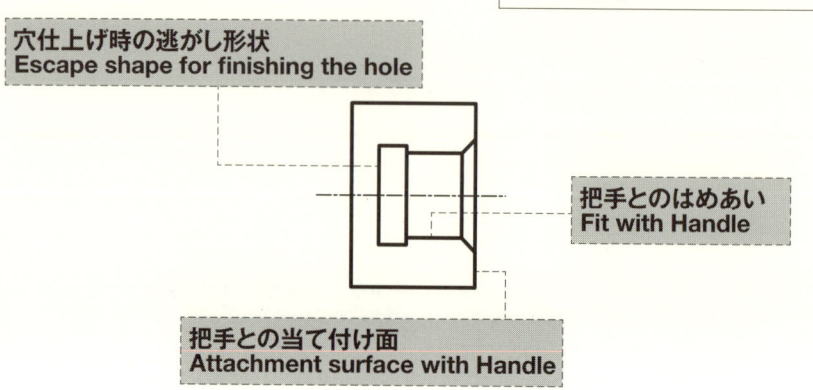

- 穴仕上げ時の逃がし形状
  Escape shape for finishing the hole
- 把手とのはめあい
  Fit with Handle
- 把手との当て付け面
  Attachment surface with Handle

キャップの投影図　View of Cap

b) 寸法と面の肌記号を指示します。
b) Dimensions and surface texture symbols are placed.

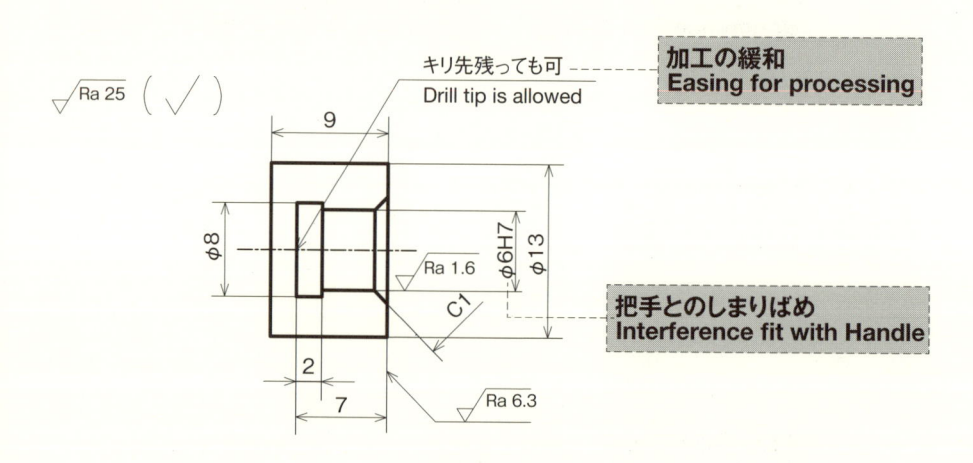

- キリ先残っても可
  Drill tip is allowed
- 加工の緩和
  Easing for processing
- 把手としまりばめ
  Interference fit with Handle

Ra 25　(√)

9, φ8, φ6H7, φ13, Ra 1.6, C1, 2, 7, Ra 6.3

寸法と面の肌記号の指示　Indication of dimensions and surface texture symbols

用語集　glossary

よく使う注記の例　Examples of notes commonly used

指示なき角隅部の丸みはR2とする
　　Unless otherwise specified, radius shall be R2.
すべての面は糸面取りのこと。
　　Break light-chamfering all edges.
特に指定がない限り、板厚は1.2mm±0.07とする。
　　Unless otherwise specified, material thickness shall be 1.2 ± 0.07mm.
特に指定がない限り、一般肉厚は5mmとする。
　　Unless otherwise specified, thickness shall be 5mm.
＊印部のエッジは、バリをとらないこと。
　　Mark (*) should not be deburred.
パーティングラインのバリは、0.1mm以下のこと。
　　Max permissible flash on parting line is 0.1mm.
抜きこう配は、1°以下のこと。こう配は体積が減る方向につけること。
　　Max permissible draft angle shall be 1°. Draft angle degrees mass.
全面にマシン油を塗布すること。
　　Coat all surface with machine oil.
二点鎖線は、展開図を示す。
　　Long-dashed double-dotted line shows development.
特に指定がない限り、塗装は白色とする。
　　Unless otherwise specified, white paint shall be used.
塗装は、仕様書No.1234による。
　　Paint shall be in accordance with specification No.1234.
硬質クロムめっきの厚さは、15～35μmのこと。
　　Thickness of hard chrome plating shall be between 15 and 35μm.
Oリングのシール部より油漏れなきこと。
　　There shall be no oil leakage from sealing such as O-ring.
溶接部は全数、浸透検査のこと。
　　All weld joint shall be examined by penetrant inspection.
指示された位置にLの文字記号を刻印すること。
　　Steel stamp letter "L" on the designation place.
ボルトの締め付けトルクは、12.5 N·mとする。
　　Torque required for bolt is 12.5 N·m.

[参考文献　Reference literature]

JIS B 0001 (日本語版 Japanese Ver.　英語版 English Ver.)

ISO Standards Handbook - Limits, fits and surface properties

ISO 468:1982 Surface roughness- parameters. Their values and general rules for specifing requirement.

ISO 4288:1996 Surface texture:Profile method- Rules and procedures for the assessment of surface texture.

Mechnical drawing problems　By:Edward Berg and Emil F. Kronquist

Basic tools for tolerance analysis of mechanical assemblies
By:Ken Chase

Gas Metal Arc Welding Handbook　　By: William H. Minnick

●著者紹介

山田　学（やまだ　まなぶ）

S38年生まれ、兵庫県出身。ラブノーツ 代表取締役。
カヤバ工業（現、KYB）自動車技術研究所にて電動パワーステアリングとその応用製品（電動後輪操舵E-HICASなど）の研究開発に従事。
グローリー工業（現、グローリー）設計部にて銀行向け紙幣処理機の設計や、設計の立場で海外展開製品における品質保証活動に従事。
平成18年4月 技術者教育を専門とする六自由度技術士事務所として独立。
平成19年1月 技術者教育を支援するため ラブノーツを設立。（http://www.labnotes.jp）
著書として、『図面って、どない描くねん！』、『設計の英語って、どない使うねん！』、『めっちゃ使える！機械便利帳』、『図面って、どない描くねん！LEVEL2』、『図解力・製図力おちゃのこさいさい』、『めっちゃ、メカメカ！リンク機構99→∞』、『メカ基礎バイブル〈読んで調べる！〉設計製図リストブック』、『図面って、どない描くねん！Plus＋』、『図面って、どない読むねん！LEVEL00』、『めっちゃ、メカメカ！2 ばねの設計と計算の作法』、『最大実体公差』、『設計センスを磨く空間認識力"モチアゲ"』 共著として『CADって、どない使うねん！』（山田学・一色桂 著）、『設計検討って、どないすんねん！』（山田学 編著）『技術士第一次試験「機械部門」専門科目 過去問題 解答と解説（第3版）』、『技術論文作成のための機械分野キーワード100解説集』（Net-P.E.Jp編著）などがある。

図面って、どない描くねん！バイリンガル
グローバルエンジニアへのファーストステップ

NDC 531.9

2014年9月19日　初版1刷発行	ⓒ著　者　山田　学
	発行者　井水 治博
	発行所　日刊工業新聞社
	東京都中央区日本橋小網町14番1号
	（郵便番号103-8548）
	書籍編集部　電話03-5644-7490
	販売・管理部　電話03-5644-7410
	FAX03-5644-7400
	URL　http://pub.nikkan.co.jp/
	e-mail　info@media.nikkan.co.jp
	振替口座　00190-2-186076
	本文デザイン・DTP——志岐デザイン事務所（矢野貴文）
	本文イラスト——小島サエキチ
	印刷——新日本印刷

定価はカバーに表示してあります
落丁・乱丁本はお取り替えいたします。
2014 Printed in Japan
ISBN 978-4-526-07292-5　C3053

本書の無断複写は、著作権法上の例外を除き、禁じられています。

日刊工業新聞社の好評図書

図面って、どない描くねん！
―現場設計者が教えるはじめての機械製図

山田 学 著
A5判224頁　定価（本体2200円＋税）

「技術者がそのアイディアを伝える唯一の方法が製図である」と信じる著者が書いた、読んで楽しい製図の入門書。著者自身が就職してはじめて図面を描いたときの戸惑いと技能検定（機械・プラント製図）を受験してはじめて知った、"製図の作法"を読者のためにわかりやすく解説した「誰もが読んで手を打ちたくなる」本。大阪弁のタイトル、めいっぱいに詰め込まれた図面やイラスト、そのすべてに製図に対する著者のストレートな愛情が詰まっています。内容はもちろん最新のJIS製図。それに現場設計者のノウハウとコツがポイントとして随所にちりばめられています。発行以来大好評で重版を重ねている、はっきり言ってお薦めの一冊です。

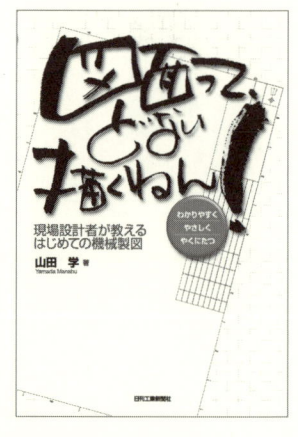

<目次>
第1章　図面ってどない描くねん！
第2章　寸法線ってどんな種類があるねん！
第3章　寸法公差ってなんやねん！
第4章　寸法ってどこから入れたらええねん！
第5章　幾何公差ってなんやねん！
第6章　この記号はどない使うねん！
第7章　こんな図面の描き方がわからへん！
第8章　図面管理ってなんやねん！

図面って、どない描くねん！LEVEL2
―現場設計者が教えるはじめての幾何公差

山田 学 著
A5判240頁　定価（本体2200円＋税）

昨今では、寸法公差だけの図面では、形状があいまいに定義されるため、幾何公差を用いたあいまいさのない図面定義が必要とされています。これについては、GPS規格としてISOでも審議されてきているのです。

本書は「幾何公差を理解することは製図を極めることである」と信じる著者による大ヒット製図入門書、第2弾。実務設計の中で戦略的に幾何公差を活用できるように、描き方から考え方、代表的な計測方法までをわかりやすく、やさしく解説しました。幾何公差をこれだけわかりやすく解説した本は他に類がありません！

<目次>
第1章　バラツキって、なんやねん！
第2章　データムって、なんやねん！
第3章　幾何特性って、なんやねん！
第4章　形状公差って、どない使うねん！
第5章　姿勢公差って、どない使うねん！
第6章　位置公差って、どない使うねん！
第7章　振れ公差って、どない使うねん！
第8章　幾何公差の相互依存って、なんやねん！
第9章　幾何公差を使ってみたいねん！

日刊工業新聞社の好評図書

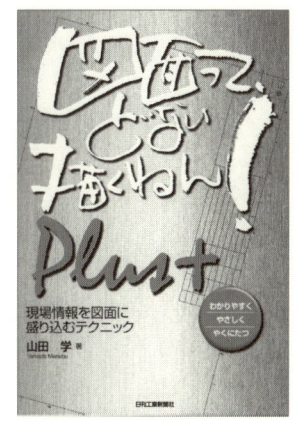

図面って、どない描くねん！Plus＋
―現場情報を図面に盛り込むテクニック

山田 学 著
A5判224頁　定価（本体2200円＋税）

　正しい製図をするためには、JIS製図の作法に則って正確に図面を描くことが必要です。ただし、本当に現場で役に立つ図面を描くためには、ルールブックには指示されていない加工や計測に配慮した現場独自の情報を図面に盛り込み、ベテラン設計者のような図面を描かなければいけません。
　そこで、本書は「図面って、どない描くねん！」シリーズのいずれの読者にも役に立つ、「ルールブックにはない現場の情報」を図面に盛り込むためのテクニックを紹介。従来描いていた図面に、「何をプラスすればベテランのような図面を描くことができるか」をやさしく、わかりやすく解説しています。本書を読んで図面を描けば、現場の作業者を唸らせることができます！

＜目次＞
第1章　設計形状と設計意図を表す寸法記入の関係
第2章　製図の手順を知り,設計の都合を図面に盛り込む
第3章　加工から図面に何を反映させるべきかを知る
第4章　図面と計測の関係から基準面の重要性を知る
第5章　加工と計測の都合を図面に盛り込む(1)
第6章　加工と計測の都合を図面に盛り込む(2)
第7章　まとめ

めっちゃ使える！機械便利帳
―すぐに調べる設計者の宝物

山田 学 編著
新書判176頁　定価（本体1400円＋税）

　著者自身が工場の現場や、CADの前でちょっとした基本的なことを調べたいときにあると便利だと思い、自作していたポケットサイズの手帳を商品化したもの。工場の現場でクレーム対応している最中や、デザインレビュー等の会議の場ですぐに利用できる手軽なデータ集です。
　記入できるメモ部分もありますので、どんどん使い込んで自分だけの便利帳にしてください。装丁は、デニム調のビニール上製特別仕立て。まさに設計現場で戦うエンジニアの宝物です。

＜目次＞
第1章　設計の基礎
第2章　数学の基礎
第3章　電気の基礎
第4章　力学の基礎
第5章　機械製図の基礎
第6章　材料の基礎
第7章　機械要素の基礎
第8章　海外対応の基礎
〈付録〉メモ帳（方眼紙）

日刊工業新聞社の好評図書

図解力・製図力 おちゃのこさいさい
―図面って、どない描くねん！LEVEL0

山田 学 著
B5判228頁（2色刷） 定価（本体2400円＋税）

　ついに登場した究極の製図入門書。ヒット作「図面って、どない描くねん！」のLEVEL0にあたるレベルでありながら、「図解力と製図力を身につけることを目的とした」ドリル形式の入門書です。「図解力が乏しいということは設計力が弱いことを意味する」と主張する著者が世界一やさしい製図本を目指して書いています。学習しやすい横レイアウト、全編2色刷の見やすい内容、豊富な演習問題(Work Shop)、従来の製図書にはなかった設計の基本的な計算問題にも対応、そして何より楽しく学習するための工夫がいっぱい詰まっています。

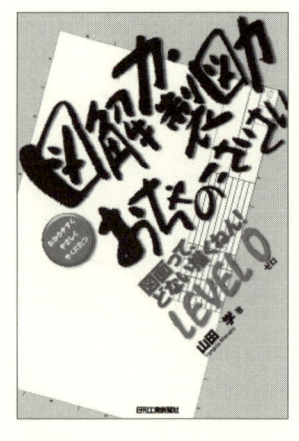

<目次>
第1章　立体と平面の図解力
第2章　JIS製図の決まりごと
第3章　寸法記入と最適な投影図
第4章　組み合せ部品の公差設定
第5章　設計に必要な設計知識と計算
第6章　Work Shop解答解説

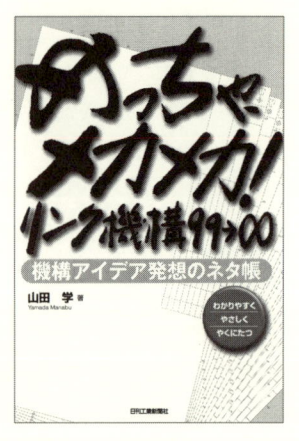

めっちゃ、メカメカ！
リンク機構99→∞
―機構アイデア発想のネタ帳

山田 学 著
A5判208頁　定価（本体2000円＋税）

　リンク機構とは、複数のリンクを組み合わせて構成した機械機構。これは、機械設計や機械要素技術の基本中の基本ですが、設計実務の中でリンク機構を考案する際、イレギュラーな機構ほど機構考案に時間がかかり、しかも、機構アイデアには経験や知識が問われます。

　本書はこのリンク機構設計の仕組みと基本がよくわかる本であり、パラパラとめくって最適な機構を探せる、あると便利なアイデア集でもあります。ぜひ、本書から無限大の発想を生み出して下さい。

<目次>
第1章　リンク機構の基本
第2章　メカトロとリンク機構
第3章　四節リンクの揺動運動
第4章　四節リンクの回転運動
第5章　四節リンクとスライド機構
第6章　その他の四節リンクの運動
第7章　多節リンクの運動

日刊工業新聞社の好評図書

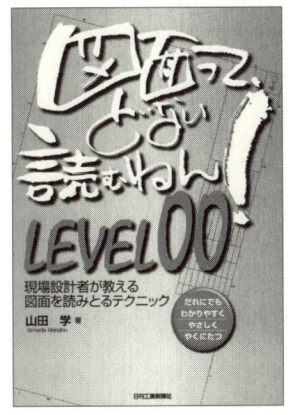

図面って、どない読むねん！
LEVEL 00
―現場設計者が教える
　図面を読みとるテクニック

山田 学 著
A5判248頁　定価（本体2000円＋税）

　図面を描く上で専門用語すら知らない「図面を読む立場の人」や、そういった相手を意識して図面を描かねばならない技術者向けの「製図＜読み／描き＞トレーニング」本。図面を見て話をする際に頻繁に出てくる用語を、具体的な図形や写真を使って解説。同時に、図面を読み描きする際に最低限必要な「LEVEL 00」相当の図解力も養います。もちろん、はじめて製図を勉強する人にもおすすめです。
　読み手の思考に合わせたページ展開で、とても読みやすく、わかりやすくなっています。

＜目次＞
第1章　正確に図形を伝える言葉を、知らなあかんねん！
第2章　投影図を読み解くとは、類推することやねん！
第3章　投影図以外の情報を、手がかりにすんねん！
第4章　投影図を読み解く、ワザがあるねん！
第5章　寸法数値以外の記号が、読み解くカギやねん！
第6章　寸法はばらつくから、公差があるねん！
第7章　幾何公差は寸法と区別して、考えなあかんねん！
第8章　溶接記号は丸暗記せんでええねん！
第9章　専門用語を知らな、読めへん図面があるねん！
第10章　図面管理に必要な記号を、見逃したらあかんねん！

めっちゃ、メカメカ！2
ばねの設計と計算の作法
―はじめてのコイルばね設計

山田 学 著
A5判218頁　定価（本体2000円＋税）

　「めっちゃ、メカメカ！」の続編として、「ばね」に焦点を当て、ばね設計を解説する本。特殊な「ばね」は割愛し、基本的なコイルばねに限定して、その設計方法を導く。実際にコイルばねを設計する際には、設計ポイントの知識をもって計算しなければいけない。本書はそのニーズに応えるわかりやすい入門書。読者に理解してもらうための、こだわりすぎなほどの著者の丁寧さが、「めっちゃ、メカメカ」の真骨頂。

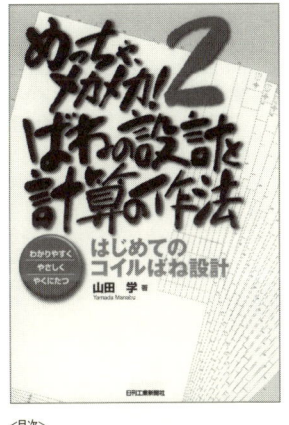

＜目次＞
第1章　ばね効果を得るための工夫ってなんやねん！
第2章　スペースや効率を考えて材料と形状を選択する
第3章　機能を考えて、コイルばねの種類を選択する
第4章　圧縮ばねを設計する前に知っておくべきこと
第5章　圧縮ばねの計算の作法（実践編）
第6章　引張りばねを設計する前に知っておくべきこと
第7章　引張りばねの計算の作法（実践編）
第8章　ねじりばねを設計する前に知っておくべきこと
第9章　ねじりばねの計算の作法（実践編）

日刊工業新聞社の好評図書

最大実体公差
―図面って、どない描くねん！LEVEL3

山田 学 著
A5判170頁　定価（本体2200円＋税）

「図面って」シリーズ最高峰のレベル3！最高難度を求める人にこそ読んで欲しい1冊。さらに進化した幾何公差、それが、「最大実体公差」。寸法公差と幾何公差の"特別な相互関係"にある最大実体公差は、論理性を持って読み解かなければ設計意図を理解できない。また同様に図面に指示することさえできない。機械製図の最高峰である「最大実体公差」をやさしく解説した本。

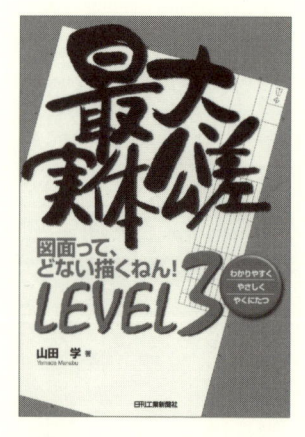

<目次>
第1章　独立の原則と相反する包絡の条件ってなんやねん！
第2章　どないしたら幾何公差だけ増やせんねん！
第3章　最大実体公差って、どの幾何公差に使ったらええねん！～形状公差・姿勢公差編～
第4章　最大実体公差って、どの幾何公差に使ったらええねん！～位置公差編～
第5章　機能ゲージって、どない設計すんねん！
第6章　最大実体公差を、もっと簡単に検査したいねん！
第7章　その他の幾何公差テクニックはどない使うねん！

設計センスを磨く
空間認識力"モチアゲ"
「勘」と「論理力」と「ポンチ絵スキル」をアップ！

山田 学 著
A5判232頁　定価（本体2000円＋税）

製図本の金字塔「図面って」シリーズ番外編。今回は、製図ではなく、その前の立体的な図面感覚や、空間としてモノを認識する能力を育てる、楽しいクイズ（演習）本。エンジニアには言葉以上に意思を的確に伝達する魔法の2つの武器、「図面」と「ポンチ絵」がある。ただし、それを駆使することができる「設計センス」が必要。本書は、その設計センスのもと「空間認識力」を育てるための本。つまり、読者の基礎設計能力を自然に「モチアゲ」ちゃう本である。

<目次>
第1章　サイズセンス　モチアゲの基本～大きさの勘を養う！
第2章　機構センス　モチアゲの基本～姿勢の変化や軌跡を読む！
第3章　レイアウトセンス　モチアゲの基本～組み合わせと分割を想像する！
第4章　投影認識センス　モチアゲの基本～投影図のルールを知る！
第5章　形状把握センス　モチアゲの基本～複数の投影図から形状を確定する！
第6章　アイデア発想センス　モチアゲの基本～オリジナルの形状を創造する！
第7章　空間認識センスSTEP1　モチアゲの基本～常に空間を意識する！
第8章　空間認識センスSTEP2　モチアゲの基本～仮想の断面を想像する！
第9章　空間認識センスSTEP3　モチアゲの基本～組立図から部品を見極める！
第10章　アイデア表現センス　モチアゲの基本～ポンチ絵は世界の共通言語！